INSTRUMENTATION IN ELEMENTARY PARTICLE PHYSICS

INSTRUMENTATION IN ELEMENTARY PARTICLE PHYSICS

The VII ICFA School

León, México   July 1997

EDITORS
G. Herrera Corral
CINVESTAV, México

M. Sosa Aquino
University of Guanajuato, México

American Institute of Physics

AIP CONFERENCE
PROCEEDINGS 422

Woodbury, New York

Editors:

Gerardo Herrera Corral
Departamento de Fisica
Centro de Investigacion y de Estudios Avanzados del I.P.N.
Apdo. Postal. 14-740
07000 Mexico, D.F.
MEXICO
E-mail: gerardo.herrera@fis.cinvestav.mx

Modesto Sosa Aquino
Instituto de Fisica
Universidad de Guanajuato at Leon
Loma del Bosque 103
Col Lomas del Campestre 37150
Leon, Guanajuato
MEXICO
E-mail: modesto@ifug4.ugto.mx

Authorization to photocopy items for internal or personal use, beyond the free copying permitted under the 1978 U.S. Copyright Law (see statement below), is granted by the American Institute of Physics for users registered with the Copyright Clearance Center (CCC) Transactional Reporting Service, provided that the base fee of $15.00 per copy is paid directly to CCC, 222 Rosewood Drive, Danvers, MA 01923. For those organizations that have been granted a photocopy license by CCC, a separate system of payment has been arranged. The fee code for users of the Transactional Reporting Service is: 1-56396-763-4/ 98 /$15.00.

© 1998 American Institute of Physics

Individual readers of this volume and nonprofit libraries, acting for them, are permitted to make fair use of the material in it, such as copying an article for use in teaching or research. Permission is granted to quote from this volume in scientific work with the customary acknowledgment of the source. To reprint a figure, table, or other excerpt requires the consent of one of the original authors and notification to AIP. Republication or systematic or multiple reproduction of any material in this volume is permitted only under license from AIP. Address inquiries to Office of Rights and Permissions, 500 Sunnyside Boulevard, Woodbury, NY 11797-2999; phone: 516-576-2268; fax: 516-576-2499; e-mail: rights@aip.org.

L.C. Catalog Card No. 98-72211
ISBN 1-56396-763-4
ISSN 0094-243X
DOE CONF- 9707140

Printed in the United States of America

CONTENTS

Organization viii
Foreword ix
Preface xi
Acknowledgements xiii

Lecture Courses

High Speed Data Acquisition 3
    P. S. Cooper
Physics of Particle Detection 14
    C. Grupen
Detector Systems for Future HEP Experiments 47
    A. Savoy-Navarro
Gaseous Wire Detectors 117
    J. Va'vra

Review Talks

Visible Light Photon Counters (VLPCs) for High Rate Tracking
Medical Imaging and Particle Astrophysics 191
    M. Atac
Perspectives in High Energy Physics 208
    A. Fernández and A. Zepeda
Cryogenic Detectors for Dark Matter 225
    D. McCammon
Neutrino Detectors 235
    K. Nakamura

Laboratory Sessions

Detection of Cosmic Ray Tracks Using Scintillating Fibers
and Position Sensitive Multi-Anode Photomultipliers 251
    M. Atac, J. Streets, and N. Wilcer
Silicon Detectors and Signal Processing 257
    P. Giubellino, M. Idzik, A. Rudge, and P. Weilhammer
Amplifiers and Electronics 283
    G. Hall and R. G. Payne
Use of De-Randomizing Buffers in a Data Acquisition System 313
    M. Johnson and M. Sheaff
Parallel Plate Avalanche Counters 324
    A. Martínez-Davalos and R. Alfaro-Molina
Muon Lifetime Measurement 333
    L. Villaseñor

X-Ray Imaging Chamber 347
       Yu. Zanevsky, S. Chernenko, L. Smykov, and G. Cheremukhina

Poster Sessions

MSGC and Fast Neutrons 367
       K. Bernier, H. Boukhal, J.-M. Denis, T. El Bardouni,
       Gh. Grégoire, O. Grégoire, and V. Tran

MSGC Tests with X-rays 369
       I. Boulogne and E. Daubie

High-Pressure Monitored Drift Tube Chambers for the ATLAS
Detector at the Large Hadron Collider 371
       U. Bratzler, T. Ferbel, A. Gabutti, H. Kroha, T. Lagouri, A. Manz,
       and M. Treichel

Order Statistics as a Tool for Analyzing Continuum Gamma Decay 375
       J. Cardona and F. Cristancho

FOCUS: A Charm Photo-Production Experiment at FERMILAB 377
       S. Carrillo and F. Vázquez

Assembly of Silicon Strip Detectors for the ATLAS
Semiconductor Tracker 382
       A. Cimmino, A. Saavedra, and G. Taylor

TILECAL-The Hadronic Calorimeter for ATLAS 385
       M. David

Can We Extract Continuum Properties from Side Feeding
Time Measurements? 387
       E. Galindo and F. Cristancho

SELEX Experiment (Fermilab E781) 389
       F. G. Garcia and J. Simon

Straw Tubes for FOCUS 392
       J. Link

Semiconductor Detectors for the ATLAS Inner Tracker 395
       D. Morgan

The Silicon Vertex Detector of HERA-B 397
       B. Moshous

The Mexican Participation at the Pierre Auger Observatory:
Recent Results 399
       S. Román, F. Alcaráz, E. Cantoral, J. Castro, A. Cordero,
       A. Fernández, R. López, C. Pacheco, M. Rubín, H. Salzar,
       J. Valdés, M. Vargas, L. Villaseñor, and A. Zepeda

DIRC, the Internally Reflecting Ring Imaging Čerenkov
Detector for BABAR: Properties of the Quartz Radiators 407
       J. Schwiening

Segmented Wire Ion Chambers (SWICs) Used As Proton Beam
Position/Profile Detectors in the Fixed Target Beamlines at Fermilab . 409
       G. Tassotto

HELIX 128S-2- A Readout Chip for the Silicon Vertex Detector and Inner Tracker Detektor of HERA-B 414
       U. Trunk, W. Fallot-Burghardt, E. Sexauer, K-T. Knöpfle,
       W. Hofmann, M. Cuje, B. Glass, M. Feuerstack-Raible, F. Eisele,
       and U. Straumann

Characterisation and Preliminary Analysis of the 1996 GaAs Keystone Test Beam Detectors 417
       S. Walsh, C. Buttar, P. Sellin, D. Morgan, C. Hardingham,
       and J. Burrage

List of Participants 419
Author Index 427

ORGANIZATION

ICFA Panel of Instrumentation Innovation and Development International Organizing Committee

Ariella Cattai	CERN/Switzerland
Boris Dolgoshein	Moscow Physics Engineering Institute/Rusia
Tord Ekelöf	Uppsala University, Sweden, (CERN/Switzerland)
Igor A. Golutvin	JINR/Rusia
Atul Gurtu	Tata Institute of Fundamental Research/India
Gerardo Herrera	CINVESTAV/Mexico
Ma JI-MAO	Inst. of IHEP/China
Takakiko Kondo	KEK/Japan
Stan Majewski	TJNAF/USA
David R. Nygren	LBL/USA
H. Okuno	Institute for Nuclear Study/Japan
Aurore Savoy-Navarro	LPNHE-Universités de Paris 6 7/France
Marleigh Sheaff	CINVESTAV/Mexico
Veniamin Sidorov	INP/Rusia
Jaroslav Va'vra	SLAC/USA
A. P. Vorobiev	IHEP/Rusia
Heinrich Walenta	Univ. of Siegen/Germany

Local Organizing Committee

Arturo Becerril Vilchis	ININ
Heriberto Castilla Valdez	CINVESTAV
Julián Felix Valdez	U. de Guanajuato (co-chair)
Arturo Fernández	FCFM, BUAP
Gerardo Herrera	CINVESTAV (chair)
Arturo Menchaca	I. de Física, UNAM
Héctor Méndez	CINVESTAV
Modesto Sosa	U. de Guanajuato
Antonio Morelos	U. de San Luis Potosí

FOREWORD

The ICFA Instrumentation School is devoted to the physics and technologies of instrumentation in elementary particle physics. Applications of these techniques to medicine and nuclear sciences as well as to the research and development in industry are also presented. An important feature of this school are laboratory courses, where students work with advanced experimental equipment that is discussed during lecture courses.

In the laboratory courses students worked in small groups which allowed them "hands-on" experience in selected experimental techniques: multiwire proportional chambers, silicon detectors, microstrip gas chambers, scintillating fibres, analog and digital circuits and data acquisition.

The School is aimed at improving the level of knowledge on instrumentation with special emphasis on applications in particle physics, medicine and industry. Participants are advanced graduate students or young researchers, the majority of them from less technologically advanced countries from all over the world, while the instructors are acknowledged experts in their fields. The selection of lectures and practical experiments has proved extremely popular among students at this as for previous Schools.

Gerardo Herrera Corral
CINVESTAV
Modesto Sosa Aquino
University of Guanajuato

PREFACE

The ICFA School on Instrumentation 1997 was held in León, Guanajuato, Mexico from July 7-19, 1997. It was organized by the Panel on Instrumentation, Innovation, and Development of the ICFA Committee, the University of Guanajuato at León, the Division of Particles and Fields of the Mexican Physical Society, and the Centro de Investigación y de Estudios Avanzados (CINVESTAV) in Mexico City.

The School in León, Guanajuato was the seventh in a highly successful series of schools of which three schools were held at ICTP in Trieste, Italy (in 1987, 1989 and 1991), one at CBPF in Rio de Janeiro, Brazil (1990), one at TIFR in Bombay, India (1993) and one at IJS in Ljubljana, Slovenia (1995).

The School has been made possible through the support of many laboratories and agencies throughout the world: CERN (Switzerland), DESY (Germany), INFN and ICTP (Italy), IN2P3 (France), RAL (UK) TRIUMF (Canada), DOE, FERMILAB, NSF (USA), UNESCO, the Centro Latinoamericano de Física (CLAF) in Brazil and México, SMF (Mexico), CINVESTAV (Mexico) and the University of Guanajuato. I wish to thank all laboratories and agencies for the generous support provided.

I also want to thank the director of the School, Dr. Gerardo Herrera, and the co-director, Dr. Julian Felix Valdez. The success of the event is mainly due to their effort and hard work as well as to the careful preparation of the laboratory sessions, lecture courses and review talks for which I wish to give special thanks to the lecturers and the laboratory instructors of the León School.

Tord Ekelöf
Chairman of the ICFA Instrumentation Panel

ACKNOWLEDGEMENTS

We would like to thank the members of the Local Organizing Committee, particularly Dr. A. Becerril Vilchis (ININ), who provided radiactive sources for the laboratories and A. Fernández (BUAP), who was in charge of the School page on the Web. Special thanks to Modesto Sosa and the staff of the University of Guanajuato, León, for their enormous contribution to the organization of the school.

We are grateful to Miguel A Pérez (chairman of the Physics Department, CINVESTAV) for his invaluable support, to Dr. Octavio Obregón (Director of IFUGto.) for providing the facilities where the laboratory sessions were held and to J. C. D'Olivo (president of the Division of Particles and Fields of the Mexican Physical Society) for his support.

We also wish to thank the engineers from IFUGto, Ing. Mario Muñoz García, Ing. Arturo González Vega and Ramón Martínez García also to Jaime García (CINVESTAV) for the excellent technical support and to the *Departmento de Importaciones* at CINVESTAV headed by Rodolfo de las Fuentes Lara. They went through the process of receiving and sending the equipment with great efficiency.

Special thanks to ConCyTEG, under the directorship of Dr. Arturo Lara López for the financial support to design and construct the lab moduli for the School. Also to the Guanajuato Government which, we recognize, is on the way to understand the important role that Science in general and Physics in particular plays in the economic and cultural development of all countries.

Thanks also to Joao dos Anjos whose support and advise was crucial.

We express here our sincere thanks to Dr. Tord Ekelöf and the members of the ICFA Instrumentation Panel for their help in defining a program and in the difficult task of obtaining financial support.

Julian Felix Valdez
Co-director of the School
Gerardo Herrera Corral
Director of the School

LECTURE COURSES

High Speed Data Acquisition

Peter S. Cooper

Fermi National Accelerator Laboratory
MS 122 P.O. Box 500 Batavia, Il 60510 USA

Abstract. A general introduction to high Speed data acquisition system techniques in modern particle physics experiments is given. Examples are drawn from the SELEX(E781) high statistics charmed baryon production and decay experiment now taking data at Fermilab.

INTRODUCTION

Data acquisition systems [DAQ] actually "do" modern particle physics experiments. We program and "train" them to select the events we want to keep. They respond on nanosecond to millisecond time scales to select the interesting data, put those data together into an event and ultimately save those events on a data tape. They also provide the command and control functions to allow the physicists to monitor what is happening on the detector and make changes.

DAQ SYSTEM FUNCTIONS

DAQ functions naturally split into three categories based upon the required response time. High speed functions are associated with the rate of events in the apparatus. Slow control functions are in real-time but with times of order seconds. On-line monitoring functions are near real-time. These operate from seconds to minutes after data is available. The sections below enumerate some of the functions required of each system.

High Speed DAQ Functions

* Select interesting interactions to keep for further study
  - With fast electronics [Trigger]
  - With fancy software [Filter]
  - With hybrids of the two

* Digitize analog detector signals into digital bytes of data
  - Event data is digitized if it is not born that way
  - Data is sparsifed - suppress the channels with zero

* Collect data for each of the detector systems
    Sub-events from each detector system (or systems) are packaged
      and usually "pipelined" to the higher levels of the DAQ.
  - Usually in high speed systems this level of data handling is
    happening in parallel in all sub-systems at the same time.

* Build events
    - Collect all the sub-events from the *same* trigger together
    - Add the necessary structures to correctly format events

* Analyze events in software
    - Some experiments run their off-line analysis code on some or all of the events.
    - If this step rejects significant numbers of events this is a software filter (Level 3 in some experiment's jargon).
    - Summaries (histograms, etc.) are built to monitor filter properties.

* Permanently record the selected events
    - Split-up up events by type (trigger?) into different files
    - usually on magnetic tape of some type
    - Newest trend is toward direct network transfer of data files to the Computer Center

A schematic of DAQ components is shown in Figure 1 below. Analog detector signals are digitized when a trigger arrives. The digitized data is "piped" to an event builder where all fragments of an event from several front end/digitizer systems come together. Built events are sent to an on-line computer where the may be filtered with software to further select interesting events. Selected events are written to data tape and/or over the network to the computer center.

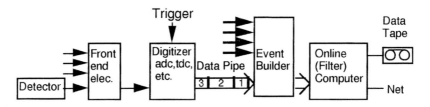

FIGURE 1. Schematic DAQ components.

Slow Control System Functions

* Control the experiment
    - Start and stop runs
    - Reload triggers, electronics, etc.

* Cold Start - reboot all or part of the DAQ itself

Monitoring Functions

* Monitor the experiment status
    - Readout non-event data (scalers, rates, etc.)
    - Log these data

* Complete sufficient analysis to display the status of the running experiment
    - Trigger rates
    - Detector status - e.g. MWPC wire maps, number of hits/plane, etc.
    - Events - Single event display (usually from the off-line)
    - Filter - pass rates, stability

TECHNOLOGIES

Data acquisition systems are built using a set of standard technologies for electronics, data transfer, computers and programming. In all of these technologies there are standards which have been developed so that all components don't have to be engineered from first principles. Very often experiments must design and engineer their own components because the functionality they require isn't available in a standard module or program. Most the manpower expended in building DAQ system is invested in this area. It is almost always advisable to use standards to the maximum degree possible, even when engineering specialized systems. The amount of effort required to replicate, in a *working* system, what already exists in an existing standard is always underestimated.

Standard Electronics and Readout Busses

Most experiments use standard electronics in different families. Most of these are busses with interfaces to computers and or other busses for readout and control. The CAMAC standard is the archetype. The big issue with these standards, and their abuses, is software support. Usual more effort is usually spent on programming a new module or system and all the things around it which aren't there, or don't work, than engineering it.

The list below gives most of the common electronics standards using in particle physics experiments. These are defined as real engineering standards so that different designers can engineer components which will work together under the definitions of a standard. (1)

- NIM   Nuclear Instrumentation modules (late 1960's)
- CAMAC   Computer Automated Measurement and Control (1970)
- Fastbus   Better, faster (harder) CAMAC (~1980)
- VME   Computer industry standard (particle physics ~1985)
- Home-brew Some people who build their own electronics don't believe
             in standards (e.g.: Transport bus(es) from Nevis Labs)
             Some systems use a standard bus, like CAMAC, in
             non-standard ways (PCOS and FERA from Lecroy,
             among many others)

Data Pipes - Systems for Long Distance (>10m) High Speed Transfers

Moving large volumes of data around is challenging. No bus *ever* achieves its maximum theoretical bandwidth. Most can't achieve 50% of the maximum in realistic applications. Some, like ethernet, can provide up to 10% to several simultaneous users but no one user can get more than 10%.

The big issues in moving data through pipes are blocking and software overhead. The maximum bandwidth is determined for one, infinitely large, block of data. The time to start and stop the transfer of a block usual dominates unless block sizes are very large and great care is taken to optimize a system.

To achieve the maximum average data transfer rates through a data pipe it is highly advisable to send data in only one direction with the largest possible block sizes to minimize the overhead in starting and finishing blocks. There is a well known example of an experiment who used the same data pipe network to send their data from detector to DAQ and their control messages from DAQ to detector. The control messages had difficulty "swimming upstream" in the large data flow and would sometime arrive too late or not arrive at all. Performance in that system suffered greatly.

- RS232   serial terminal lines (more of a straw than a pipe)
                user to download types of controllers (e.g. HV systems)
- CAMAC   branch highway (~100 Kb/sec) Allows multiple crate
                CAMAC systems
- FASTBUS   segments (~1Mb/sec) - like a CAMAC branching highway
- Ethernet   computing industry standards
- SCSI   - small Computer System Interface - disk and tape systems
- FDDI
- HPPI
- ATM
- RS 485   parallel ECL lines at 10MHz clock rate
- optical fiber up to 1 Gbit/sec on the fiber but driving electronics
                is never this fast.

Real-time Computers

Real-time computing is fundamentally different than the general purpose batch or timeshared computing. It is much harder to write and debug real time programs. The basic issue is response time - how long before an interrupt can get to it's service routine. The problem come when two things try to happen at once. (The second interrupt comes while the first is still being serviced.) The difference in real time systems is in the operating system much more than the computer hardware itself. All "big" computers treat their disk subsystems as real-time devices. They just don't treat their user's jobs with that kind of priority.

There are many different types of real-time computers and operating systems. They are typically small, complicated and for experts only. Real-time programming is much harder than usual. If you hear about a real-time problem and you are not a DAQ expert the experiment is in trouble.

These systems are often used as embedded controllers in DAQs grafting some data stream into another system. Examples are Fastbus masters and the processors at the beginning and ends of data pipes like the Selex fiber optic data paths.

General Purpose Computers

The systems we are all familiar with from our desktops and computer centers are a backbone element of high performance DAQ systems. In many ways jobs which formerly were done only in the off-line analysis phase of an experiment have migrated on-line. These include filtering events before writing them to tape and certain monitoring, calibration and alignments. These are really just doing the off-line analysis and first levels of event selection cuts before writing the data to tape. Programs like the DAQ user interface are often written using the high level graphics and programming tools available on this type of system.

These systems are typically UNIX workstations or multi-processors. Programming is done in high level languages, FORTRAN, C and recently C++. There is a strong overlap with parts of the off-line analysis programs. The unpacking routines in the off-line have much more to do with details of the DAQ than the analysis. Conversely, the single event display is usually integrated into the on-line by stealing it wholesale from the off-line codes. A filter code is often just a version of the off-line analysis rejecting events before they have been written to tape.

Programming

The majority of effort put into building a DAQ is in programming. The techniques are advanced and complicated. (Lots of queuing theory.) No system can be really tested and debugged in isolation; problem isolation is an enormous challenge. A typical problem is a filter job crashing in one event in a million because some hardware module in the experiment is dropping or corrupting bits. You cannot isolate this kind of problem without knowing a great deal about all the hardware and software between the particle detectors and your piece of analysis code.

DAQ *SYSTEMS*

The critical issue that is often missed in designing and building a new DAQ is that it is the system which has to be optimized, not a small part of the system. All new DAQ systems are, by definition, beyond the state of the art. No one designs a new one when an old system can be adapted or upgraded to do the job. DAQ's are large projects involving many people; programmers, engineers, technicians and physicists. Getting everyone to understand the big picture is hard. Getting them to all agree is impossible.

As an example, the Selex DAQ is built out of hardware and software components produced by the DART collaboration at Fermilab (2). This is a group of about 25 professional programmers, electronics engineers, and technicians plus one or more physicists from each of eight experiments. This team has produced part or all of the DAQ systems running on nearly all the Fermilab fixed target experiments now taking data. Several years were required to bring this project to its present point with many different, stable, DAQ's taking data everyday.

DEAD TIME

The most important parameter controlling the design and performance of high speed DAQ systems is dead time. Dead time occurs whenever a given step in the processing of the data in a DAQ takes a finite amount of time. If a particular step takes 100 nsec and the second event comes 50 nsec after the first there is a problem. In the simplest case the second event is just lost - the system is dead for that event. This is always enforced in electronics at the trigger level. When the DAQ cannot accept another event it asserts a logic level usual called "busy" which prevents the trigger electronics from accepting another event. When the busy is reset DAQ can continue.

One of the simplest DAQ systems possible is shown in Figure 1 below. The beam is a radioactive source and the detector a single scintillation counter consisting of a plastic scintillator and a photo multiplier tube (PMT). The trigger electronics consists of a NIM discriminator which fires whenever the inverted dynode pulse from the PMT exceeds 10 mV. With the PMT high voltage properly adjusted the trigger rate will be dominated by the rate at which beta particles from the source pass through the scintillator. There will also be a small but finite counting rate when the source is removed due to noise in the PMT.

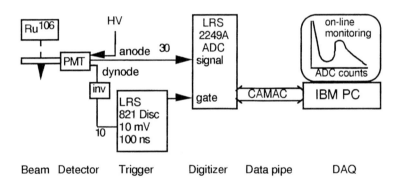

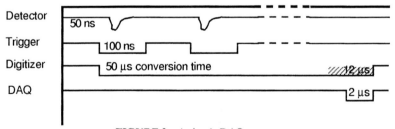

FIGURE 2. A simple DAQ system

The goal of this system is to measure the performance of the detector, specifically the pulse height spectrum of the scintillator. This is a real example in the sense that the all the parts are real devices which your experiment already has. This is an example of a "test stand"; a small and simple DAQ system used to checkout parts of an experiment.

The ADC has a finite conversion time of 50 μsec and will digitize the charge in a pulse contained with the gate. The width of the trigger pulse is set to 100 nsec to contain all the charge from a large PMT pulse. The relative timing of the gate and the anode signal are adjusted with cable delays to arrive the ADC with the trigger from a given pulse reaching the gate input just before the anode signal arrives at the signal input.

Once the ADC receives a gate it digitizes the charge within the gate to a 10 bit number with a conversion gain of 1/4 pC/count. The digitization takes 50 μsec. This type of ADC ignores subsequent gates until it is cleared with a CAMAC command (like F2 - read and clear).

The ADC interrupts the DAQ computer when the conversion is complete and the interrupt service routine reads and clears the ADC (dropping the internal ADC busy). This interrupt response time is 10 μsec plus 2 μsec for the actual CAMAC read and clear operation.

The background routine in the DAQ computer adds the new event to the pulse height histogram and updates the display of that histogram on the screen. This is on-line monitoring happening in near real time.

This DAQ system is already complicated enough to exhibit dead time. A second trigger with 62 μsec of the first is ignored. The fraction of triggers not lost to dead time is the lifetime ratio given by:

$R / R_T = 1 / [1 + R_T T_d]$

R DAQ rate
R_T Trigger rate
T_d Dead time [62 μsec in this case]

The maximum DAQ rate is $R = 1 / T_d$ independent of the trigger rate. At this trigger rate; $R_T = 1 / T_d = 16$ KHz and the livetime ratio is 50%. Half of the events are lost to dead time at this trigger rate.

Minimizing Dead Time

Consider what happens if we used two ADC's in above example instead of one with odd events going to one ADC and even events to the other. Digitization and DAQ could proceed in parallel and the dead time would be reduced. However this system would have to be more complicated in order to route odd and even events correctly and decide when both ADCs were busy. This is an example of the first technique for reducing dead time - parallelism.

A second approach is pipeline processing of the data flow. A complicated operation can be broken into several simple steps each operating on subsequent events. The dead time goes down because the time to perform each step is less than for the whole operation. Using first in first out (FIFO) buffer memories between step allows for variable times per step. The buffers will fill up while a long event is processed and empty during shorter events. A general schematic of a pipeline is shown in Figure 3 below.

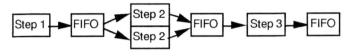

FIGURE 3. A pipeline processor.

These two techniques can be combined in ways like that shown above.

Dead time can never be completely eliminated. In the example of Figure 2 even if all dead time associated with the ADC were eliminated the trigger circuit itself has a 100 ns dead time due to the output pulse width. Since synchrotron beams are modulated with a maximum frequency by the accelerating RF a DAQ with a dead time smaller that the RF period is effectively dead timeless - it clears before the next beam particle can arrive.

The CDF and D0 collider experiments at Fermilab had DAQ systems designed for the 3.5 µsec time between bunch crossings in the Tevatron. The Tevatron collider upgrades now in progess reduce this time between crossings to ~400 nsec. All the front-end electronics in both of these experiments which exploited the 300 KHz maximum crossing rate are now being re-engineered. They are not dead timeless anymore.

A REAL DAQ SYSTEM

The schematic of my presently running Fermilab experiment is shown in Figure 4. Selex(E781) is a fixed target experiment designed to study the production and decay of charmed baryons (3). It is presently taking data in a 1MHz 600 GeV/c secondary beam in the proton center beam line at Fermilab.

The physical layout of the experiment presents a challenge to the design of the DAQ system. All of the trigger electronics and most of the DAQ systems are required to be placed close to the detectors in the radiation area. The back end of the DAQ and the Physicists are located in an upstairs counting area 100m away. Due to the requirements of radiation safety it is not possible to work on the electronics with the beam on. All changes must be remotely programmable or require a beam off access.

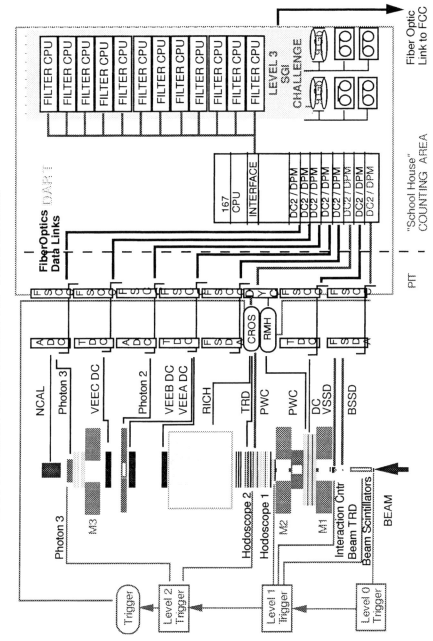

FIGURE 4. Selex (E781) DAQ system

Selex DAQ Parameters

Detector Channels
    74K   Si Strips   ADC
    17K   MWPC wires   Latches
    5K   DC wires   TDC
    2K   TRD wires   "Latches"
    2K   PBG blocks   ADC
    3K   RICH pmts   Latches
    ~100   Misc.
    ~105K   Total channels

DAQ Trigger Rate (T_2)   5000 Hz
Events Size   6500 bytes (average; heavy sparsification)
DAQ data rate   33 Mb/sec
On-line Software Filter

    Computers   20 SGI R4400 200 MHz, 144 MIP processors = 2880 MIPS
    Algorithm   Reject events with high momentum (>15 GeV/c) tracks consistent with a single vertex. Off-line tracking and vertexing code written in FORTRAN.
    Beam duty factor: 1/3 - 20 sec of beam every 60 sec
    CPU time   <12 msec/event average
    Rejection   90% - 1 trigger in 10 is written to tape.
    Taping rate   1 Mb/sec (2-8505 8mm tape drives in parallel).

Selex DAQ Architecture

The Selex DAQ has 10 parallel systems each reading out about 10% of the detector channels. The trigger strobes each of the digitizers shown in the center column of figure 4. The dead time is about 100 µsec per trigger and the livetime ratio is 50%. This is an example of parallelism on a large scale. Each data pipe is a fiber optic cable which carries one stream of event fragments from the digitizers in the radiation area to a dual ported memory upstairs in the counting area. data flow on these pipe are unidirectional and pipelined hardware inserts event headers and byte counts so that data lose or corruption can be detected. At the end of a 20 sec beam spill the ~650 Mb of data from that spill reside in those 10 memories. There are dedicated real-time processors in the data flow between the end of the optical data pipe and each memory which build a directory in each memory. This is a list of the addresses where each of the ~100K event fragments begins in that memory.

Data are transferred to the filter computers in blocks of 400 events using the event directories for each memory. The filter computers are actually two SMP (symmetric multi-processors) with 12 and 10 processors respectively. These two machines run 11 and 9 filter jobs which filter the events in parallel selecting about one in ten triggers based upon a reconstruction algorithm which looks for events with high momentum tracks which are inconsistent with a single vertex. Only some of the ten data streams are unpacked and used by this filter algorithm. The time available is about 12 CPU-msec/event. We typically require most of this time on average. Occasionally we are not done computing before then next accelerator spill comes (in 60 sec). In this case we must inhibit triggers for the next accelerator spill. This is an example of dead time at a very high level in a DAQ system.

When a trigger passes the filter criteria all its event fragments are copied to an output buffer (event building) which is written to a file on disk. These disk files are copied by a separate job to 8mm tape and renamed to a different directory on disk. There is enough disk to hold about 4 hours of data. The least recently used file is deleted to make space for new data. This allows time to recovery from tape drive problems and to look at the recent data before it is deleted.

A typical run last two hours and has about 20 such 200 Mb files. Shortly after the first files is closed during a run a set of jobs are run against this file on the on-line computers to produce on-line monitoring histograms. These allow the physicists on shift to determine that all the detectors and filters are working properly. One of the data files is copied over the network to a tape robot in the Feynman Computing Center. This makes a sample of each run available for off-line analysis on the Fermilab central computers with having to remount and read the raw data tape.

This system has worked quite well. The amount of experiment downtime due to DAQ problems has been about normal for a system of this level of complexity. We lose, at most, a few hours a week to DAQ problems.

CONCLUSIONS

We spend most of our time as experimentalists doing everything but particle physics. The ultimate goal of a DAQ system, like all the other HEP technologies, is to enable the experiment to do physics. Students, particularly, tend to forget, in the middle of all the battles to make an experiment really work, that all these techniques are merely means to and end. If the physics experiment didn't work it doesn't matter whether the DAQ did or not.

In the case of Selex there is little doubt that the experiment works. We have been taking data for about 5 months after 7 months commissioning a new apparatus. We have analyzed, off-line, about 10% of the data taken thus far and have already reconstructed the decays of about 1000 charmed particles. This is a long way from the $\sim 10^5$ charm decays we expect when a full analysis of the data is completed. This effort demonstrates that the experiment works and particularly that the relatively complicated filter algorithms succeed in keeping much of the desired signals. In this sense the Selex DAQ is a success.

REFERENCES

1. An example is the FASTBUS standard - ANSI/IEEE Std 960-1986.
2. More information on the DART collaboration can be found on the web at http://fndaub.fnal.gov:8000
3. More information on the SELEX experiment can be found on the web at http://fn781a.fnal.gov

Physics of Particle Detection

Claus Grupen

Departmend of Physics, University of Siegen
D-57068 Siegen, Germany
e-mail: grupen@sialw2.physik.uni-siegen.de

Abstract. In this review the basic interaction mechanisms of charged and neutral particles are presented. The ionization energy loss of charged particles is fundamental to most particle detectors and is therefore described in more detail. The production of electromagnetic radiation in various spectral ranges leads to the detection of charged particles in scintillation, Cherenkov and transition radiation counters. Photons are measured via the photoelectric effect, Compton scattering or pair production, and neutrons through their nuclear interactions.

A combination of the various detector methods helps to identify elementary particles and nuclei. At high energies absorption techniques in calorimeters provide additional particle identification and an accurate energy measurement.

INTRODUCTION

The detection and identification of elementary particles and nuclei is of particular importance in high energy, cosmic ray and nuclear physics [1–5]. Identification means that the mass of the particle and its charge is determined. In elementary particle physics most particles have unit charge. But in the study e.g. of the chemical composition of primary cosmic rays different charges must be distinguished.

The deflection of a charged particle in a magnetic field determines its momentum p; the radius of curvature ρ is given by

$$\rho \propto \frac{p}{z} = \frac{\gamma m_0 \beta c}{z} \tag{1}$$

where z is the particle's charge, m_0 its rest mass and $\beta = \frac{v}{c}$ its velocity. The particle velocity can be determined e.g. by a time-of-flight method yielding

$$\beta \propto \frac{1}{\tau} \quad , \tag{2}$$

where τ is the flight time. A calorimetric measurement provides a determination of the kinetic energy

$$E^{\text{kin}} = (\gamma - 1)m_0 c^2 \qquad (3)$$

where $\gamma = \frac{1}{\sqrt{1-\beta^2}}$ is the Lorentz factor.

From these measurements the ratio of m_0/z can be inferred, i.e. for singly charged particles we have already identified the particle. To determine the charge one needs another z-sensitive effect, e.g. the ionization energy loss

$$\frac{dE}{dx} \propto \frac{z^2}{\beta^2} \ln(a\beta\gamma) \qquad (4)$$

(a is a material dependent constant.)

Now we know m_0 and z separately. In this way even different isotopes of elements can be distinguished.

The basic principle of particle detection is that every physics effect can be used as an idea to build a detector. In the following we distinguish between the interaction of charged and neutral particles. In most cases the observed signature of a particle is its ionization, where the liberated charge can be collected and amplified, or its production of electromagnetic radiation which can be converted into a detectable signal. In this sense neutral particles are only detected indirectly, because they must first produce in some kind of interaction a charged particle which is then measured in the usual way.

INTERACTION OF CHARGED PARTICLES

Kinematics [2]

A particle of mass m_0 and velocity $v = \beta c$ colliding with an electron can transfer a maximum energy of

$$E^{\text{max}}_{\text{kin}} = \frac{2m_e c^2 \beta^2 \gamma^2}{1 + 2\gamma \frac{m_e}{m_0} + \left(\frac{m_e}{m_0}\right)^2} = \frac{2m_e p^2}{m_0^2 + m_e^2 + 2m_e E/c^2} \quad , \qquad (5)$$

here $\gamma = \frac{E}{m_0 c^2}$ is the Lorentz factor, E the total energy and p the momentum of the particle.

For low energy particles heavier than the electron ($2\gamma\frac{m_e}{m_0} \ll 1$; $\frac{m_e}{m_0} \ll 1$) eq. 5 reduces to

$$E^{\text{max}}_{\text{kin}} = 2m_e c^2 \beta^2 \gamma^2 \quad . \qquad (6)$$

For relativistic particles ($E_{\text{kin}} \approx E$; $pc \approx E$) one gets

$$E^{\text{max}} = \frac{E^2}{E + m_0^2 c^2 / 2m_e} \quad . \qquad (7)$$

For example, in a μ-e collision the maximum transferable energy is

$$E^{\max} = \frac{E^2}{E+11} \qquad E \text{ in GeV} \qquad (8)$$

showing that in the extreme relativistic case the complete energy can be transferred to the electron.

If $m_0 = m_e$, eq. 5 is modified to

$$E_{\text{kin}}^{\max} = \frac{p^2}{m_e + E/c^2} = \frac{E^2 - m_e^2 c^4}{E + m_e c^2} = E - m_e c^2 \quad . \qquad (9)$$

Scattering

Rutherford Scattering

The scattering of a particle of charge z on a target of nuclear charge Z is mediated by the electromagnetic interaction (figure 1).

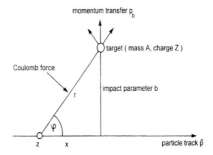

FIGURE 1. Kinematics of Coulomb scattering of a particle of charge z on a target of charge Z

The Coulomb force between the incoming particle and the target is written as

$$\vec{F} = \frac{z \cdot e \cdot Z \cdot e}{r^2} \frac{\vec{r}}{r} \quad . \qquad (10)$$

For symmetry reasons the net momentum transfer is only perpendicular to $\vec{p}$ along the impact parameter b

$$p_b = \int_{-\infty}^{+\infty} F_b \mathrm{d}t = \int_{-\infty}^{+\infty} \frac{z \cdot Z \cdot e^2}{r^2} \cdot \frac{b}{r} \cdot \frac{\mathrm{d}x}{\beta c} \quad , \qquad (11)$$

with $b = r \sin\varphi$, $\mathrm{d}t = \mathrm{d}x/v = \mathrm{d}x/\beta c$, and F_b force perpendicular to p.

$$p_b = \frac{z \cdot Z \cdot e^2}{\beta c} \int_{-\infty}^{+\infty} \frac{b\, \mathrm{d}x}{(\sqrt{x^2+b^2})^3} = \frac{z \cdot Z \cdot e^2}{\beta c b} \underbrace{\int_{-\infty}^{+\infty} \frac{\mathrm{d}(x/b)}{\left(\sqrt{1+\left(\frac{x}{b}\right)^2}\right)^3}}_{=2} \qquad (12)$$

$$p_b = \frac{2z \cdot Z \cdot e^2}{\beta c b} = \frac{2 r_e m_e c}{b\beta} z \cdot Z \quad , \qquad (13)$$

where r_e is the classical electron radius. This consideration leads to a scattering angle

$$\Theta = \frac{p_b}{p} = \frac{2z \cdot Z \cdot e^2}{\beta c b} \cdot \frac{1}{p} \quad . \qquad (14)$$

The cross section for this process is given by the well-known Rutherford formula

$$\frac{\mathrm{d}\sigma}{\mathrm{d}\Omega} = 4z \cdot Z \cdot r_e^2 \left(\frac{m_e c}{\beta p}\right)^2 \frac{1}{\sin^4 \Theta/2} \quad . \qquad (15)$$

Multiple Scattering

From eq. 15 one can see that the average scattering angle $\langle \Theta \rangle$ is zero. To characterize the different degrees of scattering when a particle passes through an absorber one normally uses the so-called "average scattering angle" $\sqrt{\langle \Theta^2 \rangle}$. The projected angular distribution of scattering angles in this sense leads to an average scattering angle of [6]

$$\sqrt{\langle \Theta^2 \rangle} = \Theta_{\mathrm{plane}} = \frac{13.6\,\mathrm{MeV}}{\beta c p} z \cdot \sqrt{\frac{x}{X_0}} \left\{ 1 + 0.038 \ln\left(\frac{x}{X_0}\right) \right\} \qquad (16)$$

with p in MeV/c and x the thickness of the scattering medium measured in radiation lengths X_0 (see **Bremsstrahlung**). The average scattering angle in three dimensions is

$$\Theta_{\mathrm{space}} = \sqrt{2}\, \Theta_{\mathrm{plane}} = \sqrt{2}\, \Theta_0 \quad . \qquad (17)$$

The projected angular distribution of scattering angles can approximately be represented by a Gaussian

$$P(\Theta)\mathrm{d}\Theta = \frac{1}{\sqrt{2\pi}\Theta_0} \exp\left\{-\frac{\Theta^2}{2\Theta_0^2}\right\} \mathrm{d}\Theta \quad . \qquad (18)$$

Energy Loss of Charged Particles

Charged particles interact with a medium by the exchange of photons (figure 2):

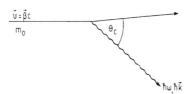

FIGURE 2. Interaction of charged particles via photon exchange

For soft collisions ($\hbar\omega \ll \gamma m_0 c^2$; $\hbar\vec{k} \ll \gamma m_0 \beta c$) conservation of energy and momentum requires [7]

$$\omega = \vec{v} \cdot \vec{k} \quad . \tag{19}$$

The behaviour of a photon in a medium is described by the dispersion relation well-known from solid state physics

$$\omega = 2\pi\nu = 2\pi \frac{c/n}{\lambda} = k \cdot \frac{c}{n} \quad , \tag{20}$$

where n is the index of refraction. Equivalently

$$\omega^2 - \frac{k^2 c^2}{\varepsilon} = 0 \quad , \tag{21}$$

where $n = \sqrt{\varepsilon}$ was used. ε is the relative permittivity or the dielectric constant of the absorber medium. Eliminating ω and k from eqs. 19 and 20 gives

$$\cos\Theta_c = \frac{1}{n\beta} = \frac{1}{\beta\sqrt{\varepsilon}} \quad , \tag{22}$$

where $\Theta_c = \sphericalangle (\vec{v}, \vec{k})$. It was assumed that the recoil of the incident particle due to photon emission is negligible (soft collisions).

Equation 22 shows that if $n\beta > 1$, there exists a real angle Θ_c for which free photons can be emitted or absorbed. This is the well-known condition for Cherenkov emission. $\beta = \frac{1}{\sqrt{\varepsilon}} = \frac{1}{n}$ is called Cherenkov threshold. At lower velocities free photons are not emitted in continuous media. In discontinuous media diffraction causes free photon emission even below Cherenkov threshold [7].

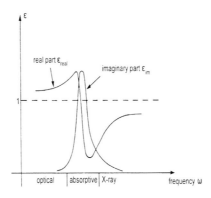

FIGURE 3. Real and imaginary part of the dielectric constant as a function of the frequency

Up to now we have assumed that ε is real. In practice this is only true below the ionization threshold of the medium. Figure 3 shows the real and imaginary (absorptive) part of the dielectric constant as a function of the frequency.

We can distinguish three distinct domains [7]:

a) The optical region: here the medium is transparent (for low frequencies); ε is real and greater than one and Cherenkov radiation is emitted above threshold.

b) The absorptive region: the dielectric constant is complex and the range of photons is short. The absorption of virtual photons constituting the field of the charged particle gives rise to ionization of the material.

c) The X-ray region: The medium becomes transparent again, but ε is still less than one. However, the emission of sub-threshold Cherenkov radiation in the presence of discontinuities can occur (transition radiation).

Ionization Energy-Loss

Bethe-Bloch Formula

This energy-loss mechanism represents the scattering of charged particles off atomic electrons, e.g.

$$\mu^+ + \text{atom} \rightarrow \mu^+ + \text{atom}^+ + e^- \quad .$$

The momentum transfer to the electron is (see eq. 13)

$$p_b = \frac{2r_e m_e c}{b\beta} z \quad ,$$

and the energy transfer in the classical approximation

$$\varepsilon = \frac{p_b^2}{2m_e} = \frac{2r_e^2 m_e c^2}{b^2 \beta^2} z^2 \quad . \tag{23}$$

The interaction probability per (g/cm^2), given the atomic cross-section σ, is

$$\phi(\mathrm{g}^{-1}\mathrm{cm}^2) = \frac{N}{A}\sigma[\mathrm{cm}^2/\mathrm{atom}] \tag{24}$$

where N is Avogadro's constant.

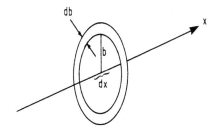

FIGURE 4. Sketch explaining the differential collision probability

The differential probability to hit an electron in the area of an annulus with radii b and $b + db$ (see figure 4) with an energy transfer between ε and $\varepsilon + d\varepsilon$ is

$$\phi(\varepsilon)d\varepsilon = \frac{N}{A} 2\pi b \, db \, Z \quad , \tag{25}$$

because there are Z electrons per target atom.

Inserting b from eq. 23 into eq. 25 gives

$$b^2 = \frac{2r_e^2 m_e c^2}{\beta^2} z^2 \cdot \frac{1}{\varepsilon}$$

$$2|b\,db| = \frac{2r_e^2 m_e c^2}{\beta^2} z^2 \cdot \frac{d\varepsilon}{\varepsilon^2}$$

$$\phi(\varepsilon)d\varepsilon = \frac{N}{A}\pi \frac{2r_e^2 m_e c^2}{\beta^2} z^2 \cdot Z \cdot \frac{d\varepsilon}{\varepsilon^2}$$

$$= \frac{2\pi r_e^2 m_e c^2 N}{\beta^2} \cdot \frac{Z}{A} \cdot z^2 \cdot \frac{d\varepsilon}{\varepsilon^2} \quad , \tag{26}$$

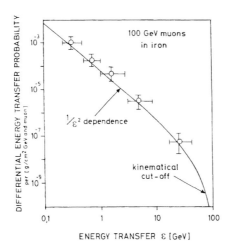

FIGURE 5. $1/\varepsilon^2$-dependence of the knock-on electron production probability [8]

showing that the energy spectrum of δ-electrons or knock-on electrons follows an $1/\varepsilon^2$ dependence (figure 5, [8]).

The energy loss is now computed from eq. 25 by integrating over all possible impact parameters [5]

$$-\mathrm{d}E = \int_0^\infty \phi(\varepsilon) \cdot \varepsilon \cdot \, dx$$
$$= \int_0^\infty \frac{N}{A} 2\pi b db \cdot Z \cdot \varepsilon \mathrm{d}x$$
$$-\frac{\mathrm{d}E}{\mathrm{d}x} = \frac{2\pi N}{A} \cdot Z \int_0^\infty \varepsilon b db$$
$$= 2\pi \frac{Z \cdot N}{A} \cdot \frac{2r_e^2 m_e c^2}{\beta^2} z^2 \int_0^\infty \frac{db}{b} \quad . \tag{27}$$

This classical calculation yields an integral which diverges for $b = 0$ as well as for $b = \infty$. This is not a surprise because one would not expect that our approximations hold for these extremes.

a) The $b = 0$ case: Let us approximate the "size" of the electron by half the de Broglie wavelength. This gives a minimum impact parameter of

$$b_{\min} = \frac{h}{2p} = \frac{h}{2\gamma m_e \beta c} \quad . \tag{28}$$

b) The $b = \infty$ case: If the revolution time τ_R of the electron in the target atom becomes smaller than the interaction time τ_i, the incident particle "sees" a more or less neutral atom

$$\tau_i = \frac{b_{\max}}{v}\sqrt{1-\beta^2} \ . \tag{29}$$

The factor $\sqrt{1-\beta^2}$ takes into account that the field at high velocities is Lorentz-contracted. Hence the interaction time is shorter. For the revolution time we have

$$\tau_R = \frac{1}{\nu_Z \cdot Z} = \frac{h}{I} \ , \tag{30}$$

where I is the average ionization potential of the target material.

The condition to see the target as neutral now leads to

$$\tau_R = \tau_i \quad \Rightarrow \quad \frac{b_{\max}}{v}\sqrt{1-\beta^2} = \frac{h}{I}$$

$$b_{\max} = \frac{\gamma h \beta c}{I} \ . \tag{31}$$

With the help of eq. 28 and 31 we can solve the integral in eq. 27

$$-\frac{dE}{dx} = 2\pi \cdot \frac{Z}{A} N \cdot \frac{2 r_e^2 m_e c^2}{\beta^2} z^2 \cdot \ln \frac{2\gamma^2 \beta^2 m_e c^2}{I} \ . \tag{32}$$

Since for long-distance interactions the Coulomb field is screened by the intervening matter one has

$$-\frac{dE}{dx} = \kappa z^2 \cdot \frac{Z}{A}\frac{1}{\beta^2}\left[\ln\frac{2\gamma^2\beta^2 m_e c^2}{I} - \eta\right] \ , \tag{33}$$

where η is a screening parameter (density parameter) and

$$\kappa = 4\pi N r_e^2 m_e c^2 \ .$$

The exact treatment of the ionization energy loss of heavy particles leads to [6]

$$-\frac{dE}{dx} = \kappa z^2 \cdot \frac{Z}{A} \cdot \frac{1}{\beta^2}\left[\frac{1}{2}\ln\frac{2m_e c^2\gamma^2\beta^2}{I^2}E_{\text{kin}}^{\max} - \beta^2 - \frac{\delta}{2}\right] \tag{34}$$

which reduces to eq. 33 for $\gamma m_e/m_0 \ll 1$ and $\beta^2 - \frac{\delta}{2} = \eta$.

The energy-loss rate in various materials is shown in figure 6 [6]. It exhibits a $\frac{1}{\beta^2}$-decrease until a minimum of ionization is obtained for $3 \leq \beta\gamma \leq 4$. Due to the $\ln\gamma$-term the energy loss increases again (relativistic rise, logarithmic rise) until a plateau is reached (density effect, Fermi plateau). If the energy loss is expressed in terms of the areal density

$$\mathrm{d}s = \rho\mathrm{d}x$$

with ρ-density of the absorber, a typical energy loss between 1 and $2\,\mathrm{MeV}/(\mathrm{g}/\mathrm{cm}^2)$ is observed for most absorber materials. For gases the Fermi-plateau, which saturates the relativisitc rise, is about 60% higher compared to the minimum of ionization. Figure 7 shows the measured energy loss rates of electrons, pions, kaons, protons, deuterons and tritons in the OPAL jet chamber [9].

Landau Distributions

The Bethe-Bloch formula describes the average energy loss of charged particles. The fluctuation of the energy loss around the mean is described by an asymmetric distribution, the Landau distribution [10,11].

The probability $\phi(\varepsilon)\mathrm{d}\varepsilon$ that a singly charged particle loses an energy between ε and $\varepsilon + \mathrm{d}\varepsilon$ per unit length of an absorber was (eq. 26)

$$\phi(\varepsilon) = \frac{2\pi N e^4}{m_e v^2} \frac{Z}{A} \cdot \frac{1}{\varepsilon^2} \quad . \tag{35}$$

Let us define

$$\xi = \frac{2\pi N e^4}{m_e v^2} \cdot \frac{Z}{A} x \quad , \tag{36}$$

where x is the areal density of the absorber:

$$\phi(\varepsilon) = \xi(x)\frac{1}{x\varepsilon^2} \quad . \tag{37}$$

Numerically one can write

$$\xi = \frac{0.1536}{\beta^2}\frac{Z}{A} \cdot x \quad [\mathrm{keV}] \quad , \tag{38}$$

where x is measured in $\mathrm{mg}/\mathrm{cm}^2$.

For an absorber of 1 cm Ar we have for $\beta = 1$

$$\xi = 0.123\,\mathrm{keV} \quad .$$

We define now

$$f(x,\Delta) = \frac{1}{\xi}\omega(\lambda) \tag{39}$$

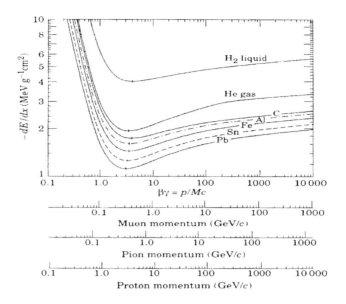

FIGURE 6. Rate of energy loss in various materials [6]

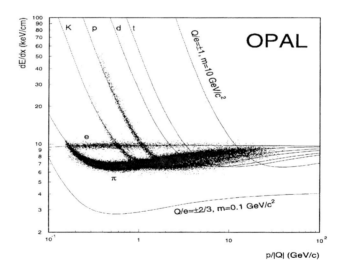

FIGURE 7. Measured ionization energy loss of electrons, pions, kaons, protons, deuterons and tritons in the OPAL jet chamber [9]

as the probability that the particle loses an energy Δ on traversing an absorber of thickness x. λ is the deviation from the most probable energy loss $\Delta^{\text{m.p.}}$

$$\lambda = \frac{\Delta - \Delta^{\text{m.p.}}}{\xi} \quad . \tag{40}$$

The most probable energy loss is calculated to be [10]

$$\Delta^{\text{m.p.}} = \xi \left\{ \ln \frac{2 m_e c^2 \beta^2 \gamma^2 \xi}{I^2} - \beta^2 + 1 - \gamma_E \right\} \quad , \tag{41}$$

where $\gamma_E = 0.577\ldots$ is Euler's constant.

Landau's treatment of $f(x, \Delta)$ yields

$$\omega(\lambda) = \frac{1}{\pi} \int_0^\infty e^{-u \ln u - \lambda u} \sin \pi u \, du \quad , \tag{42}$$

which can be approximated by [12]

$$\Omega(\lambda) = \frac{1}{\sqrt{2\pi}} \exp \left\{ -\frac{1}{2} (\lambda + e^{-\lambda}) \right\} \quad . \tag{43}$$

Figure 8 shows the energy loss distribution of 3 GeV electrons in an Ar/CH_4 (80:20) filled drift chamber of 0.5 cm thickness [13] and figure 9 demonstrates the particle identification power of the OPAL jet chamber based on the ionization loss measurement [14].

The asymmetric property of the energy-loss distribution becomes obvious for thin absorbers. For larger absorber thicknesses or truncation techniques applied to thin absorbers the Landau distribution gets more symmetric.

Scintillation in Materials

Scintillator materials can be inorganic crystals, organic liquids or plastic and gases. The scintillation mechanism in organic crystals is an effect of the lattice. Incident particles can transfer energy to the lattice by creating electron-hole pairs or taking electrons to higher energy levels below the conduction band. Recombination of electron-hole pairs may lead to the emission of light. Also electron-hole bound states (excitons) moving through the lattice can emit light when hitting an activator center and transferring their binding energy to activator levels, which subsequently deexcite. In thallium doped NaI-crystals about 25 eV are required to produce one scintillation photon. The decay time in inorganic scintillators can be quite long (1μs in CsI (Te); 0.62 μs in BaF_2).

In organic substances the scintillation mechanism is different. Certain types of molecules will release a small fraction ($\approx 3\%$) of the absorbed energy as optical photons. This process is expecially marked in organic substances which

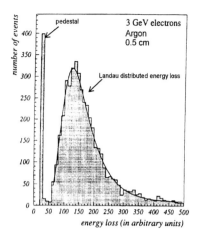 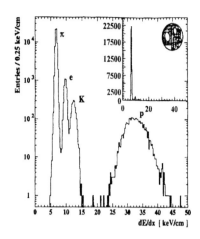

FIGURE 8. Energy-loss distribution of 3 GeV electrons in a thin-gap multiwire drift chamber [13]

FIGURE 9. Energy-loss distribution of electrons, pions, kaons and protons in the momentum interval 0.45 - 0.48 GeV/c in the OPAL jet chamber [14]

contain aromatic rings, such as polystyrene, polyvinyltoluene, and naphtalene. Liquids which scintillate include toluene or xylene [6].

This primary scintillation light is preferentially emitted in the UV-range. The absorption length for UV-photons in the scintillation material is rather short: the scintillator is not transparent for its own scintillation light. Therefore, this light is transferred to a wavelength shifter which absorbs the UV-light and reemits it at longer wavelengths (e.g. in the green). Due to the lower concentration of the wavelength shifter material the reemitted light can get out of the scintillator and be detected by a photosensitive device. The technique of wavelength shifting is also used to match the emitted light to the spectral sensitivity of the photomultiplier. For plastic scintillators the primary scintillator and wavelength shifter are mixed with an organic material to form a polymerizing structure. In liquid scintillators the two active components are mixed with an organic base [2].

About 100 eV are required to produce one photon in an organic scintillator. The decay time of the light signal in plastic scintillators is substantially shorter compared to inorganic substances (e.g. 30 ns in naphtalene).

Because of the low light absorption in gases there is no need for wavelength shifting in gas scintillators.

Plastic scintillators do not respond linearly to the energy-loss density. The number of photons produced by charged particles is described by Birk's semiempirical formula [6,15,16]

$$N = N_0 \frac{\mathrm{d}E/\mathrm{d}x}{1 + k_B\,\mathrm{d}E/\mathrm{d}x} \quad , \tag{44}$$

where N_0 is the photon yield at low specific ionization density, and k_B is Birk's density parameter. For 100 MeV protons in plastic scintillators one has $\mathrm{d}E/\mathrm{d}x \approx 10\,\mathrm{MeV}/(\mathrm{g/cm^2})$ and $k_B \approx 5\,\mathrm{mg}/(\mathrm{cm^2 MeV})$, yielding a saturation effect of $\sim 5\%$.

For low energy losses eq. 44 leads to a linear dependence

$$N = N_0 \cdot \mathrm{d}E/\mathrm{d}x \quad , \tag{45}$$

while for very high $\mathrm{d}E/\mathrm{d}x$ saturation occurs at

$$N = N_0/k_B \quad . \tag{46}$$

There exists a correlation between the energy loss of a particle that goes into the creation of electron-ion pairs or the production of scintillation light because electron-ion pairs can recombine thus reducing the $\mathrm{d}E/\mathrm{d}x|_{\mathrm{ion}}$-signal. On the other hand the scintillation light signal is enhanced because recombination frequently leads to excited states which deexcite yielding scintillation light.

Figure 10 shows this correlation for lanthanum ions passing through a liquid-argon scintillator [17]. The two lines are caused by La-ions travelling through different parts of the scintillator and depositing different amounts of energy.

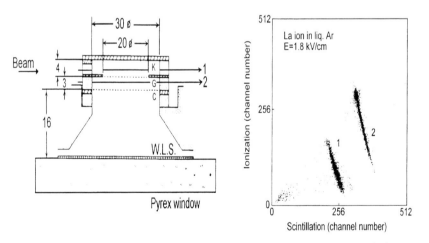

FIGURE 10. Correlation of signals due to scintillation and ionization [17]

Cherenkov Radiation

A charged particle traversing a medium with refractive index n with a velocity v exceeding the velocity of light c/n in that medium, emits Cherenkov radiation. The threshold condition is given by eq. 22

$$\beta_{\text{thres}} = \frac{v_{\text{thres}}}{c} \geq \frac{1}{n} \quad . \tag{47}$$

The angle of emission increases with the velocity reaching a maximum value for $\beta = 1$, namely

$$\Theta_c^{\max} = \arccos \frac{1}{n} \quad . \tag{48}$$

The threshold velocity translates into a threshold energy

$$E_{\text{thres}} = \gamma_{\text{thres}} m_0 c^2 \tag{49}$$

yielding

$$\gamma_{\text{thres}} = \frac{1}{\sqrt{1 - \beta_{\text{thres}}^2}} = \frac{n}{\sqrt{n^2 - 1}} \quad . \tag{50}$$

The dependence of $\Theta_c = f(\beta)$ for various Cherenkov radiators is shown in figure 11.

The number of Cherenkov photons emitted per unit path length dx is

$$\frac{dN}{dx} = 2\pi \alpha z^2 \int \left(1 - \frac{1}{n^2 \beta^2}\right) \frac{d\lambda}{\lambda^2} \tag{51}$$

for $n(\lambda) > 1$, z – electric charge of the incident particle, λ – wavelength, and α – fine structure constant. The yield of Cherenkov radiation photons is proportional to $1/\lambda^2$, but only for those wavelengths where the refractive index is larger than unity. Since $n(\lambda) \approx 1$ in the X-ray region, there is no X-ray Cherenkov emission. Integrating eq. 51 over the visible spectrum ($\lambda_1 = 400\,\text{nm}$, $\lambda_2 = 700\,\text{nm}$) gives

$$\begin{aligned}\frac{dN}{dx} &= 2\pi \alpha z^2 \frac{\lambda_2 - \lambda_1}{\lambda_1 \lambda_2} \sin^2 \Theta_c \\ &= 490 \cdot z^2 \cdot \sin^2 \Theta_c \,[\text{cm}^{-1}] \quad .\end{aligned} \tag{52}$$

Figure 12 shows the photon yield for various materials [2]. The Cherenkov effect can be used to identify particles of fixed momentum by means of threshold Cherenkov counters or by devices which can also measure the Cherenkov angle or the ring of Cherenkov photons emitted in a radiator.

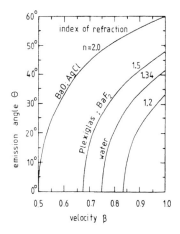 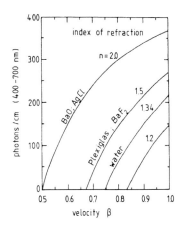

FIGURE 11. Velocity dependence of the emission angle for Cherenkov radiation in various materials [2]

FIGURE 12. Cherenkov radiation yield for various radiators [2]

Reprinted with the permission of Cambridge University Press.

Transition Radiation

Transition radiation is emitted when a charged particle traverses a medium with discontinuous dielectric constant. A charged particle moving towards a boundary, where the dielectric constant changes, can be considered to form together with its mirror charge an electric dipole whose field strength varies in time. The time dependent dipole field causes the emission of electromagnetic radiation. This emission can be understood in such a way that although the dielectric displacement $\vec{D} = \varepsilon\varepsilon_0\vec{E}$ varies continuously in passing through a boundary, the electric field does not.

The energy radiated from a single boundary (transition from vacuum to a medium with dielectric constant ε) is proportional to the Lorentz-factor of the incident charged particle [6,15,18]:

$$S = \frac{1}{3}\alpha z^2 \hbar\omega_p\gamma \quad , \tag{53}$$

where

$$\omega_p = \sqrt{4\pi N_e r_e^3 \frac{m_e c^2}{\alpha\hbar}} = \sqrt{\frac{N_e e^2}{\varepsilon_0 m_e}} \tag{54}$$

is the plasma frequency, which depends on the electron density N_e (ε_0 – permittivity of free space). The dielectric constant is related to N_e by

$$\varepsilon(\omega) = 1 - \frac{N_e e^2}{\varepsilon_0 m_e} \cdot \frac{1}{\omega^2} = 1 - \left(\frac{\omega_p}{\omega}\right)^2 \quad . \tag{55}$$

The plasma energy can be expressed as

$$\hbar\omega_p = \sqrt{4\pi N_e a_\infty^3}\, \alpha^2 m_e c^2$$
$$= 27.2 \sqrt{4\pi N_e a_\infty^3} \quad [\text{eV}] \quad , \tag{56}$$

where $a_\infty = r_e/\alpha^2$ is the Bohr radius. For plastic radiators (styrene or similar materials) $\sqrt{4\pi N_e a_\infty^3} \approx 0.8$, so that

$$\hbar\omega_p \approx 20\,\text{eV} \quad .$$

The typical emission angle of transition radiation is proportional to $1/\gamma$. The radiation yield drops sharply for frequencies

$$\omega > \gamma \omega_p \quad . \tag{57}$$

The γ-dependence of the emitted energy originates mainly from the hardening of the spectrum rather than from the increased photon yield. Since the radiated photons also have energies proportional to the Lorentz factor of the incident particle, the number of emitted transition radiation photons is

$$N \propto \alpha z^2 \quad . \tag{58}$$

The number of emitted photons can be increased by using many transitions (stack of foils, or foam). However, the foils or foams have to be of low Z material to avoid absorption in the radiator. Interference effects for radiation from transitions in periodic arrangements cause an effective threshold behaviour at a value of $\gamma \approx 1000$. These effects also produce a frequency dependent photon yield. The foil thickness must be comparable to or larger than the formation zone

$$D = \gamma c/\omega_p \tag{59}$$

which in practical situations ($\hbar\omega_p = 20\,\text{eV}$; $\gamma = 5 \cdot 10^3$) is about $50\,\mu\text{m}$.

Figure 13 shows the energy loss spectra of energetic electrons and the average energy of transition radiation photons in their dependence on the electron momentum [19]. Transition radiation detectors are mainly used for e/π-separation.

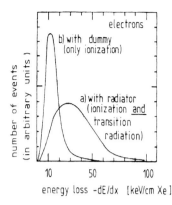 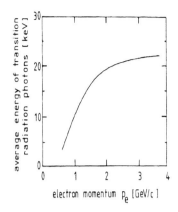

FIGURE 13. Energy-loss spectra of energetic electrons and average energy of transition radiation photons in their dependence on the electron momentum [19]

Bremsstrahlung

If a charged particle is decelerated in the Coulomb field of a nucleus a fraction of its kinetic energy will be emitted in form of real photons (bremsstrahlung). The energy loss by bremsstrahlung for high energies can be described by [2]

$$-\frac{dE}{dx} = 4\alpha N_A \frac{Z^2}{A} \cdot z^2 r^2 E \ln \frac{183}{Z^{1/3}} \quad , \tag{60}$$

where $r = \frac{1}{4\pi\varepsilon_0} \cdot \frac{e^2}{mc^2}$. Bremsstrahlung is mainly produced by electrons because

$$r_e \propto \frac{1}{m_e} \quad . \tag{61}$$

Equation 60 can be rewritten for electrons

$$-\frac{dE}{dx} = \frac{E}{X_0} \quad , \tag{62}$$

where

$$X_0 = \frac{A}{4\alpha N_A Z(Z+1) r_e^2 \ln(183\, Z^{-1/3})} \tag{63}$$

is the radiation length of the absorber in which bremsstrahlung is produced. Here we have included also radiation from electrons ($\sim Z$, because there are Z electrons per nucleus). If screening effects are taken into account X_0 can be more accurately described by [6]

$$X_0 = \frac{716.4\, A}{Z(Z+1)\ln(287/\sqrt{Z})} \quad [\text{g/cm}^2] \quad . \tag{64}$$

The important point about bremsstrahlung is that the energy loss is proportional to the energy. The energy where the losses due to ionization and bremsstrahlung for electrons are the same is called critical energy

$$\left.\frac{dE_c}{dx}\right|_{\text{ion}} = \left.\frac{dE_c}{dx}\right|_{\text{brems}} \quad . \tag{65}$$

For solid or liquid absorbers the critical energy can be approximated by [6]

$$E_c = \frac{610\,\text{MeV}}{Z + 1.24} \quad , \tag{66}$$

while for gases one has [6]

$$E_c = \frac{710\,\text{MeV}}{Z + 0.92} \quad . \tag{67}$$

The difference between gases on the one hand and solids and liquids on the other hand comes about because the density corrections are different in these substances, and this modifies $\left.\frac{dE}{dx}\right|_{\text{ion}}$.

The energy spectrum of bremsstrahlung photons is $\sim E_\gamma^{-1}$, where E_γ is the photon energy.

At high energies also radiation from heavier particles becomes important and consequently a critical energy for these particles can be defined. Since

$$\left.\frac{dE}{dx}\right|_{\text{brems}} \propto \frac{1}{m^2} \tag{68}$$

the critical energy e.g. for muons in iron is

$$E_c = \frac{610\,\text{MeV}}{Z + 1.24} \cdot \left(\frac{m_\mu}{m_e}\right)^2 = 960\,\text{GeV} \quad . \tag{69}$$

Direct Electron Pair Production

Direct electron pair production in the Coulomb field of a nucleus via virtual photons ("tridents") is a dominant energy loss mechanism at high energies. The energy loss for singly charged particles due to this process can be represented by

$$-\left.\frac{dE}{dx}\right|_{\text{pair}} = b(Z, A, E) \cdot E \quad . \tag{70}$$

It is essentially – like bremsstrahlung – also proportional to the particle's energy. Because bremsstrahlung and direct pair production dominate at high energies this offers an attractive possibility to build also muon calorimeters [2]. The average rate of muon energy losses can be parametrized as

$$\frac{dE}{dx} = a(E) + b(E) \cdot E \tag{71}$$

where $a(E)$ represents the ionization energy loss and $b(E)$ is the sum of direct elektron pair production, bremsstrahlung and photonuclear interactions. The energy dependence of the energy-loss parameters ("b-parameters") for muons in iron is shown in figure 14 [6,30]. Figure 15 shows the various contributions to the energy loss of muons in iron and also the total losses for uranium and hydrogen [6].

The linear dependence of the energy loss on the energy for energetic muons is verified experimentally (figure 16, [20]). Also the observed spectrum of secondaries (δ-electrons, bremsstrahlung photons, direct electron-pairs) is in agreement with expectation (figure 17, [21]).

Nuclear Interactions

Nuclear interactions play an important role in the detection of neutral particles other than photons. They are also responsible for the development of hadronic cascades. The total cross section for nucleons is of the order of 50 mbarn and varies slightly with energy. It has an elastic (σ_{el}) and inelastic part (σ_{inel}). The inelastic cross section has a material dependence

$$\sigma_{\text{inel}} \approx \sigma_0 A^\alpha \tag{72}$$

with $\alpha = 0.71$. The corresponding absorption length λ_a is [2]

$$\lambda_a = \frac{A}{N_A \cdot \rho \cdot \sigma_{\text{inel}}} [\text{cm}] \tag{73}$$

(A in g/mole, N_A in mole^{-1}, ρ in g/cm^3, and σ_{inel} in cm^2).
This quantity has to be distinguished from the the nuclear interaction length λ_w, which is related to the total cross section

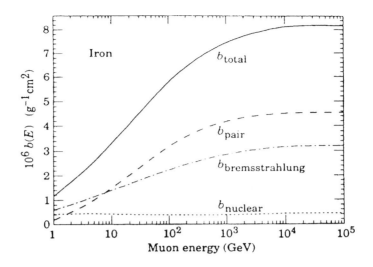

FIGURE 14. Energy-loss parameters due to direct electron pair production, bremsstrahlung and nuclear interactions for muons in iron [6,30]

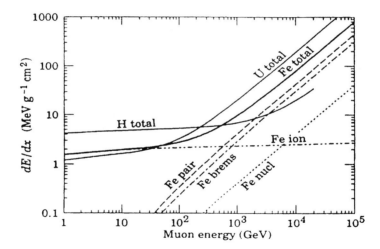

FIGURE 15. Average energy losses for muons in hydrogen, iron and uranium. Contributions of the individual energy losses in iron are also shown [6]

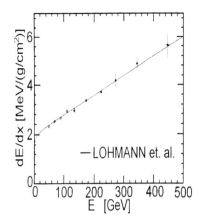
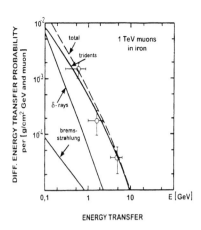

FIGURE 16. Average energy loss of muons in iron vs. the muon energy [20]

FIGURE 17. Energy spectrum of secondaries produced by 1 TeV cosmic ray muons in iron by direct electron pair production (tridents), bremsstrahlung und δ-electrons [21]

$$\lambda_w = \frac{A}{N_A \cdot \rho \cdot \sigma_{\text{total}}} [\text{cm}] \quad . \tag{74}$$

Since $\sigma_{\text{total}} > \sigma_{\text{inel}}$, $\lambda_w < \lambda_a$ holds.

Strong interactions have a multiplicity which grows logarithmically with energy. The particles are produced in a narrow cone around the forward direction with a typical transverse momentum of $p_T = 350\,\text{MeV/c}$, which is responsible for the lateral spread of hadronic cascades.

A useful relation for the calculation of interaction rates per (g/cm^2) is

$$\phi((\text{g/cm}^2)^{-1}) = \sigma_N \cdot N_A \tag{75}$$

where σ_N is the cross section per nucleon and N_A Avogadro's number.

INTERACTION OF PHOTONS

Photons are attenuated in matter via the processes of the photoelectric effect, Compton scattering and pair production. The intensity of a photon beam varies in matter according to

$$I = I_0\, e^{-\mu x} \quad , \tag{76}$$

where μ is mass attenuation coefficient. μ is related to the photon cross sections σ_i by

$$\mu = \frac{N_A}{A} \sum_{i=1}^{3} \sigma_i \quad . \tag{77}$$

Photoelectric Effect

Atomic electrons can absorb the energy of a photon completely

$$\gamma + \text{atom} \to \text{atom}^+ + e^- \quad . \tag{78}$$

The cross section for absorption of a photon of energy E_γ is particularly large in the K-shell (80% of the total cross section). The total cross section for photon absorption in the K-shell is

$$\sigma_{\text{Photo}}^K = \left(\frac{32}{\varepsilon^7}\right)^{1/2} \alpha^4 Z^5 \sigma_{\text{Thomson}} [\text{cm}^2/\text{atom}] \quad , \tag{79}$$

where $\varepsilon = E_\gamma/m_e c^2$, and $\sigma_{\text{Thomson}} = \frac{8}{3}\pi r_e^2 = 665$ mbarn is the cross section for Thomson scattering. For high energies the energy dependence becomes softer

$$\sigma_{\text{Photo}}^K = 4\pi r_e^2 Z^5 \alpha^4 \cdot \frac{1}{\varepsilon} \quad . \tag{80}$$

The photoelectric cross section has sharp discontinuities when E_γ coincides with the binding energy of atomic shells. As a consequence of a photoabsorption in the K-shell characteristic X-rays or Auger electrons are emitted [2].

Compton Scattering

The Compton effect describes the scattering of photons off quasi-free atomic electrons

$$\gamma + e \to \gamma' + e' \quad . \tag{81}$$

The cross section for this process, given by the Klein-Nishina formula, can be approximated at high energies by

$$\sigma_c \propto \frac{\ln \varepsilon}{\varepsilon} \cdot Z \tag{82}$$

where Z is the number of electrons in the target atom. From energy and momentum conservation one can derive the ratio of scattered (E'_γ) to incident photon energy (E_γ)

$$\frac{E'_\gamma}{E_\gamma} = \frac{1}{1 + \varepsilon(1 - \cos\Theta_\gamma)} \quad , \tag{83}$$

where Θ_γ is the scattering angle of the photon with respect to its original direction.

For backscattering ($\Theta_\gamma = \pi$) the energy transfer to the electron E_{kin} reaches a maximum value

$$E_{\text{kin}}^{\max} = \frac{2\varepsilon^2}{1 + 2\varepsilon} m_e c^2 \quad , \tag{84}$$

which, in the extreme case ($\varepsilon \gg 1$), equals E_γ.

In Compton scattering only a fraction of the photon energy is transferred to the electron. Therefore, one defines an energy scattering cross section

$$\sigma_{cs} = \frac{E'_\gamma}{E_\gamma} \sigma_c \tag{85}$$

and an energy absorption cross section

$$\sigma_{ca} = \sigma_c - \sigma_{cs} = \sigma_c \frac{E_{\text{kin}}}{E_\gamma} \quad . \tag{86}$$

At accelerators and in astrophysics also the process of inverse Compton scattering is of importance [2].

Pair Production

The production of an electron-positron pair in the Coulomb field of a nucleus requires a certain minimum energy

$$E_\gamma \geq 2m_e c^2 + \frac{2m_e^2 c^2}{m_{\text{nucleus}}} \quad . \tag{87}$$

Since for all practical cases $m_{\text{nucleus}} \gg m_e$, one has effectively $E_\gamma \geq 2m_e c^2$.

The total cross section in the case of complete screening $\left(\varepsilon \gg \frac{1}{\alpha Z^{1/3}}\right)$; i.e. at reasonably high energies ($E_\gamma \gg 20\,\text{MeV}$), is

$$\sigma_{\text{pair}} = 4\alpha r_e^2 Z^2 \left(\frac{7}{9} \ln \frac{183}{Z^{1/3}} - \frac{1}{54}\right) \quad [\text{cm}^2/\text{atom}] \quad . \tag{88}$$

Neglecting the small additive term 1/54 in eq. 88 one can rewrite, using eq. 60 and eq. 63,

$$\sigma_{\text{pair}} = \frac{7}{9} \frac{A}{N_A} \cdot \frac{1}{X_0} \quad . \tag{89}$$

The partition of the energy to the electron and positron is symmetric at low energies ($E_\gamma \ll 50\,\text{MeV}$) and increasingly asymmetric at high energies ($E_\gamma > 1\,\text{GeV}$) [2].

Mass-Attenuation Coefficients

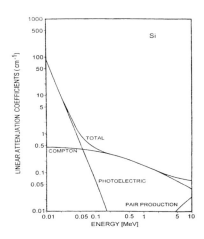

FIGURE 18. Mass attenuation coefficients for photon interactions in silicon [23]

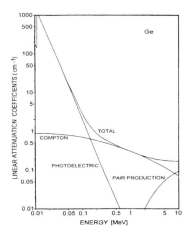

FIGURE 19. Mass attenuation coefficients for photon interactions in germanium [23]

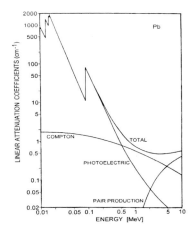

FIGURE 20. Mass attenuation coefficients for photon interactions in lead [23]

The mass-attenuation coefficients for photon interactions are shown in figures 18-20 for silicon, germanium and lead [23]. The photoelectric effect dominates at low energies ($E_\gamma < 100\,\text{keV}$). Superimposed on the continuous photoelectric attenuation coefficient are absorption edges characteristic of the absorber material. Pair production dominates at high energies ($> 10\,\text{MeV}$). In the intermediate region Compton scattering prevails.

INTERACTION OF NEUTRONS

In the same way as photons are detected via their interactions also neutrons have to be measured indirectly. Depending on the neutron energy various reactions can be considered which produce charged particles which are then detected via their ionization or scintillation [2].

a) Low energies ($< 20\,\text{MeV}$)

$$\begin{aligned} n + {}^6\text{Li} &\to \alpha + {}^3\text{H} \\ n + {}^{10}\text{B} &\to \alpha + {}^7\text{Li} \\ n + {}^3\text{He} &\to p + {}^3\text{H} \\ n + p &\to n + p \end{aligned} \tag{90}$$

The conversion material can be a component of a scintillator (e.g. LiI (Tl)), a thin layer of material in front of the sensitive volume of a gaseous detector (boron layer), or an admixture to the counting gas of a proportional counter (BF_3, ^{3}He, or protons in CH_4).

b) Medium energies ($20\,\text{MeV} \leq E_{\text{kin}} \leq 1\,\text{GeV}$)
The (n,p)-recoil reaction can be used for neutron detection in detectors which contain many quasi-free protons in their sensitive volume (e.g. hydrocarbons).

c) High energies ($E < 1\,\text{GeV}$)
Neutrons of high energy initiate hadron cascades in inelastic interactions which are easy to identify in hadron calorimeters.

Neutrons are detected with relatively high efficiency at very low energies. Therefore, it is often useful to moderate neutrons with substances containing many protons, because neutrons can transfer a large amount of energy to collision partners of the same mass. Figure 21 shows the cross sections for neutron induced reactions in the energy range from $1\,\text{eV}$ to $1\,\text{MeV}$ [2]. In some fields of application, like in radiation protection at nuclear reactors, it is of importance to know the energy of fission neutrons, because the relative biological effectiveness depends on it. This can e.g. be achieved with a stack of plastic detectors interleaved with foils of materials with different threshold energies for neutron conversion [22].

INTERACTIONS OF NEUTRINOS

Neutrinos are very difficult to detect. Depending on the neutrino flavor the following inverse beta decay like interactions can be considered:

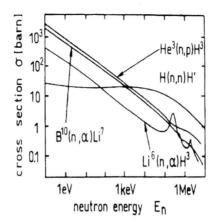

FIGURE 21. Cross sections for neutron induced reactions [2]
Reprinted with the permission of Cambridge University Press.

$$\nu_e + n \to p + e^-$$
$$\bar{\nu}_e + p \to n + e^+$$
$$\nu_\mu + n \to p + \mu^-$$
$$\bar{\nu}_\mu + p \to n + \mu^+$$
$$\nu_\tau + n \to p + \tau^-$$
$$\bar{\nu}_\tau + p \to n + \tau^+$$

The cross section for ν_e-detection in the MeV-range can be estimated as [24]

$$\sigma(\nu_e N) = \frac{4}{\pi} \cdot 10^{-10} \left(\frac{\hbar p}{(m_p c)^2} \right)^2$$
$$= 6.4 \cdot 10^{-44} \text{cm}^2 \text{ for } 1 \text{ MeV} \quad . \tag{91}$$

This means that the interaction probability of e.g. solar neutrinos in a water Cherenkov counter of $d = 100$ meter thickness is only

$$\phi = \sigma \cdot N_A \cdot d = 3.8 \cdot 10^{-16} \quad . \tag{92}$$

Since the coupling constant of weak interactions has a dimension of $1/\text{GeV}^2$, the neutrino cross section must rise at high energies like the square of the center-of-mass energy. For fixed target experiments we can parametrize

$$\sigma(\nu N) = 0.67 \cdot 10^{-38} E_\nu [\text{GeV}] \quad \text{cm}^2/\text{nucleon}$$
$$\sigma(\bar{\nu} N) = 0.34 \cdot 10^{-38} E_\nu [\text{GeV}] \quad \text{cm}^2/\text{nucleon} \tag{93}$$

This shows that even at 100 GeV the neutrino cross section is lower by 11 orders of magnitude compared to the total proton-proton cross section.

ELECTROMAGNETIC CASCADES

The development of cascades induced by electrons, positrons or photons is governed by bremsstrahlung of electrons and pair production of photons. Secondary particle production continues until photons fall below the pair production threshold, and energy losses of electrons other than bremsstrahlung start to dominate: the shower multiplicity decays exponentially.

Already a very simple model can describe the main features of particle multiplication in electromagnetic cascades (figure 22): A photon of energy E_0 starts the cascade by producing an e^+e^--pair after one radiation length. Assuming that the energy is shared symmetrically between the particles at each multiplication step, one gets at the depth t

$$N(t) = 2^t \tag{94}$$

particles with energy

$$E(t) = E_0 \cdot 2^{-t} \quad . \tag{95}$$

The multiplication continues until the electrons fall below the critical energy E_c

$$E_c = E_0 \cdot 2^{-t_{\max}} \quad . \tag{96}$$

From then on ($t > t_{\max}$) the shower particles are only absorbed. The position of the shower maximum is obtained from eq. 96

$$t_{\max} = \frac{\ln E_0/E_c}{\ln 2} \propto \ln E_0 \quad . \tag{97}$$

The total number of shower particles is

$$S = \sum_{t=0}^{t_{\max}} N(t) = \sum 2^t = 2^{t_{\max}+1} - 1 \approx 2^{t_{\max}+1}$$

$$S = 2 \cdot 2^{t_{\max}} = 2 \cdot \frac{E_0}{E_c} \propto E_0 \quad . \tag{98}$$

If the shower particles are sampled in steps t measured in units of X_0, the total track length is obtained as

$$S^* = \frac{S}{t} = 2\frac{E_0}{E_c} \cdot \frac{1}{t} \quad , \tag{99}$$

which leads to an energy resolution of

$$\frac{\sigma}{E_0} = \frac{\sqrt{S^*}}{S^*} = \frac{\sqrt{t}}{\sqrt{2E_0/E_c}} \propto \frac{\sqrt{t}}{\sqrt{E_0}} \quad . \tag{100}$$

In a more realistic description the longitudinal development of the electron shower can be approximated by [6]

$$\frac{dE}{dt} = \text{const} \cdot t^a \cdot e^{-bt} \quad , \tag{101}$$

where a, b are fit parameters.

Figure 23 shows a muon induced electromagnetic cascade in a multiplate cloud chamber [25]. The longitudinal development of electron cascades in a streamer tube calorimeter is shown in figure 24, [26].

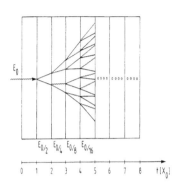

FIGURE 22. Simple model of particle multiplication in an electromagnetic cascade

FIGURE 23. Cloud chamber photograph of a muon induced electromagnetic cascade [25]
Reprinted with the permission of Taylor and Francis.

The lateral spread of an electromagnetic shower is mainly caused by multiple scattering. It is described by the Molière radius

$$R_m = \frac{21\,\text{MeV}}{E_c} X_0 \, [\text{g/cm}^2] \quad . \tag{102}$$

95% of the shower energy in a homogeneous calorimeter is contained in a cylinder of radius $2R_m$ around the shower axis.

Figure 25 demonstrates the interplay of the longitudinal and lateral development of an electromagnetic shower [2].

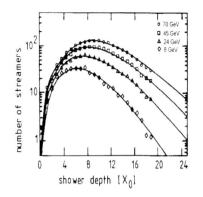 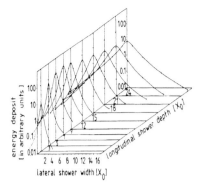

FIGURE 24. Longitudinal development of electron cascades in a streamer-tube calorimeter [2]

Reprinted with the permission of Cambridge University Press.

FIGURE 25. Sketch of the longitudinal and lateral development of an electromagnetic cascade in a homogeneous absorber [2]

Reprinted with the permission of Cambridge University Press.

HADRON CASCADES

The longitudinal development of electromagnetic cascades is characterized by the radiation length X_0 and their lateral width is determined by multiple scattering. In contrast to this, hadron showers are governed in their longitudinal structure by the nuclear interaction length λ and by transverse momenta of secondary particles as far as lateral width is concerned. Since for most materials $\lambda \gg X_0$, and $\langle p_T^{\text{interaction}} \rangle \gg \langle p_T^{\text{multiple scattering}} \rangle$ hadron showers are longer and wider.

Part of the energy of the incident hadron is spent to break up nuclear bonds. This fraction of the energy is invisible in hadron calorimeters. Further energy is lost by escaping particles like neutrinos and muons as a result of hadron decays. Since the fraction of lost binding energy and escaping particles fluctuates considerably, the energy resolution of hadron calorimeters is systematically inferior to electron calorimeters.

The longitudinal development of pion induced hadron cascades is plotted in figure 26 [27]. Figure 27 shows a comparison between proton and photon induced cascades in the earth's atmosphere [28]. The punch-through and sail-through probabilities for pions in iron are described in figure 28 (after [29]).

The different response of calorimeters to electrons and hadrons is an undesireable feature for the energy measurement of jets of unknown particle composition. By appropriate compensation techniques, however, the electron to hadron response can be equalized.

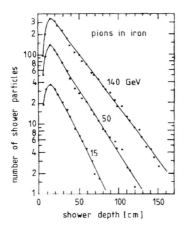
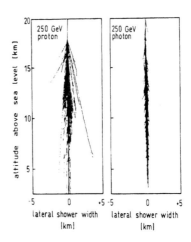

FIGURE 26. Longitudinal development of pion induced hadron cascades [27]

FIGURE 27. Comparison of the structure of photon and proton induced showers in the Earth's atmosphere [28]

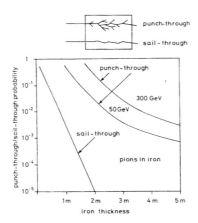

FIGURE 28. Punch-through and sail-through probabilities for pions in iron (after [29])

CONCLUSION

Basic physical principles can be used to identify all kinds of elementary particles and nuclei. The precise measurement of the particle composition in high energy physics experiments at accelerators and in cosmic rays is essential for the insight into the underlying physics processes. This is an important ingredient for the progress in the fields of elementary particles and astrophysics aiming at the unification of forces and the understanding of the evolution of the universe.

ACKNOWLEDGEMENTS

The author thanks Mrs. U. Bender (text), C. Hauke (figures) and Mr. G. Prange (layout) for their help in preparing the manuscript.

REFERENCES

1. K. Kleinknecht, *Detectors for Particle Radiation*, Cambridge University Press 1986
2. C. Grupen, *Particle Detectors*, Cambridge University Press 1996
3. R. Fernow, *Introduction to Experimental Particle Physics*, Cambridge University Press 1989
4. W.R. Leo, *Techniques for Nuclear and Particle Physics Experiments*, Springer, Berlin 1987
5. B. Rossi, *High Energy Particles*, Prentice-Hall (1952)
6. Particle Data Group, R.M. Barnett et al., *Phys.Rev.* **D54** 1 (1996)
7. W.W.M. Allison, P.R.S. Wright, *Oxford Univ. Preprint* **35/83** (1983) and W.W.M. Allison, J.H. Cobb, *Ann.Rev.Nucl.Sci.* **30** 253 (1980)
8. C. Grupen, *Ph.D. Thesis*, University of Kiel 1970
9. OPAL Collaboration, R. Akers et al., *CERN-PPE* **95-021** (1995) and *Z.Phys.* **C67** 203 (1995)
10. D.H. Wilkinson, *Nucl.Instr.Meth.* **A 383** 513 (1996)
11. L.D. Landau, *J.Exp.Phys. (USSR)* **8** 201 (1944)
12. S. Behrens, A.C. Melissinos, *Univ. of Rochester Preprint* **UR-776** (1981)
13. K. Affholderbach et al., *ALEPH-Note* **96-146**, *Physics* **96-133** (1996)
14. OPAL-Collaboration, R. Akers et al., *CERN-PPE* **94-49** (1994) and *Z.Phys.* **C63** 181 (1994)
15. R.C. Fernow, *Brookhaven Nat.Lab. Preprint* **BNL-42114** (1988)
16. J. Birks, *Theory and Practice of Scintillation Counting*, MacMillan 1964
17. H.J. Crawford et al., *Nucl.Instr.Meth.* **A256** 47 (1987)
18. S. Paul, *CERN-PPE* **91-199** (1991)
19. C.W. Fabjan, H.G. Fischer, *CERN-EP* **80-27** (1980)
20. P.S. Auchincloss et al., *Nucl.Instr.Meth.* **A343** 463 (1994)
21. C. Grupen, *Fortschr. der Physik* **23** 127 (1976)

22. E. Sauter, *Grundlagen des Strahlenschutzes*, Thiemig, München 1982
23. Harshaw Chemical Company (H. Lentz, L. van Gelderen, Th. Courbois) 1969
24. D.H. Perkins *Introduction to High Energy Physics*, Addison-Wesley, 1986
25. G.D. Rochester, K.E. Turver,*Contemp. Physics* **22** 425 (1981)
26. R. Baumgart et al., *Nucl.Instr.Meth.* **A256** 254 (1987)
27. M. Holder et al., *Nucl.Instr.Meth.* **151** 69 (1978)
28. T.C. Weekes, *Phys.Rep.* **160** 1 (1988)
29. M. Aalste et al., *Z.Phys.* **C60** 1 (1993)
30. W. Lohmann et al., *CERN* **85-03** (1985)

DETECTOR SYSTEMS FOR FUTURE HEP EXPERIMENTS

Aurore Savoy-Navarro
LPNHE-Universites de Paris 6&7, IN2P3-CNRS-France

1 INTRODUCTION

In HEP, the third millenium will start with, as a real challenge, *to go beyond the standard model*. To be able to discover this possible new world, there is a world-wide effort to build a new machine, the Large Hadron Collider (LHC), at CERN and the associated experiments. In order to face the harsh LHC environment and the very demanding physics issues at these energies in pp collisions, people have been working hard since the beginning of the nineties to develop new detector systems. Important technological advances have been achieved.

Besides the LHC and just at the start of the year 2000, the TEVATRON at FNAL will run at 2 TeV, with a luminosity of 2×10^{32} cm^{-2} s^{-1}, reaching at least 5×10^{32} cm^{-2} s^{-1} in 2003. The bunch crossing time will be shortened from 3.5 μs (the value at the end of the RUN I) to 396 ns and then to 132 ns in the RUN II. These TeVII and TeV 33 new phases of the TEVATRON require a substantial upgrade of the detector systems for both the CDF and D0 experiments.

Finally, and to complete the game, the next millenium will see the "comeback" of the "sky-experiments" renamed astroparticle experiments. These also demand large and challenging detection systems. It will not be possible to cover this topic in these lectures[1].

Multipurpose experiments at colliders must be able to achieve simultaneous, reliable-high-efficiency, identification and measurement of ALL the objects that make up an "event". Objects mean: leptons, jets, "ν-like", i.e. missing

---

[1] *An example of future astroparticle experiment, the Auger Observatory, has been presented in one lecture given at this School.*

energy, charged tracks, secondary vertices, etc., all measured with high precision implying understood calibrations, high resolutions and low fake rates. These detectors have therefore to be 4π hermetic and to comprise the following different pieces or detector systems: *the tracking system, the μ-detector system, the calorimetry*. An impressive variety of different detector techniques are used in each case and the purpose here is not to describe them in detail [2]. These lectures will concentrate on some striking points that show the evolution and at the same time the most appealing features of each detector system. The emphasis is put also on the aspects that are real new technical challenges and even that lead to important steps forward in the conception of these devices.

The course will be organized in three sections. The first of these sections is devoted to the muon-detector system. Two points of view are discussed: the ATLAS and the CMS ones. The type and the role of the magnet are the focus in this section, as they determine the architecture, the functioning and the performances of the muon-detector system. The second of these sections presents the tracking systems. After a brief description of the ATLAS, CMS and CDF tracking systems, this section concentrates on the impressive advances driven by the physics needs, which have been achieved in the CDF experiment. They concern the introduction, for the first time in a pp collider environment, of a highly performing microvertex detector and of the insertion of the tracking information in the full trigger architecture, including the level 1. The use of the microvertex is of primary importance for the b-tagging in the LEP experiments and it has been instrumental in the discovery of the top quark by CDF. The introduction of the tracking system in the trigger architecture of CDF, including level 1 will allow this experiment to strongly refine the physics selection already at this early stage in the triggering. Besides, CDF is building a processor for the level 2, which will include the microvertex information. This will allow the incorporation of the b-tagging in the triggering system. Both types of tracking information will be very powerful tools to record samples of data fitting well with the physics to be studied.

The last section will describe the calorimetry, and in particular two types of e.m. calorimeters: the liquid-argon calorimetry in ATLAS and the crystal e.m. calorimetry in CMS. Of these two techniques, one is well established, but it will have to be pushed to its limits to be adapted to the LHC requirements, whereas the other one is more innovative. Both will be shown to be technically very challenging, when having to operate in the LHC environment.

---

[2] *The reader is referred to the lectures given at this School, on many different topics such as multiwire chambers, silicium detectors, MSGCs, front-end electronics, data acquisition systems, scintillating fibre tracking, and so on.*

2 THE MUON-DETECTOR SYSTEM

The muon-detector system must fulfil three basic tasks: *the muon identification, the muon momentum measurement, and the triggering on muons*. The tools to achieve these tasks are: the muon chambers (fast ones for triggering and precise ones for momentum measurement), plus a magnet for momentum measurement and, of course, enough material (absorber) to ensure the muon identification.

Two different *"philosophies"* have been developed by the LHC experiments for the μ-detection system. In the CMS (Compact Muon Solenoid) experiment, the muon detection function is more integrated as a whole in the overall detector architecture, whereas in the ATLAS (Air Toroidal LHC Apparatus), the muon-detection function is rather autonomous. This is due to the use of a dedicated air-core toroidal magnet in ATLAS, whereas in CMS, a 4 teslas solenoid serves as magnet for both the central tracking system and for the μ-detector (see Figs. 1 and 2).

In this section are discussed:

- *The magnets:* the 4 T solenoidal magnet in CMS and the 2 T solenoid plus the air-core toroid of ATLAS.

- *The muon chambers:* the trigger chambers and the muon momentum measurement chambers.

- *The muon spectrometer of ATLAS:* as an example of muon detector system.

2.1 *MAGNETS*

Magnets are crucial devices for detector systems that aim to measure the momentum of charged particles, i.e. the tracking systems or the muon-detector systems. ATLAS and CMS experiments have different approaches when considering the momentum measurement of charged particles, including the muons. CMS uses a unique device to perform the momentum measurement of both the muons and the central tracking, namely a *long 4 T solenoid* that covers up to ± 2.4 in pseudorapidity η range. In ATLAS, the μ-momentum measurement is achieved almost independently by the air-core toroidal magnet plus the dedicated muon chambers, without a real need for the central tracking

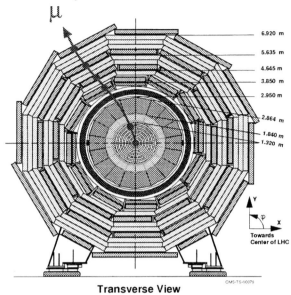

Figure 1: Transverse view of the CMS experiment.

system that includes a 2 T solenoid.

The primordial importance of the magnets in the overall design of these experiments is such that the name of the experiments reflects the type of magnet they use: CMS for Compact Muon Spectrometer and ATLAS for Air Toroidal LHC ApparatuS. Both the air-core toroid of ATLAS and the 4 T solenoid of CMS are impressive and very challenging devices.

The magnets are made of two parts: the magnet yoke and the cold mass, which are now described.

2.1.1 The magnet yoke and the field map

The magnetic forces generated in the magnet yoke are characterized by the flux distribution lines recorded as field maps. This flux distribution depends on the geometrical arrangement of the different parts constituting the magnet

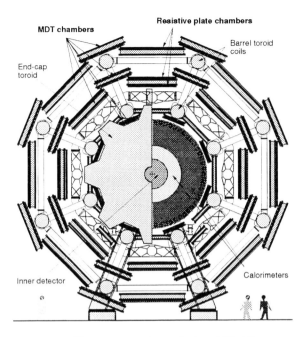

Figure 2: Transverse view of the ATLAS muon spectrometer.

yoke.

In the case of the solenoids of both ATLAS and CMS the situation is quite different. In the CMS solenoid the return yoke covers the space at both ends of the magnets. This ensures a uniform field, i.e. the flux distribution lines stay parallel in the overall η-range covered by the solenoid, as shown in Fig. 3.

On the contrary, the 2 T solenoid of ATLAS has almost no return yoke at both ends. This leads to a non-uniformity of the flux distribution lines already at η around ± 0.8, as shown in Fig. 4.

The field integral length inside the tracking volume, defined by:

$$I_1 = \int B \sin\theta (d\vec{l} \cdot \vec{B}) dl$$
$$I_2 = \int \int B \sin\theta (d\vec{l} \cdot \vec{B}) dl dr$$

are the relevant parameters for the track-momentum measurement. A study

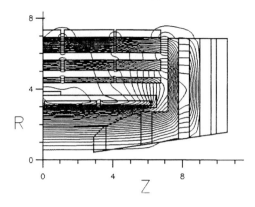

Figure 3: The magnetic flux line distribution inside and outside the CMS solenoid.

of these parameters emphasizes this point. A comparison has been performed between the ATLAS solenoid and an ideal solenoid of the same size, based on the two quantities I_1 and I_2 as a function of the η extent of the magnet. I_1 and I_2 stay at a constant value, namely 2.1 Tesla m for I_1 and 1.1 Tesla m^2 for I_2, up to $\eta = 1.6$ in the case of the ideal solenoid, whereas it keeps these values only up to $\eta = 0.8$ in the case of the actual ATLAS solenoid. However I_2 drops slowly from $\eta = 0.8$ to $\eta = 1.6$ for the ATLAS solenoid. From $\eta = 1.6$ to $\eta = 2.8$, I_2 falls quickly down to 0.1 Tesla m^2 for both the actual and ideal cases. Another way to compare the ATLAS solenoid with its ideal case is by using the ratio I_2(ATLAS) $/I_2$(ideal) (see Fig. 5). This parameter measures the capability of momentum measurement with respect to the ideal case. This ratio stays at a value around 1 up to $\eta = 0.8$. From $\eta = 0.8$ to $\eta = 1.6$, this ratio drops down to 0.9, and then it remains constant at this value, from $\eta = 1.6$ to $\eta = 3.2$. Thus, at $\eta = 0.8$, the ATLAS solenoid diverges from the ideal case, which would be able to cover the full η-range of this magnet (namely, to $\eta = 1.6$).

However, it should be noted that this does not affect so much the momentum measurement in ATLAS, as the central tracking extends essentially up to ± 0.8 in η and the μ-momentum measurement in ATLAS will be achieved by the air-core toroid. Moreover, for the region from 0.8 to 1.6 in η, the detailed field map allow us to correct for a non-uniformity of the field in the off-line estimate of the charged particle momentum.

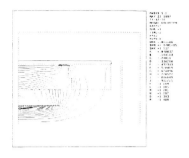
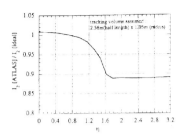

Figure 4: The magnetic flux line distribution inside and outside the ATLAS solenoid.

Figure 5: Comparison between the ideal and the ATLAS cases of the capability of momentum measurement: $I_2(\text{ATLAS}) / I_2(\text{ideal})$.

In the case of the ATLAS μ spectrometer, the μ-momentum measurement is based on the magnetic deflection of muon tracks in a system of three large supraconducting air-core toroidal magnets (see Fig. 6).

In $|\eta| \leq 1.0$, the magnetic bending is provided by a large barrel magnet constructed from 8 coils surrounding the hadron calorimeter. For $1.4 \leq \eta \leq 2.7$, μ tracks are bent in two smaller end-cap magnets. In $1.0 \leq \eta \leq 1.4$, i.e. the transition region, the $\vec{B}$-field is provided by a combination of the $\vec{B}$-field of the barrel magnet and the $\vec{B}$-field of the end-cap magnet. This gives origin to a non-uniform field in this transition region as shown in Figs. 7 and 8. . This magnet configuration provides a field that is mostly orthogonal to the μ trajectories, while minimizing the degradation of the resolution due to the multiple scattering.

2.1.2 The magnet cold mass

The cold mass is the part of the magnet that operates at LHe temperature. It consists essentially of the superconducting winding and the quench-back external cylinder to which the LHe cooling is attached.

a) The supraconducting coil

First let's have a reminder about supraconductors. A perfect conductor

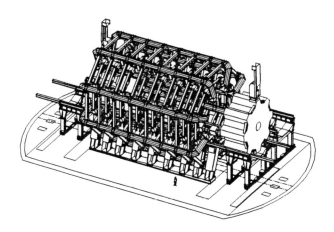

Figure 6: General view of the ATLAS muon spectrometer with the air-core toroid.

($R{=}0$) cannot have variations of flux of magnetic inductance, i.e. $dB/dt = 0$, so B is a constant. *A supraconductor is a perfect conductor but in addition it satisfies: B=0.* The supraconductor ejects the magnetic flux, its susceptibility $\chi = 1$. It behaves as a perfect diamagnetic.

A keyword for supraconductor is **stability**. A supraconductor cable, even well below the critical conditions of temperature, field or heat by Joule effect (T_c, B_c, J_c), can quench locally or over its full length under the effect of **mechanical or thermal or electromagnetic** perturbations. The risk is the dissipation of heat by Joule effect, leading to irreversible damage of the cable by fusion of part of this cable.

To remedy these instabilities, there are a certain number of tricks to play with the structure of the conductor itself, and thus various ways to improve the conductors. Let us discuss a few important ones:

- *Importance of increasing the dimension of the cable*
  Initially, the diameter of a suprafilament is very small (a few tens of μm). In order to increase its section, a supraconducting cable is made of several strands, each strand being made of many suprafilaments. For instance, in the ATLAS solenoid, the diameter of the suprafilament is 20 μm, the strands are made of 2050 filaments and have a diameter of 1.22 mm. The corresponding supraconductor cable has a rectangular section of 2.2 x 7.4 mm^2 and includes 12 strands (see Fig. 9).

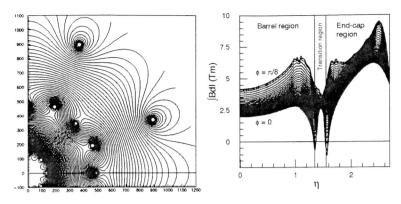

Figure 7: Magnetic flux line distribution for the ATLAS air core toroid

Figure 8: Magnetic flux line distribution for the ATLAS air core toroid, in the transition region between barrel and end caps.

- *Cryostability*

  The need for stability in temperature requires cooling of the supraconductor. The heat dissipated by Joule effect must be evacuated by the cooling liquid, before the supraconductor reaches the critical temperature T_c.

  Moreover the dissipated energy being very large, it is necessary to surround the supraconductor with a low-resistivity material (Cu and Al). The current density is therefore reduced and so the heat dissipation by Joule effect. A multifilamentary wire is therefore made of supraconduc-

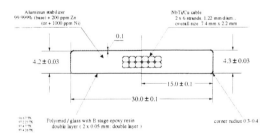

Figure 9: Design of aluminium stabilized conductor as done for the ATLAS solenoid.

tor strands with interspaces filled with Al or Cu stabilizer, as sketched in Fig. 9.
An increase in the surface of exchange with the cryofluid allows a good evacuation of the heat. Another way to ensure a good cryostability is by circulating the cryofluid by "forced convection", inside the cable.

- *Thickness of the supraconducting filament*
  By minimizing the thickness of the supraconducting material, the risk of quench is decreased. Therefore supraconductors are made of very thin filaments. For instance, in the case of the NbTi filament, the risk is minimized under 6 T, if the diameter of the filament is less than 100 μm. In ATLAS, the filaments are 20–30 μm, for each type of magnet; in CMS the solenoid has 50 μm diameter filaments.

- *Dynamical stability*
  The Cu or Al stabilizers around the filaments ensure this stability. Indeed if a quench occurs, there is a redistribution of the field and the Al or Cu material detects this field variation and tends to oppose it.

- *Losses in supraconductors*
  In the case of a variation of the magnetic field or of the current, there are losses in the supraconductors by Foucault currents in between the filaments as sketched in Fig. 10.
  By twisting the filaments with a regular twist pitch, this effect is cancelled (see Fig. 10). For example the twist pitch in the ATLAS solenoid is of 20 mm while it is of 50 mm in the CMS solenoid.

Another important point for the supraconductors is the stored energy and the protection against quench. In case of a brutal transition of the bobine from $R=0$ to $R \neq 0$, i.e. **a quench**, the large current that circulates in the supraconductor must be discharged. This is achieved as in the circuit of Fig. 11, by using a resistor r that has to have a very low value (in the CMS quench system the r-value is of 0.050 mΩ). The mass of the resistor must be large to absorb all the energy. For example, in the ATLAS toroid, the energy reaches 1000 MJ so the mass of the resistor is 8 tons!

Another parameter is the discharge time $\tau = \frac{L}{r}$. It must not be too short in order to avoid breakdowns. On the other hand, the bobine, becoming resistive, has to be protected against an increase of heat by Joule effect. The Al

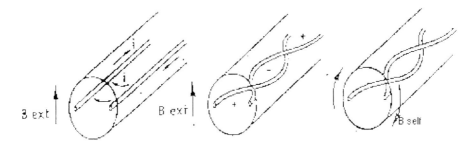

Figure 10: Losses in the supraconductors by Foucault's currents in between the filaments and after twisting the filament.

stabilizer around the supraconductive bobine provides additional matter that allows for a better repartition of the current.

The supraconductor coils of ATLAS and CMS are now described. In Table 1 are summarized the main magnetic and overall geometrical parameters of each of the three magnets. There are striking differences between the now "conventional" 2 T solenoid of the inner tracking in ATLAS and the 4 T solenoid of CMS. The total stored energy is 44.2 MJ in the ATLAS solenoid compared to 2.7 GJ in the CMS solenoid. The operating current is 19.5 kA in CMS compared to 7.6 kA in ATLAS. For the air-core toroid (barrel device) these two parameters are 1.5 GJ for the stored energy and 20 kA as operating current.

A useful parameter to define and characterize the relative transparency of superconducting coils, is the ratio of the stored energy of the coil to the cold mass = E/M. The energy E is given by: $E = \frac{1}{2} \int (\vec{H} \cdot \vec{B}) dv = \frac{1}{2\mu_0} B_0^2 V$, where V is the cylinder volume of the solenoid.

So $E/M = 44$ MJ $/2 \pi r l \rho$, for the ATLAS solenoid, with ρ = effective density of the coil, l = coil length and r = coil radius. Therefore, the ATLAS coil thickness is fixed by this formula to a value of 44 mm. This parameter E/M is of the order of 4 to 6 in most of the known solenoids (DELPHI, H1, ALEPH and CDF). In the case of the ATLAS solenoid it will be of the order of 7 to 8; and for the CMS solenoid it will be above 12.

The CMS solenoid is 12.4 m length, with an external diameter (without cooling tubes) of 6.98 m and an internal diameter of 6.36 m and a total mass of 220 tons (without the supports). The ATLAS solenoid is 5.3 m length and 1.22 m inner radius and a total cold mass of 5.4 tons. The ATLAS air-core

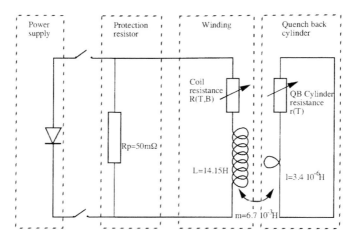

Figure 11: Quench circuitry for the CMS 4 T solenoid.

toroid has an inner bore of 9.4 m, 19.5 m as outer diameter. It has 26 m of axial length and the total weight, excluding the external loads, is 1000 tons.

b) The coil cryogenic system

Another aspect of the magnet is the coil cryogenic system. As an example, we describe here that of the CMS solenoid. It is composed of two parts: the cryogenic plant that stands outside the coil vacuum vessel and the coil cryogenic system that sits inside. The cryogenic plant at the surface includes the surface cryogenic equipment, helium compressors and storage. The coil cryogenic system is composed of the 220 ton superconducting coil working at the liquid helium temperature (4.4 K), the thermal shield system cooled by He to 60–80 K and two chimneys crossing the Fe yoke.

A fair amount of engineering work in the magnets is also involved in the studies of horizontal and vertical displacements that can go up to about ± 5 mm and even more, due to the applied forces. An impressive gantry crane of 2000 tons capacity will be used to lift these large and heavy pieces. Mechanical studies on these aspects, including detailed studies of the suspension system with mock-up are also part of the design work for these magnets.

Table 1: Main parameters of the three considered magnets

Main parameters	4 T CMS solenoid	2 T ATLAS solenoid	Air-core toroid	
			Barrel	End-cap
Geometrical dimensions:				
- Coil length (m)	12.4	5.3	26	5.6
- Inner radius (m)	3.2	1.2		
Mass:				
- Total cold mass (tons)	220	5.7	450	120
Electrical parameters:				
- Peak field (T)	4.6	2.6	4.2	4.4
- Current (A)	19500	7600	20000	20000
- Stored energy (MJ)	2690	44.2	1490	238
- E/M (kJ/kg)	≥ 12	8.1		

2.2 MUON CHAMBERS

The muon chambers have two main functions: one is to identify the muon and to measure its momentum; the other one is to trigger on muon tracks. Besides, as for the calorimetry, the μ detector must be **hermetic**, i.e. muons should not escape detection, since they enter also in the estimate of the total missing energy of an event. There are two types of chambers:

- *The precision chambers* that allow a good recognition of the track and the measurement of its momentum thanks to the bending of the magnet. The precision measurement of the μ-tracks are made in the *(r–z)* projection, in a direction parallel to the bending direction of the $\vec{B}$-field. Two types of chambers are used:

  – *The Monitored Drift Tubes (MDTs)* are used over a large extent of the solid angle. They exhibit a single-wire resolution of about 80 μm at a pressure of 3 bars operation.

  – *The Cathode Strip Chambers (CSCs)* are used to provide a finer granularity to cope with the demanding rates and background conditions at $|\eta| \geq 2$ and even $|\eta| \geq 1.4$ for the first station.

- *The trigger chambers* are of two kinds:
  - *The Resistive Plate Chambers (RPCs)* provide a typical space-time resolution of 1 cm × 1 ns
  - *The Thin Gap Chambers (TGCs)* are used for the end-cap regions because they better match with the demanding rate capability and spatial resolution in this case.

The μ chambers are gas chambers working in proportional or drift mode. For the theory of these detectors, I refer to the lectures given at this school on the subject. Here are summarized the main characteristics of functioning and fabrication of these detectors.

2.2.1 The precision chambers

The specifications to be fulfilled by these chambers in terms of accuracy is 50 μm per chamber to achieve $\Delta p_T/p_T$ of order 10% for a p_T of 1000 GeV.

In the central barrel the precision-chamber system are made of MDTs arranged as sketched in Fig. 12.

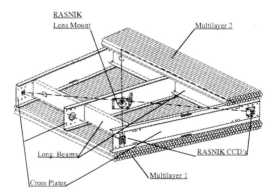

Figure 12: The monitored drift tube μ-chamber arrangement.

The MDT principle is based on the proportional drift tube, where the position is determined from the measured drift time (Fig. 13).

Figure 13: The MDT principle:schematic view of a monitored drift tube

The tube diameter is 30 mm (+ 0, − 30 μm), the wall thickness is 400 μm (± 15 μm) and the tube length is 70 to 630 cm. The operating point is characterized by a mixture of $Ar/N_2/CH_4$ (91%/4%/5%) at a pressure of 3 bars and a gas gain of 2×10^4. The performances are 500 ns maximal drift time and 80 μm resolution per tube. The MDT chambers are equipped with a *RASNIK alignment monitor* (see subsection 2.3.2).

The MDT chamber fabrication is now based on a large experience. The fabrication of each tube is realized by assembling the different components: tube, wire, end plugs and wire locators. The wiring of a single tube takes few minutes: about 40 tubes are produced per day.

The tubes are stacked individually; the layers are glued one at a time and there is on-line quality assurance, allowing to build the large number of chambers that are required (1194 chambers in total). The gas parameters are crucial to the proper MDT operation. The chamber gas volume is 800 m^3 at 3 bars. It functions in closed loop with one volume exchange every day. There is a complete volume renewal every ten days. The signal from the drift tubes is read out by ASD circuits consisting of a preamplifier, a shaper and a discriminator. It is then digitized by a TDC.

In the end-plug regions the precision chambers are made by CSCs of which the principle is sketched in Fig. 14 .The wire pitch is 2.54 mm, the strip readout pitch is 5.08 mm and the anode to cathode distance is 2.54 mm. The CSCs are multiwire proportional chambers in which the position is given from induced charge distribution in etched cathodes and the timing is obtained from the arrival time of signal on the wires.

The operating conditions are defined by a gas mixture of $Ar/CO_2/CF_4$ in the proportions (30%/50%/20%) at 1 bar gas pressure and 4×10^4 gas gain.

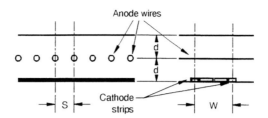

Figure 14: The CSC principle.

Contrary to the MDTs, precise gas parameters do not affect CSC operation. The chamber gas volume is about 1.1 m^3; it works in closed loop with one volume exchange every 6 hours. There is a complete volume renewal every 2.5 days.

The performances are 30 ns maximal drift time, 7 ns timing per plane and 60 μm resolution per plane.

The readout is made by an ASP circuit (i.e. amplifier, shaper and capacitive storage array), followed by an ADC that digitizes the induced charge that gives the position.

2.2.2 The trigger chambers

In the barrel region, the trigger chambers are made by RPCs. These chambers, sketched in Fig. 15 are similar to parallel-plate chambers but with highly resistive electrodes (bakelite). They operate with a gas mixture of Tetrafluoroethane-Isobutane in proportional mode. Indeed, to avoid strong rate dependence and ageing problems it is necessary to avoid streamers. They give excellent time resolution with a σ of 1.5 ns.

In the end-cap region, RPCs are replaced by TGCs. The principle of the TGC is shown in Fig. 16. It is a MWPC-like arrangement, but the anode-to-anode distance is larger than the cathode-to-anode distance. It operates with a highly quenching gas (CO_2-n-pentane), which results in no streamers. The operation in a saturated mode leads to a small sensitivity to mechanical deformations and to variations in primary ionization. The operating voltage is 3.1 ± 0.1 kV. The presence of wires permits small granularity (given by the wire distance) and high rate capabilities.

Both LHC experiments use these μ-chamber techniques. As an example,

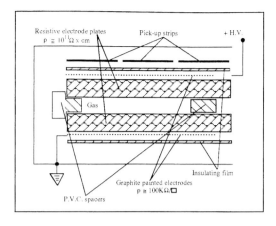

Figure 15: The RPC principle.

the ATLAS μ spectrometer is now described.

2.3 THE ATLAS MUON SPECTROMETER

The ATLAS μ spectrometer is based on an air-core toroid that we discussed in Section 2.1. Thanks to this toroid, the μ detector of ATLAS is acting as an *autonomous piece* of the detector, able to fulfil all the functions (identification, momentum measurement and triggering) by itself. In particular it does not need the central tracking, contrary to the CMS μ detector, which fully relies on the central tracker plus the 4 T solenoid to measure the μ momentum precisely.

Besides, the choice of the air-core toroid is justified by the fact that it fits within the cavern and it maximizes the coverage (for multi-μ states and hermeticity issues). As shown in Fig. 2 the projectivity is ensured by large and small sectors that overlap. The μ chambers installed in the magnet structures lead to a combined optimization. It accommodates services, supports, optical alignment corridors. It balances optimal use of the field in terms of Bl^2 with the material distribution. It minimizes the number of different chamber types.

2.3.1 Momentum measurement in the ATLAS μ spectrometer

The momentum measurement is based on using three stations, with $\sigma \sim 50$ μm ($\sigma_\theta \sim 0.4$ mrad) per station, that measure z-coordinate in the central

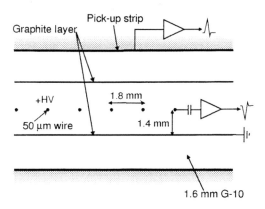

Figure 16: The TGC principle.

barrel or R- coordinate in the forward detector. The second coordinate in the "non-bending" plane is measured by the axial (resp. radial) strips in the trigger chambers with a resolution $\sigma_W \leq 5$ to 10 mm.

The pattern recognition is presently based on studies using a full GEANT simulation. It is guided by the information provided by the trigger chambers that determine the zones in which to look for muons. The way the pattern recognition is achieved using the drift tubes hits is sketched in Fig. 17. It exploits both the presence of hits and the "justified" absence of a hit due, for instance, to tube boundaries or electronics deadtime. It distinguishes hits used in the reconstruction from background hits.

The μ momentum resolution depends on various factors. One is the magnet configuration. An interesting feature of the used toroidal configuration is that it provides a flat resolution, over a large range in η, of typically 2 to 3%, up to η of 2.5, apart from the transition region around 1.5 in η where the resolution is degraded up to 5%. Other facts influencing the μ-momentum resolution are the precision chamber accuracy (intrinsic resolution, alignment), the multiple scattering in the magnet and chamber structures, the energy loss fluctuations, the momentum tails from the pattern recognition errors. Figure 18 shows how each of these contributions affects the overall resolution as a function of the transverse momentum of the μ. Moreover, in order to improve the momentum measurement one can implement the inner tracker measurement.

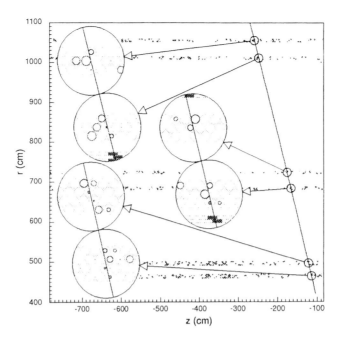

Figure 17: Pattern recognition using the drift tubes of the MDT in the ATLAS muon spectrometer.

Figures 19 and 20 show that the μ momentum, as determined by the μ spectrometer only, is quite good and that adding the inner tracking would give only slightly better results, particularly for p_T below 100 GeV; for p_T above, there should not be a significant gain. This result demonstrates that the ATLAS μ spectrometer is an autonomous system.

2.3.2 The alignment system of the ATLAS μ spectrometer

An important element in the measurement of the μ is the good alignment of the chambers. As the ATLAS spectrometer is an autonomous system, the alignment layout we are describing is a **relative positioning** of each chamber within the triplet of chambers, and not an absolute positioning, i.e. with respect to the overall tracking system. The alignment system must monitor the in-plane alignment for chamber deformations, the projective alignment for relative positions of the triplets (within a required precision of 20 μm) and the axial alignment to limit the number of projective rays (20 μm). The in-plane

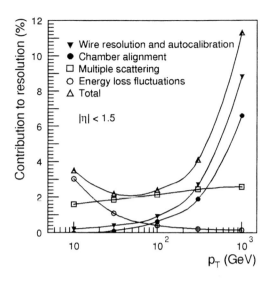

Figure 18: Contribution of various geometrical and physics parameters to the overall resolution in μ momentum (ATLAS μ spectrometer).

alignment for chamber deformations and the projective alignment for relative positions of "triplets" are done using the *RASNIK* system method (i.e. Relative Alignment System NIKhef). The system has already been used for the μ chambers of the L3 experiment at LEP.

It allows the monitoring of the relative positioning of each chamber of the triplet relative to the two others (i.e. the second one versus the first and the third chamber in the triplet). The RASNIK system alignment monitor is sketched in Figs. 21 and 22.

It is made of a CCD sensor (400 000 pixels with a pixel size of 7×7 μm^2), an arbitrarily large mask (range) infra-red light source and lens adapted to the layout. The alignment information is obtained from the image pattern recognition. The performances are about 1 μm r.m.s. for the perpendicular coordinate and about $0.025\times f$ μm r.m.s (where f is the focal distance in mm) for the parallel coordinate. The rotations around the perpendicular axis are measured with $20f$ μrad r.m.s (f in mm).

After verifying the position of chamber 2 versus the other two at each corner of the chamber, one measures track segments on each chamber and corrects

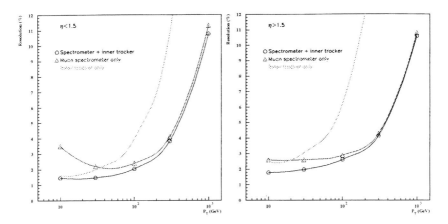

Figure 19: Performances of the ATLAS μ spectrometer for the μ momentum measurement with and without the tracking system, for $\eta \leq 1.5$.

Figure 20: Performances of the ATLAS μ spectrometer for the μ momentum measurement with and without the tracking system, for $\eta \geq 1.5$.

the track segment if there is a mispositioning of 2 with respect to 1 and 3. The needed frequency of this monitoring is not yet defined. This is under study in the test set-up installed at CERN. The first indications tend to show that one monitoring per hour should be good enough.

There is also a multipoint alignment monitor (ALMY) to measure the alignment of the chambers with respect to the beam. It consists of two-sided transparent silicon strip detectors of sensitive area 20×20 to 30×30 mm^2 (300 + 12 μm pitch), a semiconductor laser diode, single-mode optical fibres and collimator optics to obtain about 2 mm diameter Gaussian beam profile. The alignment information is obtained from the charge distribution on the strips. The beam deflections are measured (without calibration) to 10 μrad r.m.s and to 3 μrad r.m.s after calibration. The perpendicular coordinates are measured within at most 5 μm r.m.s.

System tests called DATCHA have been installed at both CERN and Saclay Laboratories. The DATCHA objectives at CERN are to gain experience with a complete sector of barrel μ spectrometer, including the MDT chambers, the alignment system, the RPC chambers, the data acquisition and detector control. Moreover this test system aims to validate the alignment concept and to study the thermal effects. The aim of the DATCHA set-up installed at Saclay is to verify the in-plane alignment system (RASNIK). Cross-plate shifts known from readings of mechanical sensors ("dial gauge" readings) are

Figure 21: The RASNIK alignment monitor arrangement for the MDT chambers.

monitored with the RASNIK system. In-plane deviations of ±1 mm are corrected by this monitoring system to ± 10 μm. Thus the monitoring system meets the specifications.

2.3.3 The muon level 1 trigger of the ATLAS μ spectrometer

The level 1 μ trigger requires sharp threshold and minimization of accidental triggers, mainly due to neutrons and photon backgrounds. It is therefore based on high granularity and fast chambers. The level 1 trigger concept is shown in Fig. 23.

The trigger decision is derived from three trigger stations (RPCs in the barrel and TGCs in the end-cap regions). The trigger is based on a coincidence, in both projections, between a strip (or a group of wires) in the first station and a range of "windows" of strips or wires in the second and possibly the third region. Such coincidences define the "region of interest" of the trigger, which extends in the space $(\Delta \eta \times \Delta \phi)$ over (0.1×0.1). The low-p_T trigger (i.e. $p_T^\mu \geq 6$ GeV) requires 3 or 4 coincidences in RPC 1 and 2 (barrel)

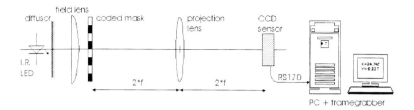

Figure 22: The RASNIK alignment monitor system for the ATLAS μ spectrometer.

or TGC2 and 3 (end-caps).

The high-p_T trigger (i.e. $p_T^\mu \geq 20$ GeV) requires 3 or 4 coincidences in RPC 1 and 2 and 1 or 2 coincidences in RPC3(in the barrel). In the end-cap it is formed by 3 or 4 coincidences in TGC2 and 3 and 2 or 3 coincidences in TGC1.

The high granularity is crucial in achieving the sharp thresholds needed for low trigger rates. Figure 24 shows the computed trigger efficiencies both in the barrel and end-cap regions for low-p_T and high-p_T muons. It indicates the particularly good (sharp) threshold of the low-p_T muon trigger in the barrel region thanks to the high granularity of the MDTs.

Located at the periphery of the experiments, the μ detector plays a crucial role as μ's are among the main actors in the events interesting for physics. The performances demanded at LHC for the μ identification and their momentum measurement require very large and precise μ-chambers and highly performing magnets. This makes μ detectors challenging detector systems not only to build but also to operate when having to face the LHC issues.

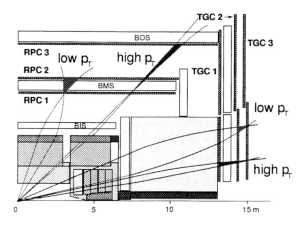

Figure 23: Principle of the muon momentum measurement and muon triggering in the ATLAS muon spectrometer.

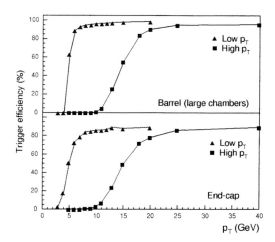

Figure 24: Trigger efficiencies in the barrel and in the end-cap ATLAS μ spectrometer.

3 TRACKING SYSTEM

The tracking system is the detector system that is situated nearest to the interaction point. It allows the detection of all the charged particles by visualizing the tracks they produce when passing through it. The main functions of this system are therefore:

- *To detect charged particles.*

- *To measure with accuracy their momentum.*
  And also to cross-check with accuracy the energy measurement in the calorimeter with the ratio E/p.

- *To do particle identification by the dE/dx measurement.*

- *To tag heavy flavours: c's, b's and top.*
  And to accurately define the coordinates of the primary vertex (interaction point) and of secondaries.

- *To be part of the full triggering architecture, including level 1.*

This last point is quite an innovative issue, where the CDF experiment at the TEVATRON is a precursor.

These lectures will particularly focus on the two last issues mentioned in the above list, after describing a few typical complete tracking systems designed for future experiments. The tracking systems of the two LHC experiments and that of the CDF experiment at the TEVATRON for RUN II will be taken as examples.

For thirty years, a large variety of tracking detector techniques have been developed. For a long time there has been an "hegemony" of the multiwire proportional chambers. Some years ago, new techniques were developed, based on silicon tracking detectors, or a sort of "hybrid" detectors that combine the silicon technique and the multiwire chamber technique as MSGCs, or also tracking detectors based on the use of scintillating fibres technique. We will now show how they are used in various experiments[3].

Typical tracking systems include a $\vec{B}$-field, i.e. a magnet and a set of tracking devices. The momentum resolution is proportional to the product Bl^2 where l is the lever arm of the tracking system. The main components of the tracking system are:

- *The microvertex near the interaction point*

- *High resolution tracking devices*

- *The magnet*

The main motivation of the tracking system is to visualize the charged tracks that correspond to the charged particles passing through this detector system. Thanks to the development of the detector techniques and of the associated readout electronics, it is now possible, even in the demanding environment of pp collisions at the LHC, to comprehend in great detail the full charged particle content of an event. Figure 25 gives, as an example, a "realistically" simulated Higgs event in the ATLAS experiment. This picture demonstrates that this detector system, will be of primary importance when looking, for instance, for low mass Higgs decaying into ZZ^* going then into 4 charged leptons (electrons or μ's).

Let us describe now a few typical tracking systems for future experiments at the LHC and at the new TEVATRON upgrade.

3.1 Examples of tracking systems at future experiments

Certainly the tracking systems that will be facing the pp collisions at the upgraded TEVATRON or *a fortiori* at the LHC will be very challenging.

---

[3]*For the detailed description of these different tracking techniques, their principle and theory of functioning, I refer to the dedicated lectures on these various subjects given at this School.*

ATLAS Barrel Inner Detector
$H \to ZZ^* \to \mu^+\mu^-e^+e^-$ (m_H = 130 GeV)

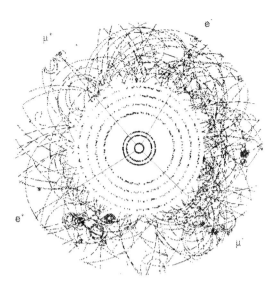

Figure 25: Picture of a simulated ZZ^* event in the ATLAS tracking system.

3.1.1 *The ATLAS tracking system*

The main parameters of this tracking system are: the magnet, i.e. the 2 T solenoid that we already described in detail in Section 2, its lever arm l of 1.05 m and its η-coverage down to η of 2.5.
As shown in Fig. 26 below, it is organized into three parts that use mainly two different tracking techniques, namely semi-conductor techniques, i.e. *pixels* and *silicium microstrips*, and *straw tubes*.

$\Longrightarrow$ *First part: the microvertex detector*

It is made of very fine granular pixels (a total of 140 million channels), which will provide resolutions of about 10 μm in $(r - \Phi)$ transverse plane and 50 μm along the z-beam axis. It will permit the pattern recognition and the reconstruction of the track origin and of the secondary decay vertices[4].

---
[4] *For details on the semi-conductor trackers, see the dedicated lectures given on the subject*

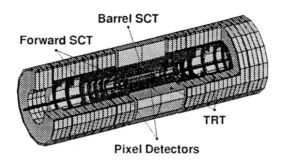

Figure 26: ATLAS Tracking system

The ATLAS microvertex is arranged in three barrel layers, situated respectively at a radius of 4.7 cm, 10.5 cm and 13.7 cm from the beam axis, and it covers ±38.5 cm in the z–direction (i.e. the beam direction). The forward disks are situated at the following z–points: 49 cm, 60.8 cm, 75.9 cm and 103.5 cm, on each side of the interaction point.

The basic component of the pixel silicium detectors is the *pixel sensor*. The *radiation environment at the LHC will cause damage to the sensor substrate*, which may eventually require that the sensors be operated partially depleted. This fact is the dominant factor in the choice of *n-type implants in an n-type substrate*. Radiation damage to the operating system will *increase the operating voltage* during the detector's lifetime. The ability to operate the sensors at high voltage without electrical breakdown or microdischarge also influences the detector design. Another important point is the ability to test the sensors before attaching the readout integrated circuits.

To build the pixel modules, pixels of dimension for instance 400 × 50 μm^2, are grouped into a tile, which is a "large" array of pixels organized in n rows × N columns. Likewise, the pixel detector is composed of modular units. The readout integrated circuits are mounted on the detector substrate, i.e. the silicium sensor, using bump bonding techniques. An additional integrated circuit for control and clock distribution and data compression is mounted on each module; flexible cables connect each module to data transmission/control circuitry, located within the detector volume (see Fig. 27).

The pixel front-end electronics should be able to cope with very small signals ($\geq$ 6000 e$^-$) to allow operation with highly irradiated and thinned detector

---

at this School.

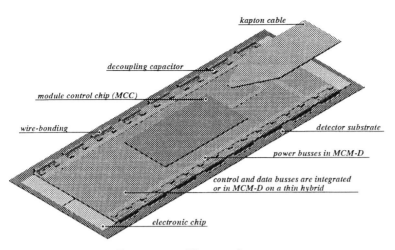

Figure 27: The pixel sensor arrangement in the ATLAS microvertex

substrates. This is possible through the state-of-the-art integrated circuit design due to low noise arising from the small (about 200 fF) individual sensor capacitance.

This part of the tracking system, namely the pixel micorvertex, is still in the conceptual phase and R & D is still in progress, as it is a very difficult technological issue.

$\Longrightarrow$ *Second part: the semi-conductor tracker (SCT)*

The microvertex is followed by the semi-conductor tracker, made of silicium strips, measuring the tracks with small stereo angle. It will achieve the precision measurement with resolutions of about 20 μm in $(r - \Phi)$ and 700 μm in the z–direction[5].

It extends from 30 cm to 52 cm with respect to the beam axis in the central barrel, made of four layers. The forward parts are made of nine layers each, situated along the z-beam direction from 83.5 cm to 279 cm from the interaction point, and its extent in radius is 56 cm for the outer part and varies from 25 cm to 44 cm in the inner part, depending on the z-position. The SCT is based upon silicon microstrip detector technology, used with success over the past 8 years in the LEP experiments and in the CDF experiment at the

---

[5] *For details on the semi-conductor trackers, see the dedicated lectures given on the subject at this School.*

TEVATRON (see next two subsections).

However the main striking points of this detector for the LHC is that it will have an active silicon area of about 63 m^2, more than 50 times that of any existing silicon vertex detector. In addition, precision mechanical and cooling techniques, used successfully in the vertex detectors, must be engineered on a cylindrical scale of diameter 1 m and length 6 m, a factor of 10 larger than before.

$\implies$ *Third part: the transition radiation tracker (TRT)*

At the outer part of the tracking system there is the precision tracking device for achieving a good pattern recognition. It is based on the *straw tube technique*[6].

A straw functions as a drift tube for measuring the position of a track that passes through it and as an ionization chamber for recording the amplitude of the signals.
The straws are made from a coated polyimide film. The bare material consists of 25 μm thick kapton film. They have a diameter of 4 mm, with a wall thickness of 72 μm. The active length in the barrel is 1.5 m and 39 cm or 55 cm in the end-caps. The anode wire positioned in the centre of the straw has a diameter of 30 μm. The ionization gas mixture is composed of 70% Xe + 20% CF_4 + 10% CO_2. This is the result of many years of dedicated studies. Electron identification capability is added by employing xenon gas to detect transition-radiation photons created in a radiator between the straws. The transition radiator material that completely surrounds the straws, consists of propylene-polyethylene fibres (see Fig. 28).

The total charge collection time is a crucial parameter for straw operation at the LHC. It is determined by the electron drift velocity; its behaviour is a function of the electric field in the straw and of the relationship to the direction of the magnetic field. The presence of a magnetic field generally leads to an increase of the total charge collection time, which depends on the relative orientation of the electric and magnetic fields. As a result of various studies, the straws will be placed either parallel to the solenoidal magnetic field in the central barrel or orthogonal to the dominant longitudinal component of this field in the end-cap.

---

[6] *For details on the TRT detector, see ATLAS Inner Detector Technical Report Volumes 1 and 2, CERN/LHCC/97-17 ATLAS TDR 4 and 5, 30 April 1997.*

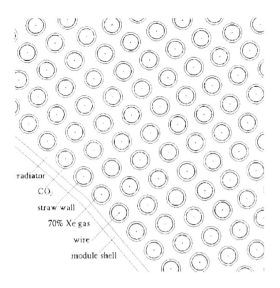

Figure 28: The straws and the transition radiator arrangement in the ATLAS TRT

Another crucial parameter for the TRT detector is the *material budget*. A lot of effort has been put into diminishing its value as much as possible.

Each track will pass through more than 30 straw tubes in the TRT, ensuring a rather robust pattern recognition (there will be of the order of 36 hits on average per track).

It has a total length of 7 m and an outer diameter of 2.16 m. Its weight is 1.5 tons. There is a total of the order of 425 000 electronic channels. Each channel provides a drift time measurement, giving a spatial resolution of 170 μm per straw and two independent thresholds. The total power dissipation of this TRT system is 15 kW.

The role of the TRT, besides ensuring the pattern recognition, is to provide a fast and efficient level 2 trigger for high-p_T leptons at high luminosity and B-physics at relatively low luminosity.

Another important role of the TRT is to provide *electron identification thanks to the transition radiation effect* they produce. The following performances are anticipated in this respect:

- *For isolated electrons with $p_T \geq 20\ GeV$, the TRT electron identification,* combined with the electromagnetic calorimeter and the E/p matching between it and the Inner tracking detector, provides a sample with 85% purity, with an overall efficiency of about 70% and an overall rejection against hadronic jets of about 10^5.

- A rejection of photon conversion and π^0/η Dalitz decays

- Combined with the electromagnetic calorimeter and the E/p matching, the TRT electron identification performance provides efficient tagging of b-jets containing soft electrons with $p_T \geq 1.0~GeV$. The overall b-tagging efficiency obtained is about 10% (including the inclusive branching ratio of 17% to electrons) and the rejection against gluon jets is larger than 100

- Extraction of the signal from $J/\Psi \to e^+e^-$ decays

- The combination of tracking near the vertex (pixel detectors) and of the TRT pattern recognition and electron identification provides the required veto against electrons from $Z^0 \to e^+e^-$ decays in the search for $H^0 \to \gamma\gamma$ with $m_H \sim m_Z$. Even at the LHC design luminosity, isolated high-p_T electrons can be rejected by a factor larger than 500 with an efficiency for isolated photons better than 95%.

The anticipated global pattern recognition performance is indicated in Fig. 29 below. This figure shows a view in the transverse plane of the barrel detector for a simulated $B_d^0 \to J/\Psi K_s^0$ decay for a luminosity $\geq 5 \times 10^{33}$ cm^{-2} s^{-1}. The low-p_T electrons are identified, using the transition radiation information, and the large radius $K_s^0 \to \pi^+\pi^-$ are also easily recognized.

3.1.2 The CMS tracking system

The CMS tracking system is characterized by a high B-field of 4 T that was described in detail in Section 2. Moreover it has a lever arm of 1.2 m and an η-coverage of 2.4. As the ATLAS tracking system it is organized into three parts as shown in Fig. 30 below.

The first two parts use the same technique as the ATLAS tracker, namely a pixel microvertex followed by a silicium microstrip tracker. As for ATLAS, these semi-conductor trackers that are situated very near the beam will be

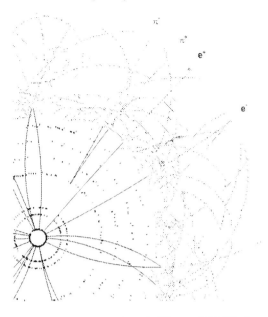

Figure 29: A simulated $B_d^0 \to J/\Psi K_s^0$ decay with $K_s^0 \to \pi^+\pi^-$ as seen by the ATLAS tracking system

submitted to very high radiation levels. Various tests on radiation hardness are made to study, for instance, the effect on the spatial resolution provided by the Si detectors. After a fluence of 3.6×10^{13} n/cm^2 at a bias voltage of 330 V, plus a heating implying an effective fluence larger than 10^{14} n/cm^2, the spatial resolution stays at a σ of 12 µm (resp. 14 µm) for 50 µm (resp. 100 µm) readout pitch[7].

The particularity of the CMS tracker are the *microstrip gas chambers* (MSGC)[8] used as outer tracker to provide the precise pattern recognition. The principle of this device is sketched in Fig. 31.

[7] *See the dedicated lectures on silicium tracker at this School, for more details on the very important issue of radiation hardness.*

[8] *See the dedicated lectures on MSGCs, given at this School, for a complete description of this tracking detector.*

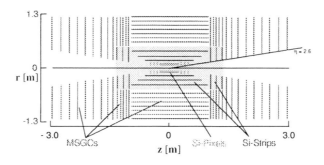

Figure 30: Schema of the CMS tracking system

The construction of these devices is based on the "rod" concept, namely a carbon fibre structure with cooling and gas supply channels; in these, the MSGC modules that are made of 250 mm long strips are placed. To emphasize the type of performances of these devices, Figure 32 shows the track finding efficiency in jets, as a function of the signal-to-noise ratio. It also includes the evolution of this ratio as a function of the applied voltage, as obtained at a test beam or as expected at CMS.

The track finding efficiency in jets of $E_t \geq 300$ GeV/c for a track with a $p_T \geq 2$ GeV/c will be above 97-98% when the signal-to noise-ratio is of the order of 20.
This very challenging tracking technique, especially when applied to an LHC tracker, now seems to converge to a realistic working solution.

3.1.3 *The CDF tracking system*

To work at the upgraded TEVATRON, which will start running year 2000, both experiments CDF and D0 are making important upgrades. This is in order to be able to cope with the new machine conditions, and to yet improve their detector performances. In particular an increase of at least a factor 10 for the instantaneous luminosity and a decrease of the bunch crossing time from 3.5 μsec to 396 nsec, and then 132 ns, are the two main issues that are influencing the upgrade of these detectors, and in particular of their tracking systems.

The upgrade of the CDF tracking system is performed in order to adapt

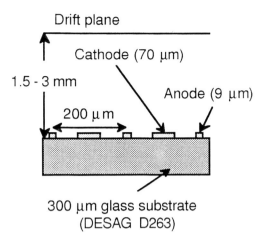

Figure 31: Principle of the MSGC

it to these new machine conditions, while keeping the same *"philosophy"* as in the earlier stage of the experiment. This is what has made its present success. It will at the same time keep in line with its *precusor and quite innovative way* to introduce the tracking information in the overall trigger architecture. CDF was the first pp experiment to succesfully use a microvertex. CDF will be the first one to introduce the tracking information in the level 1 trigger and the microvertex information (b-tagging) in the level 2 trigger.

The upgrade of the D0 tracking system follows quite different lines. D0 was so far a non-magnetic detector. For RUN II, D0 makes a drastic change, by introducing a 2 T solenoid and by completely changing the tracking system. It will comprise now a microvertex, followed by a central tracker based on a totally new technique, i.e. using 800 μm diameter *scintillating fibres* read out by VLPCs[9].

In these lectures we will concentrate more on the CDF tracker, which has more similarities to both the ATLAS and the CMS trackers. Moreover, it presents a very rich past experience and, at the same time, it is quite innovative and precursor for the next runs at the TEVATRON. This will be of highly *"educative" interest* for the LHC experiments.

The architecture of the CDF tracking system for TEVATRON II, is shown

---

[9]*I refer to the dedicated lectures on scintillating fibre tracking given at this School, for more details.*

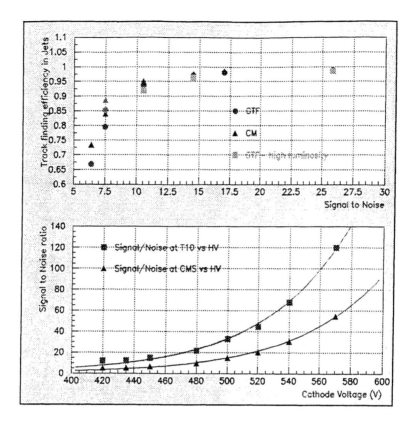

Figure 32: Performances of the MSGCs in CMS for the track finding in jets

in Fig. 33. It is characterized by a 1.4 T solenoid, a rather large lever arm of 1.4 m and a total η-coverage going down to $\eta = 2$. It comprises a new microvertex made of silicium microstrips extending up to 10 cm in radius, a new intermediate tracker also made of Si-microstrips and a new version of the central tracking chamber, keeping the same technique but modifying it in order to cope with the new machine conditions. This is the central outer tracker (COT). It covers from 46 cm to 1.35 m in radius.

The extension of the microvertex and of the intermediate Si tracker to $\eta = 2$, provides a precise momentum measurement, namely $\delta p_T/p_T = \sqrt{0.0066^2 + 0.0009 p_T^2}$
in the end-plug region down to $\eta = 2$. This will also permit a good E/p cross-check, in this end-plug region, improving in this way the precision in the energy

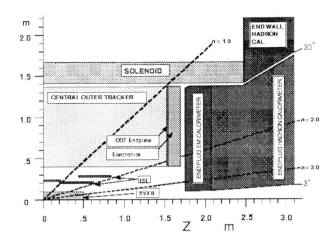

Figure 33: The CDF tracking system

measurement.
This tracking system is also characterized by a redundancy in the number of measured points per track (of order 80 to 90 hits on average) and in the fact that the COT on the one hand, and the set constituted by the SVX plus the ISL on the other hand, form two autonomous systems. The redundancy is another good point of this system, especially useful for the harsh high luminosity conditions.

Two main challenging aspects of this system are now described in more detail in the following subsection. It concerns the use of the microvertex, and consequently the b-tagging at the $p\bar{p}$ colliders, as well as the insertion of the tracking information in the overall trigger architecture.

3.2 Microvertex and b-tagging at pp colliders

CDF was the first experiment to successfully introduce a microvertex in the harsh $p\bar{p}$ environment and therefore to achieve b-tagging for the first time at a hadron collider. It has been quite instrumental to the discovery of the top quark. Indeed Figure 34 shows a picture of one of the first $p\bar{p} \to t\bar{t}$, with each top decaying into Wb. One of the W decays into the hadronic mode (i.e. two jets) and the other one into the leptonic mode i.e. into $e\nu_e$. The double b-tagging as achieved by the microvertex is clearly seen from this picture. This technique of b-tagging is widely used and also of primordial importance in the LEP experiments. Although not straightforward, b-tagging is easier to achieve in this e^+e^- environment than it was at the TEVATRON phase I. This will of course be even harder at TEVATRON II.

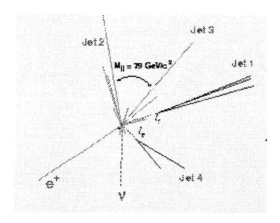

Figure 34: One of the first top event discovered by CDF, with the double b-tagging

Withstanding the conditions of TEVATRON II means an overall factor of 20 more in luminosity with:

- *Interbunch down to 132 ns*: so a faster front-end electronics with an analogue pipeline; this is achieved with the SVX3 new front-end circuit.

- *Increased radiation damage*: more than 1 Mrad are expected for the inner silicon layers.

- *More overlapping minimum bias events*: so more layers in the microvertex detector and the addition of the intermediate silicon tracker (ISL).

The upgrades are performed not only to face the new machine conditions, but also to yet improve the performances of the detector. In this sense, CDF now has double-sided silicium instead of single-sided as it was in the previous versions of the microvertex. This will give a 3D information instead of 2D. CDF will also improve the acceptance of the detector by building a longer vertex detector and by adding the ISL.

The microvertex is a key component of an integrated tracking system for a wide range of physics:

- *High statistics top studies*

- *Precision electroweak measurements*

- *Searches for supersymmetry*

- *QCD studies*

- *Measurement to constrain the CKM matrix*

The principal features of this microvertex for RUN II includes five layers of double-sided silicon that allow a 3D track measurement and the vertex reconstruction with an impact parameter resolution ≤ 30 μm in $(r - \Phi)$ and ≤ 60 μm in *(r–z)* longitudinal plane. It will be able to sustain a radiation hardness greater than 1 Mrad for a delivered luminosity of more than 3 fb^{-1} (more than the overall integrated luminosity expected for RUN II). Table 2 summarizes the main characteristics of this detector and Fig. 35 shows a view of this device in the transverse plane.

The silicium detectors are mounted on ladders that are arranged into 12 wedges with 5 layers each (see Fig. 36). A ladder has four silicon sensors with the readout electronics (i.e. the SVX3 chips) mounted at both ends.

The power and control cables and the cooling system are inserted in the overall detector structure, as shown in Fig. 35.

The microvertex is surrounded and in some sense completed by the intermediate silicon layers situated at a radius between 20 and 30 cm from the beam pipe. The silicon detectors are 112 μm pitch both sides, 4 chips per side. They are 1.2° stereo angle n^+ strips. They also are mounted on ladders.

Both detectors are read out by the SVX3 new chip set that consists of analogue front-end and digital back-end chips. The front-end has a low noise

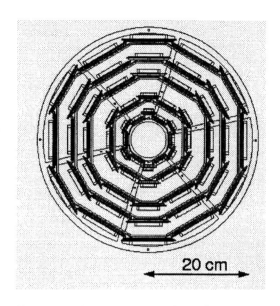

Figure 35: Tranverse view of the CDF II Microvertex

integrator and 42 cell analogue pipeline with four buffer cells. The back-end has a comparator, an on-board 7-bit ADC with data sparcification. It is optimized for 396 nsec or 132 nsec crossing interval. It comprises 128 channels per chip and performs simultaneous analogue and digital operation. It provides data within 6 to 7 μsec to the level 2 trigger processor (SVT). It has a highly parallel fibre readout system that follows the wedge geometry described in Fig. 35. This will be a very powerful tracking device, able to handle in a quite autonomous way the tracking at the vicinity of the interaction point. However, after 2003, there will be another phase in the upgrade of the TEVATRON that will lead to an increase in luminosity up to at least 5×10^{32} cm^{-2} s^{-1}. It will then be necessary to replace the actual microvertex by a pixel detector comparable to the ones built for LHC experiments.

3.3 Tracking information in the overall trigger architecture

Traditionally, the level 1 trigger includes both the calorimeter and the μ-detector informations. The information from the tracking is introduced at the

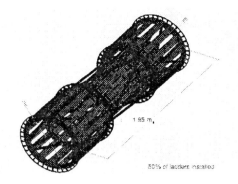

Figure 36: The CDF II intermediate silicon layers (ISL) detector

Table 2: Comparison of the CDF Microvertex for CDF I (SVX ') and the one for CDF II (SVX II)

Parameter	SVX' Detector	SVX II Detector
Readout coordinates	$(r - \phi)$	$(r - \phi)$ $(r-z)$
Number of barrels	2	3
Number of layers per barrel	4	5
Combined barrel length	51.0 cm	87.0 cm
Layer geometry	3 degree tilt	staggered radii
Radius innermost layer	3.0 cm	2.44 cm
Radius outermost layer	7.8 cm	10.6 cm
Total number of channels	46,080	405,504

earliest at the level 2 trigger. CDF has been pursuying for a long time and quite successfully, the integration of the tracking information in the overall trigger architecture. This has been driven by the physics needs, especially in the demanding case of pp collisions. A very high rate of events are produced, from which those interesting for the physics to be studied have to be extracted. A refinement of the selection is made necessary even at an early stage (i.e. level 1 trigger) and a fortiori when the luminosity increases. Moreover, the search for new physics, as for instance SUSY, or the high statistics top physics, or the B-physics, will require an enhanced b-tagging selection. The b's are produced in large quantities at pp colliders but one has to be able to select them, otherwise they are usually lost by most of the non-dedicated triggers.

Taking into account all these physics requirements, CDF is developing two main challenging devices for RUN II:

- **A very fast tracker processor (XFT)** that uses the information from the COT detector and allows **the insertion of this information at the level 1 trigger**.

- **A processor for the information from the microvertex** that will act at **the level 2 trigger**, therefore allowing the inclusion of the b-tagging at this trigger stage.

These are the two main issues that are discussed below.

Figure 37 gives the block diagram of the CDF II trigger system.

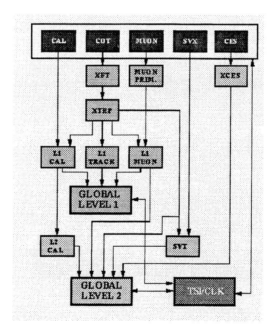

Figure 37: Block diagram of the CDF II trigger system

The key parameters for the triggering system of CDF II are:

- A crossing rate of 7.6 MHz

- *At level 1*: the latency is fixed at 5.5 μsec, requiring a storage capability of 42 BCO's at the front-end. An acceptance rate of 50 kHz for a dead-time ≤ 10% is foreseen.

- *At level 2*: the processing time is anticipated to be 20 μsec.

- *At level 3*: a processor farm will achieve the full event reconstruction and record ≥ 50 Hz on permanent storage.

Introducing the tracking information will greatly help in refining the selection at all the trigger levels.

3.3.1 The eXtremely Fast Tracker (XFT) at level 1 trigger: tracking information at the level 1

CDF I already used, during the run of 1988-1989, a fast tracker processor (CFT) that was able to trigger on high-p_T tracks using the information from the central tracking chamber. It was fast enough to be part of the level 2 trigger. For RUN II, the information of the COT chamber will be used to trigger on "high-p_T" tracks at level 1.

The design specifications are:

- The track-finding efficiency will be greater than 96% when the single-hit efficiency of the central tracker is greater than 92%.

- The momentum resolution reported to the trigger will be $\Delta p_T/p_T^2 \leq 2\%$.

- The resolution on ϕ_0 will be better than 6 mrad.

- The fake-track rejection will be at least twice as good as in the CFT.

- The minimum track p_T will be 1.5 GeV/c.

These different design specifications are set by various needs, including for instance the efficiency for high p_T physics and better trigger capabilities on high p_T muons.

The block diagram that describes the XFT trigger is made essentially of three parts. First the TDC information is multiplexed 4:1, so that prompt hits (i.e. hits that occur in the time window of 33 nsec) and delayed hits (i.e. hits that are falling in the window of 33 to 100 nsec, assuming the maximum drift time in the COT is of the order of 100 nsec) information from 4 neigbouring COT cells comes in on a single cable (see Fig. 38).

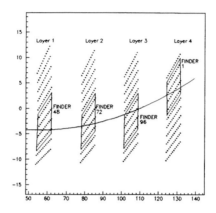

Figure 38: Superlayer hit informations from the COT used to define the track segments in the eXTRemely fast Processor for the level 1 trigger of CDF II.

A first processor called *finder logic* is designed to look for valid track segments in each of these 4 cells. It searches for high-p_T track segments in each of the outer-four axial superlayers of the COT. Each found segment is characterized by a mean ϕ position in the axial superlayer. A second processor, *linker logic*, searches for a four-out-of-four match among segments in the 4 layers, consistent with a prompt high-p_T track. This information is then distributed by another unit, the eXTRaPolation (XTRP) unit, which distributes the tracks or the information derived from the tracks to the level 1 and the level 2 subsystems. This will enhance the quality of the selection as provided from the calorimetry and the muon systems already at this first stage of the triggering.

3.3.2 *Microvertex and b-tagging at the level 2 trigger*

The ability to use impact parameter information in the trigger to detect secondary vertices can substantially increase the physics reach of a hadron collider experiment. This is the purpose of the silicon vertex tracker (SVT) that will be installed at the level 2 trigger in the CDF II detector.

The SVT combines the information coming from the COT and the one provided by the microvertex (SVX) in order to reconstruct the tracks in the transverse plane $(r - \phi)$. Three parameters are computed for each reconstructed track: the transverse momentum, the ϕ-angle and the impact parameter d. The impact parameter is defined as the distance of minimum approach of the track to the beam axis, as seen in the transverse plane projection (see Fig. 39). The aim is to achieve, at the level 2 (i.e. within less than 20 μsec), quasi offline performances. This means that for tracks with a p_T above 2 GeV/c, the track parameters will be computed by the SVT with the following resolutions: $\sigma_d = 35$ μm, $\sigma_\phi = 1$ mrad and $\sigma_{p_T} = 0.3\% \times p_T$ GeV/c.

To achieve this, the SVT system mainly relies on the use of performant and fast processor farms and associative memories, following the architecture sketched in Fig. 40. As indicated there, the raw SVX data flow from the front-end to the "hit-finders". The task of the hit finders is to find clusters of strips with a significant energy deposit and compute, for each of them, the coordinate of the centroid. The results of this processing are the SVX hits. The data from the central tracking chamber (COT) are fed to the XFT processor that finds, for each beam crossing, all the tracks in the COT with a minimum p_T of 1.5 GeV/c. The transverse momentum and the ϕ-angle of these COT tracks, as computed by the XFT, are fed both to the associative memory and the hit buffers together with all the SVX hits. The associative memory performs the first stage of the pattern recognition process by finding track candidates using a coarse spatial resolution in the SVX (250 μm). These track candidates are called *roads*. For each *road* output by the associative memory, the full tracking information, i.e. the hits and the COT track belonging to that *road*, are retrieved in the hit buffers and delivered to the track fitters. The track fitter is a processor farm where each processor receives one *road*, performs

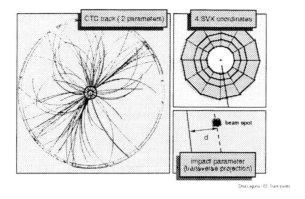

Figure 39: Three parameters defining the tracks (p_T, ϕ, d) as processed by the SVT in CDF II.

the final quality cuts and estimates track parameters using the full available spatial resolution. Different processors in the farm work in parallel on different track candidates. The track fitters run a special algorithm designed to perform the fitting procedure very efficiently and without compromising the accuracy. It is based on the classical method of *linear digital filtering*. The track parameters are expressed in terms of a first-order linear expansion as follows: $d \approx c_0 + \sum_{i=1}^{6} c_i x_i$.

A lab test bench has been mounted to simulate the SVT processing system, in the real environment of RUN II. It is based on the use of technology available today (i.e. Motorola 96000) and a VME-type set-up, with real data recorded from RUN I. This test set-up permits to realistically evaluate the performances achievable with the SVT. For instance, with the present technology, the overall processing time is of 13 μsec. This is already well within the level-2 time constraints.

This set-up is also used to compute and histogram the impact parameters of all the tracks found by the SVT and to compare them with the same quantities as computed by the standard offline CDF analysis. As a result, the σ of

SVT architecture

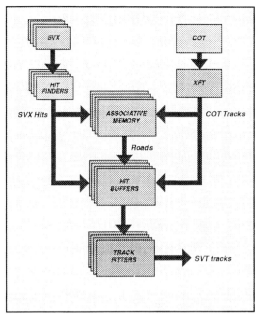

Figure 40: Block diagram of the SVT processing for the level 2 trigger of CDF II.

the distribution is of the order of 50 μm and the widths of the two distributions are virtually equal, ensuring that the results are quite comparable (see Fig. 41).

As already emphasized, the impact of the SVT processor on many of the main aspects of the physics reach at the pp collider will be of major importance.

Because of its location at the heart of the interaction, and of its crucial impact in the physics studies, the tracking system is in several aspects the most technologically difficult detector system. The overall story of the development of the tracking and its continuous improvement with quite innovative ideas, all driven by physics motivations, have led to major progresses in the use of this detector system in a running experiment such as CDF. This will have a highly educative value and a strong impact, especially for the LHC experiments.

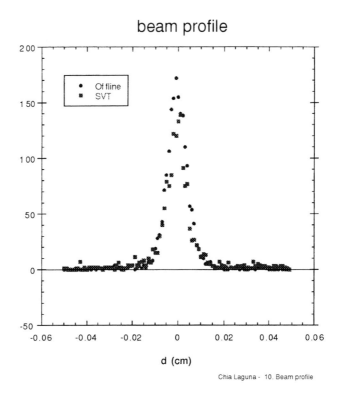

Figure 41: Comparison of the impact parameter distribution as computed by the offline and by the SVT processor for CDF II.

4 CALORIMETRY

The functions of the calorimeter system are:

- *To identify electrons and γ's in the electromagnetic part, and τ's and π's in both the electromagnetic and hadronic parts, and to measure their energy.* The benchmark at the LHC for the e.m. calorimetry is the diphoton mass resolution, to be able to identify the decay of H^0 into two photons for a Higgs mass of 100 GeV.

- *To identify jets and measure their energy*

- *To identify ν's or ν-like particles by accurately measuring the missing energy in the events.* This requires a good missing E_t resolution and a hermetic overall calorimeter system.

- *To be part of the trigger.* Traditionally the level 1 trigger mainly relies on the calorimeter fast information.

The calorimeters include two sectors, sometimes physically different, namely the e.m. calorimetry and the hadronic calorimetry.

Several arguments as for instance homogeneity, non-discontinuity between the hadronic and the e.m. parts and hermeticity, would favour the use of the same calorimeter technique for the two components of the calorimeter system. Various reasons are used to do the contrary [10]. ATLAS and CMS use different calorimeter techniques to build the e.m. and the hadronic components. For ATLAS, this is indeed only the case in the central region, since the e.m. part is made with liquid-argon whereas the hadronic part is made with tiles. But, in the end-caps, the overall calorimetry is all made with liquid-argon sampling devices.

In this section, the emphasis is on the e.m. calorimetry, where ATLAS and CMS have chosen different approaches. One is a "conservative" approach in the sense that it uses the liquid-argon technology which is a well established technology. The other one is more innovative, as it is based on the use of

---

[10] *This is discussed, for instance, in " Calorimetry at LEP a critical point of view", by H. Videau in the Proceedings of the VIth Conference on Calorimetry in High Energy Physics, Frascati 1996.*

totally new crystals and new associated front-ends.

But as we will see, both, at the end, are quite challenging calorimeter techniques when facing the LHC environmental constraints.

If one takes as benchmark the process $H^0 \to \gamma\gamma$ at the LHC for the e.m. calorimetry, to define the required performances of these devices, it sets the scenario. Indeed, the mass resolution is given by the following quadratic sum:

$$\sigma_M/M = 1/2(\sigma_{E_1}/E_1 \otimes \sigma_{E_2}/E_2 \otimes \sigma_\theta/\tan\theta/2)$$

and it therefore depends on the resolution in energy (E_1, E_2) of the two photons expressed in GeV and their angular separation (θ) expressed in radians. The energy resolution is usually parametrized by the following quadratic sum:

$$\sigma_E/E = a/\sqrt{E} \otimes b \otimes \sigma_N/E$$

where a is the stochastic term, b is the constant term and σ_N is the noise term that includes two sources of noise: the electronic noise and the pile-up noise.

It is difficult to achieve a stochastic term $\leq 10\%/\sqrt{E}$ in a large volume, without demanding strict mechanical tolerances. This sets the scale for the requirement of the constant term to $\leq 1\%$ and the noise term to ≤ 500 MeV equivalent noise. Presently running e.m. calorimeters achieve around $15\%/\sqrt{E}$ and 1% constant term (TEVATRON or HERA calorimeters). The LHC e.m. calorimeters aim to do better, in a harsher environment. This is quite a challenging task.

4.1 *An innovative approach: the CMS crystal calorimetry*

The CMS crystal calorimetry is meant to reach a resolution in energy of about 3% to $4\%/\sqrt{E} \otimes 0.5\%$. In order to reach it, CMS has decided to use crystals for the e.m. calorimetry. The CMS e.m. calorimeter covers down to 3 degrees with respect to the beam axis and corresponds to 25.8 X_0. It is made of PbWO$_4$ crystals (82 000 in total) of trapezoidal shape: 22 mm × 20 mm × 230 mm. The final design shown in 42 demonstrates an almost perfect overlap between the central and end-caps parts. This means that this device will have a particularly good hermeticity even in the space devoted to the services. The crystals are arranged in a projective way in sets of baskets as indicated in Fig. 43. They are read out by avalanche photodiodes (APDs). The details of the capsule containing the photosensor are also shown in Fig. 43.

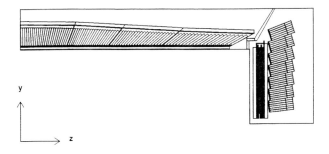

Figure 42: Overview of the CMS crystal e.m. calorimetry

Two key issues for this e.m. calorimeter are the choice of **crystals** and of **APDs** as photosensor. Let us discuss these two challenging aspects.

4.1.1 Issues on crystals for the LHC e.m. calorimetry

The specifications for the choice of the crystal are:

- *The density (g/cm^3):*
  It must be high in order to have a compact e.m. calorimeter. This implies a short radiation length (X_0) to contain the e.m. shower in length and a small Moliere radius to contain the e.m. shower in width.

- *The light yield or photostatistics:*
  The light yield is of course the number of photons per MeV, but it is also a function of the emission spectrum (i.e. the wavelength in nm of the emitted light) with respect to the photodetector response moreover, it depends on the efficiency of the optical transmission and on the light collection.

- *The radiation hardness:*
  The radiation hardness for γ's (50 kRad/year) and for neutrons fluence (2×10^{13} n / cm^2 for 10 years of LHC).

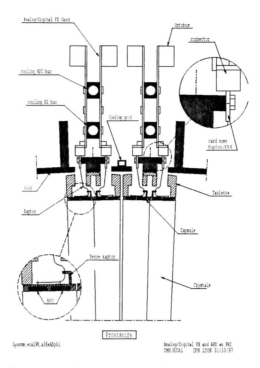

Figure 43: Detailed view of the basket where are located a set of crystals with their very front-end electronics and monitoring system

- *The decay time:*
  The scintillation decay time must be smaller than 25 nsec which is the bunch crossing time at the LHC.

- *Temperature dependence of scintillation light yield:*
  The crystal must have a good light yield at low temperature so that this does not decrease too much when the temperature grows.

- *Mechanical resistance:*
  The crystals must present a not too bad mechanical resistance to cuts, polishing and mounting on the detector.

Table 3: Comparison of the properties of BGO, CsI, CeF$_3$ and PbWO$_4$ crystals

Properties	BGO	CsI	CeF$_3$	PbWO$_4$
X_0 (cm)	1.12	1.85	1.68	0.89
Moliere radius (cm)	2.3	3.5	2.6	2
Density (g/cm^3)	7.13	4.51	6.16	8.28
Light yield (photons/MeV)	4000	1500	2000	120
Decay time (ns)	300	35	30	100 (*)
% of emitted light in 25 ns	10%	85 %	50%	80%
Emission spectrum (nm)	480	310	300	450-500
Radiation hardness	NO	NO	$\geq$ 1 MRad	$\geq$ 1 MRad
Temperature dependence (%/K)	$-$ 1.6	$-$ 0.6	0.15	$-$ 2.0

- *Cost:*
  Last, but not least, the crystal must be cheap.

Four types of crystals have been investigated: **BGO, CsI, CeF$_3$** and **PbWO$_4$**. Table 3 shows a comparison of the main properties of these crystals.

The lead-tungstate crystal (PbWO$_4$) has been chosen as the best compromise. Indeed the tunsgtate crystals have been known for a long time, and progress with large crystals has been made recently. As seen from Table 3 the following reasons have guided the choice:

- *Its small radiation length and small Moliere radius*

- *Its fast scintillation decay time constant ((*)namely: 30% of the signal in 5 ns, the next 60% in 15 ns and the remaining 10% in 100ns)*

- *Its good radiation hardness on full length crystals doped with niobium*

- *The substantial production capacity that already exists.* It is relatively easy to grow from readily available raw materials.

The drawback of the light yield is partly overcome by recent developments on relatively large area Si APDs. However there is on-going R&D work on the crystals, to optimize the scintillation light yield (by reducing the impurities on the basic materials: PbO and WO_3), to optimize the light collection, to overcome the radiation effects.

The study of the radiation effects on the crystals is an R&D area particularly active where progress has recently been made. A problem that has been especially investigated is the radiation damage, i.e. the decrease of light transmission at low dose. The following experimental facts have been established:

- The scintillation mechanism is not affected by irradiation. The loss of light output is due to the absorption by colour centres (defects in the crystal) produced by the irradiation.

- The level of light reaches an equilibrium at LHC doses between the creation of defects and the self annealing.

- The level of saturation and the recovery time is a function of the dose rate. At LHC dose rates (≤ 100 rad/h), the loss for crystals produced in 1996 was 10% to 20%, and the recovery time of the order of two weeks. Improvements in the crystal production in 1997, have brought these levels down to 5%.

To overcome these problems, CMS is, on the one hand, developing harder crystals both in Russia (using the Czochralski method) and in China (using the Bridgeman method). On the other hand, the residual calibration fluctuations are monitored and corrected using the light injection system (see Fig. 44).

To get a better radiation hardness, two ways have been explored; one is by decreasing the concentration of defects by stoechiometry or by annealing. The second one is by compensating the remaining defects, by the purity control of the raw material or by specific doping: pentavalent on W site and trivalent on Pb site.

Work still remains to be done, but substantial progress has been made lately on the understanding of the problems and consequently on the ways to overcome them.

4.1.2 Issues on the photodetector for the CMS e.m. calorimetry

The required properties of the photodetector are:

- *Its ability to operate in the 4 T $\vec{B}$ field*

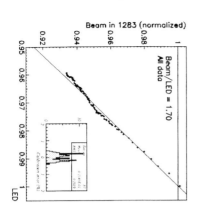 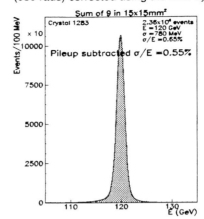

Figure 44: Calibration system results to correct for the effect of radiations

- A high quantum efficiency over the range of the emission wavelength ($350 \leq \lambda_{nm} \leq 600$)

- The gain stability to meet the requirement on the constant term of the energy resolution.

- The radiation tolerance: it must be able to meet the specifications even after an irradiation with an integrated neutron fluence of a few $\times \ 10^{13}$ n/cm^2, in the central part after 10 years of running. Phototetrodes (VPTs) are used instead of APDs in the end-caps because of the higher doses.

- The effective input capacitance and leakage current should be small enough to avoid excessive noise.

- The response to charged particles (the nuclear counter effect) should be small.

Table 4: Present conclusions on APDs (March 1997)

Parameter	Hamamatsu APDs	EG&G APDs
Area (mm^2)	25	25
Capacitance (pF/mm^2)	6.5	1.2
Excess noise factor at a Gain of 50	2.0	2.2
$\Delta gain/\Delta V$ at gain of 50	$\geq 5\% \times$ M	$\geq 2\% \times$ M
$\Delta gain/\Delta T$ at gain of 50	$- 2\% \times$ M	$- 2\% \times$ M
$\sigma(V_{operating})$ at gain of 50	$\sim \pm 1$V	$\sim \pm 50$V

All these conditions have lead to the choice of *avalanche photodiodes (APDs)* as photodetectors. Two firms are at present considered: Hamamatsu and EG&G. Table 4 summarizes the present state of the art of the various devices proposed by these two companies. The R&D continues on both types of APDs, as both present pros and cons.

4.1.3 Front-end readout challenges

Figure 45 sketches the overall chain of transmission of the information delivered by the detector to the data acquisition system (DAQ). The information of the calorimeter (i.e. the energy deposited in the calorimeter), given as a light in the crystal (scintillating material), is transformed in a current by the photodetector (APDs); then the electronic readout chain transforms this current into voltage in the preamplifier stage and then the voltage into bits in the digitization stage (i.e. the ADCs). The bits are then sent to the later stage (DAQ chain) by optical fibres through light transmission.

The challenges that have to be overcome by this overall chain are also clearly stated in Fig. 45: the relatively low light yield of the crystals, the new type of photodetector that is used, i.e. APDs functioning in a 4 T field, the 16- to 18-bit dynamic range, implying a 12 bit ADC working at 40 MHz. All this in a radiation-hard environment. An overall electronic chain has been developed and tested. It is a prefiguration of what the readout of this calorimeter will look like. It is based on fast digitization at the front-end.

The readout prototype consists of a set of chips: full custom preamplifier and shaping circuits made in Bi-CMOS and commercial 12 bit–40 MHz ADC from Analog Device, the AD9042 Bipolar XFCB device. At the present time a XILINX operates as a pipelined memory and readout circuit. Very encouraging results have been obtained in the test beams. The crucial point here is the

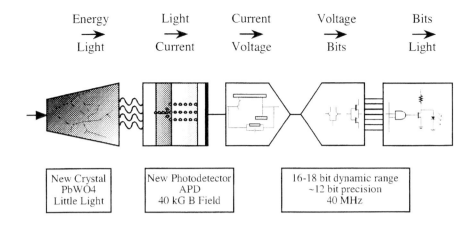

Figure 45: CMS readout chain

impressive progress made on the side of fast and large dynamic range ADCs. This progress is triggered, in particular, by the development of the market of cellulars.

4.2 A well established approach: the ATLAS liquid-argon calorimetry

ATLAS's aim is to reach a resolution in energy for the e.m. part of the calorimetry of slightly better than $10\%/\sqrt{E} \otimes 1\%$, which makes it a little bit less demanding than CMS. In order to achieve this resolution, the liquid-argon (LAr) calorimetry has been chosen.

Liquid-argon calorimeters are sampling calorimeters. They are based on the use of lead, iron or uranium as absorber and liquid-argon or krypton or xenon as ionizing medium with electrodes (usually made of kapton and copper) that record the signal corresponding to the deposition of energy. This signal is due to the produced electrons that drift to the electrode where they are registered. The sampling is made as shown in Fig. 46 by alternating absorber plates with an ionizing gap and the electrode. Likewise, the development of an e.m. shower originated by photons or electrons is followed step by step in this device.

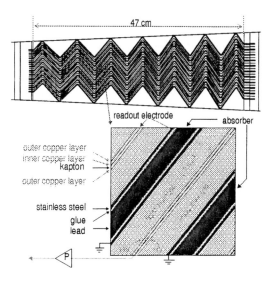

Figure 46: Sketch of the structure of a sampling calorimeter: the ATLAS liquid argon e.m. calorimeter

The analogue electric signal has a triangular shape. The recorded charge is directly proportional to the deposited energy in the corresponding compartment (sampling) of the calorimeter. The maximum length of the signal corresponds to the maximum possible distance of drift of the electrons from the point they are created to the electrode, i.e. the dimension of the ionizing gap.

Various running experiments are using such calorimeters. This is the case for instance of H1 at HERA, D0 at the TEVATRON, which both have a liquid-argon calorimetry and NA48 at the SPS-CERN, which uses Xe-calorimetry. Although well established, this type of calorimetry, presents the following challenges:

- It is a relatively slow device ($\geq$ 400 ns signal duration). So the question was the way in which to transform it in a fast detector as required by the LHC

- The cryostat system used to cool down the argon has to overcome a certain number of difficult issues (mechanical issues, feedthroughs that have to handle a very large number of signals, power consumption and cool-down time).

- The stringent constraints on the mechanics of the detector, as they strongly influence the performance of the device in terms of energy resolution (constant term).

- The readout system.

These different aspects are now reviewed.

4.2.1 *A fast accordion liquid-argon calorimetry*

The typical LAr response is a triangular signal; because of the dimension of the gap, this has a duration of order 400 nsec. This is by far too slow for the LHC standards, where the basic time is 25 nsec. To shorten it, one applies the traditional technique, i.e. a bipolar shaping such as to have a peaking time of order 40 nsec (see Fig. 47). In this way the LAr calorimeter becomes a fast-response detector. But the main issue is to have a very low noise front-end electronics, i.e. preamplifier and shaper, in the harsh environment in which they will be located.

Another point is the design of the absorber plates. Usually the absorber plates in sampling calorimeters are positioned perpendicularly to the incoming particles. To obtain the granularity required at the LHC, this type of geometry would imply many dead spaces due to the high number of needed connexions. This large amount of connexions, needed to link between them the absorber plates in depth, has a large contribution to the inductance L and the capacity C_d of each cell. The time to transfer the signal from the electrodes to the preamplifiers is proportional to $4\sqrt{LC_d}$. It has of course to be kept as small as possible. Thus, in order to reduce the dead space between the electrodes, the accordion shape has been adopted. The accordion waving is placed orthogonally to the incident particles, as shown in Fig. 48.

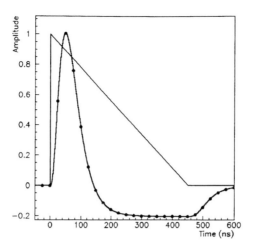

Figure 47: Signal shape as produced in the liquid Argon calorimeter (triangle), and after shaping (curve with dots). The dots represent the position of the successive bunch crossings.

4.2.2 The cryostat issues

The cryostat that houses the central e.m. calorimeter of ATLAS (i.e. a weight of 110 tons for the detector itself plus 60 tons for the liquid argon) has, in addition, to support and provide the vacuum for the superconducting solenoid coil (5.5 tons) that is described in Section 2 and serves the inner tracker. Furthermore, the cryostat supports the inner tracker (5 tons including the cables) and is itself supported by the hadronic calorimeter. It is made of a warm vessel (12 tons), which forms a hollow cylinder and contains a similarly shaped cold vessel filled with liquid argon. The e.m. calorimeter is inside the cold vessel, as well as approximately 100 000 signal cables and 2000 HV cables that are part of the detector. The most critical interface is with the superconducting solenoid and its power leads.

This cryostat is thus no ordinary pressure vessel, since it has to support a detector of 170 tons (including the liquid argon), which dominates the stress distribution and deformations in both the warm and the cold vessels. In addi-

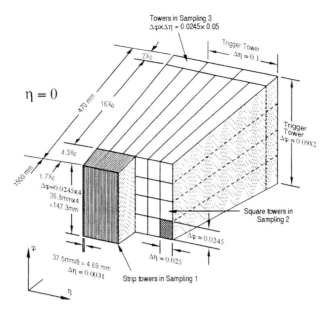

Figure 48: Sketch of the accordion structure of the ATLAS e.m. calorimeter

tion the location of all the critical components has to be known to an accuracy of 1 mm. This overview of the cryostat, its content and its environment (see Figs. 49 and 50) already sets the mechanical engineering difficulties that have to be overcome and how crucial this piece of the liquid-argon calorimetry is.

To these environmental constraints have to be added other design requirements. One is that cryogenic calorimetry, and particularly e.m., has to minimize the material between the interaction region and the calorimeter. This requirement has several consequences in the design of the cryostat. The cryostat is made of aluminium instead of stainless steel. The walls of the cryostat (warm and cold walls) are made as thin as possible. However the thickness must be a compromise with the pressure and load constraints to which they are submitted. Still in order to minimize the amount of material, the barrel LAr calorimeter and the superconducting solenoid share the cryostat. Consequently, it eliminates two vacuum walls between the solenoid and LAr calorimeter, reducing the radiation length in front of the calorimeter and saving space. However this also introduces various designing as well as manufacturing constraints on both devices, and it restricts their operation.

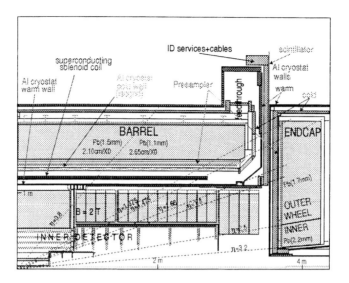

Figure 49: Quadrant view of the LAr cryostat and its environment

Another design constraint is the requirement that the barrel e.m. calorimeter is constructed azimuthally in a "seamless way" (see Fig. 50), with no gaps between individual modules. Thus all the cables must follow this azimuthal symmetry, so that all signal delays from individual cells in the calorimeter are identical. This in turn constrains the location of the feedthroughs.

The feedtroughs are themselves another challenging aspect of the cryostat. The detector located inside it requires about 120 000 signal and calibration lines and, in addition, a large number of HV leads. There will be a total of 64 signal feedthroughs of 1920 signal lines each, connecting the inside of the cold vessel to the outside. Besides the impressive number of signals handled by each feedthrough (a factor order of 10 higher than usual), the specifications that drive the technical design of the feedthroughs are quite complex. They involve geometrical and space constraints in the cryostat design, physical limitations on the space that is also allocated to the feedthroughs, signal transmission quality, vacuum integrity, access and reliability issues (see subsection 4.2.4).

All this sets the high level of difficulty in the design, building and operation (in particular cool-down time and power-consumption considerations also

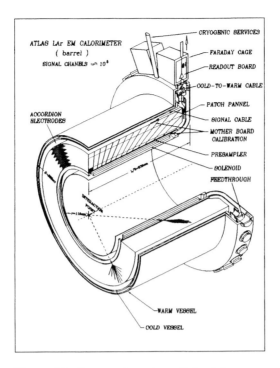

Figure 50: Perspective view of the LAr cryostat and its environment

have to be taken into account) this very fundamental piece of the liquid-argon calorimetry.

4.2.3 *Quality controls and detector mechanics*

The precision on the mechanics commands the value of the constant term in the energy resolution when aiming to go below 1%. Thus very severe constraints are imposed on the mechanics of the detector. To cope with that, a number of quality controls are set-up all along the chain of production of both the absorbers and the electrodes.

For instance non-uniformities in the thickness of the lead plates used to make the absorbers lead to local variations of the calorimeter response. An increase of 10 μm on the lead thickness implies a decrease of 0.4% of the measured signal. Two systems are used to select in two steps the lead plates within the required ± 10 μm tolerance. One is based on an X-ray measurement. The radiation passes through the lead sheet just being rolled and is then detected by NAI scintillators. A more precise measurement is performed afterwards, based on a detailed scan of each manufactured lead plate by an ultrasonic probe. Once the absorber is mounted and bent in an accordion shape, its geometry is carefully measured and verified with a 3D machine, with an expected accuracy better than 20 μm.

The electrodes are submitted to systematical complete electrical tests at the different stages of their fabrication.

4.2.4 The readout challenges

The readout is another crucial issue to ensure that the calorimeter is able to function in a satisfactory way in the LHC environment. This is especially true for the front-end part. *The primary duty of any readout system is to introduce a minimal degradation to the signal delivered by the detector when processing it.*

Among other things, the front-end readout must be fast (40 MHz), with a large dynamic range ($\geq$ 16 bits for the e.m. calorimetry), pipelined (to be compatible with the level-1 latency of order 2.5 μsec), compact (because it will be located on the detector and a lot of channels to handle), radiation-hard, reliable (it is not easy to access to it), easy to standardize (large number of channels), to maintain and to upgrade.

$\implies$ *Digital or analogue front-end readout*

Two types of front-end have been studied in ATLAS. One approach is totally digital, as it is based on an immediate digitization of the signal at the front-end, right after the shaping. The pipelining is then done using a pipelined memory. The other one stays analogue by doing the pipelining first using swith capacitor array (SCA) and then digitizing the signal with a relatively slow ADC. The digital front-end solution relies on the recent progress of commercial, fast and of large dynamic range ADCs. This progress has been achieved thanks in particular to the development of the industry of cellulars. As already seen (section 4.1), the digital solution has been adopted by the CMS calorimetry. Figure 51 shows how a prototype version of this digital readout is able to reproduce in great detail (on-line imaging capability) the shape of the incoming signal.

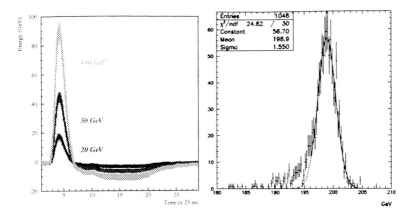

Figure 51: Online imaging capability of a digital readout on signals at test beam from 20 to 100 GeV electrons

Figure 52: Resolution in energy obtained with a SCA prototype bipolar shaped LAr calorimeter for 200 GeV electrons

The second approach, which has been adopted by the liquid-argon AT-LAS calorimetry, is based on a well established technique, the SCAs, pushed to work at its limits in the sense that it must be able to cover a dynamic range of order 12 to 13 bits. Figure 52 shows the obtained resolution in a test beam with 200 GeV electrons with a prototype version of this electronic device.

The readout chain comprises, after the low-noise preamplifier, a linear shaping with 3 gain scales, an analogue pipeline to store the data during 2.5 μsec. It is made of a 144-cell-deep SCA followed by a 12 bit–6 MHz ADC.
The calibration and the integration of the readout system within the overall experiment are particularly touchy points.

$\Longrightarrow$ *Calibration of the readout chain*

The calibration has two main tasks: it is part of the absolute calibration, i.e. the correct determination of the energy, and it is used to handle the functioning of the overall readout system and to detect its possible failures. It is achieved in a quite usual way, i.e. using a test pulse system. But the unusual aspect of it is the use of an injection resistor instead of a traditional capacitor to apply the voltage pulse. It makes it easier and inexpensive to obtain a

1% precision. It is also robust and radiation-hard. This pulse simulates the physics signal and is applied directly on the electrodes. The pulsing pattern allows us to measure the cross-talk between neighbouring channels.

$\implies$ *Integration of the readout in the experiment*

The integration of this readout system in the overall architecture of the experiment addresses several points. This is not the case for this detector system only. It is of course true for all the different detector systems we have been discussing, and it is even more difficult for obvious reasons in the case of the tracking system. However we will discuss the integration issues for the readout of the calorimetry in more detail, just as an example.

Space, dead material (therefore "compacity"), easiness to be incorporated in the allocated space, modularity and capability to be standardized (important because of the high number of channels), accessibility, power consumption and cooling, radiation hardness, maintenance, capability to be upgraded and extended, cost, are among the important integration issues.

We will now review some of these points.

- *Space, dead material and "compacity"*
  In the case of the liquid argon calorimeter, the front-end will be installed in crates sitting by the feedthroughs (see Fig. 53). Front-end boards will contain the electronics to fully process the signals coming from the detector to send the primitives to the level-1 processing on the one hand, and on the other hand the digitized pre-processed data to the DAQ system. A total of 128 channels will be treated per board. The overall space taken by this electronics is $80 \times 49 \times 41$ cm^3. It therefore takes a fair amount of the space in the gap due to the LAr cryostat. Other detectors are also demanding some space in this strategic region; this is in particular the case of the tracking system as well as the solenoid plus cryostat services. Thus every mm^3 is crucial as hermeticity and no dead material are strongly required by physics.

- *Radiation hardness*
  Although the level of radiation in this region where the electronics will be located is relatively not too high (namely 20 kGy for charged radiation and 10^{12} n/cm^2 for neutron fluence), it demands that every element of this readout system be sufficiently radiation-tolerant.

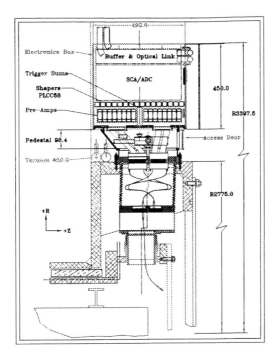

Figure 53: Side-view of the enclosure containing on-detector readout electronics on the barrel cryostat

- *Power consumption and cooling*
  The proposed electronic chain is foreseen to have a power consumption of 750 mW per channel. This means 96 W per board, 3.5 kW per crate, 700 A current per crate and a total power in crates of 145 kW. This large amount of power has to be removed by water cooling. The needed water flow in crates will be about 25 m^3/h. Therefore every additional 100 mW in the front-end circuits translates into 30 kW of additional power and 3 m^3/h of cooling water. These numbers show the level of concern one might have about this particular point.

As a conclusion of this section on calorimetry for future HEP experiments, we have reviewed two different calorimeters: the crystal calorimetry from CMS and the LAr calorimetry of ATLAS. One is based on new technologies, namely a new type of scintillating crystals, new front-end (APDs), new readout, based on a digitization at the very front-end using commercial, new, fast and large dynamic range ADCs. The other one is based on well established technologies: liquid-argon sampling calorimetry for the detector and SCA pipeline plus relatively slow ADC for the readout. But in order to face the requirements of LHC these well-established technologies must be revisited and pushed to their limits.

Thus for both detectors it will be quite challenging to build them and make them work at the LHC. Let us wait ten years to know the end of and/or the true story.

5 CONCLUSIONS

The impressive advances achieved in the development of the various detector systems for future high energy physics experiments are the consequences of:

- A world-wide collaborative effort,
- An active R&D especially in view of LHC experiments,
- The upgrade of challenging running experiments.

They are primarily driven by physics needs and they are made in conjunction with the progresses in "high technologies", in the domains of interest for HEP (microelectronics, computing, telecoms,...). This fact stresses how important it is for fundamental research and high tech industry to collaborate even more closely. This is particularly true in our field.

This leads to an impressive potentiality in terms of detection capabilities, making it possible to perform the very challenging experiments that will be running in the (near) future, i.e. at the turn of the next millenium.

Acknowledgments

I would like to thank all the colleagues from ATLAS, CDF and CMS experiments who provided me with useful informations: In particular I am indebted to P. Besson, A. Cattai, J. David, M. Della Negra, L. Dobrzynski, D. Froidevaux, F. Kapusta, W. Kozanecki, G. Martin, T. Mouthuy, L. Ristori, P. Siegrist, K. Tanaka, P. Vincent and J. Wotschak. A special thanks also to C. Fabjan and J.L. Faure for a critical reading of the manuscript. I am grateful to S. Vascotto for proof-reading.

Bibliography

These lectures are based on:

- [1] G.L. Bayatian et al., CMS Collaboration, *The Compact Muon Solenoid Technical Proposal*, CERN/LHCC/94-38, LHCC/P1, 15 December 1994. And references therein.

- [2] W.W.Amstrong et al., ATLAS Collaboration, *Technical Proposal for a General-Purpose pp Experiment at the Large Hadron Collider at CERN*, CERN/LHCC/94-43, LHCC/P2, 15 December 1994. And references therein.

- [3] R. Blair et al., CDF Collaboration, *The CDF II Detector Technical Design Report*, FERMILAB-Pub-96/390-E, November 1996. And references therein.
  L. Ristori, *The SVT Microvertex Second-Level Trigger for CDF II*, private communication.

- [4] ATLAS Collaboration, *ATLAS Calorimeter Performance Technical Design Report*, CERN/LHCC/97-40, ATLAS TDR 1, 15 December 1996. And references therein.

- [5] ATLAS Collaboration, *ATLAS Liquid Argon Calorimetry Technical Design Report*, CERN/LHCC/97-41, ATLAS TDR 2, 15 December 1996. And references therein.

- [6] ATLAS Collaboration, *ATLAS Tile Calorimeter Technical Design Report*, CERN/LHCC/97-42, ATLAS TDR 3, 15 December 1997. And references therein.

- [7] ATLAS Collaboration, *ATLAS Inner Detector Technical Design Report, Volume I*, CERN/LHCC/97-16, ATLAS TDR 4, 30 April 1997. And references therein.

- [8] ATLAS Collaboration, *ATLAS Inner Detector Technical Design Report, Volume II*, CERN/LHCC/97-17, ATLAS TDR 5, 30 April 1997. And references therein.

- [9] ATLAS Collaboration, *ATLAS Barrel Toroid Technical Design Report*, CERN/LHCC/97-19, ATLAS TDR 7, 30 April 1997. And references therein.

- [10] ATLAS Collaboration, *ATLAS End-Cap Toroid Technical Design Report*, CERN/LHCC/97-20, ATLAS TDR 8, 30 April 1997. And references therein.

- [11] ATLAS Collaboration, *ATLAS Central Solenoid Technical Design Report*, CERN/LHCC/97-21, ATLAS TDR 9, 30 April 1997. And references therein.

- [12] ATLAS Collaboration, *ATLAS Muon Spectrometer Technical Design Report*, CERN/LHCC/97-22, ATLAS TDR 10, 31 May 1997. And references therein.

- [13] CMS Collaboration, *The Magnet Project Technical Design Report*, CERN/
  LHCC/97-10, CMS TDR 1, 2 May 1997. And references therein.

- [14 CMS Collaboration, *CMS Crystal Calorimeter Technical Design Report*, in preparation.

- [15] CMS Collaboration, Presentation to the LHCC Meeting 5th of June 1997 by M. Della Negra, *CMS Status Report*.

- [16] ATLAS Collaboration, Presentation to the LHCC Meeting 5th of June 1997 by W. Kozanecki, *ATLAS Muon Spectrometer: Introduction and Overview*, Copies of the transparencies are available on Web.

- [17] ATLAS Collaboration, Presentation to the LHCC Meeting 5th of June 1997 by F. Linde, *ATLAS Muon Spectrometer: Precision Chambers*, Copies of the transparencies are available on Web.

- [18] ATLAS Collaboration, Presentation to the LHCC Meeting 5th of June 1997 by G. Mikenberg, *ATLAS Muon Spectrometer: Trigger Chambers*, Copies of the transparencies are available on Web.

- [19] Proceedings of the Sixth International Conference on *Calorimetry in High Energy Physics*, Frascati, June 1996, Frascati Physics Issues, editors: A. Antonelli, S. Bianco, A. Calcaterra, F.L. Fabbri. See in particular:

H. Videau, *Calorimetry at LEP, A Critical Point of View*, P 905.

P. Denes, *New Crystal Calorimeters for Colliders*, P 85.

A. Savoy-Navarro, *Realization and Test of a Fast Digital Readout for LHC calorimeters: present performances*, P 883.

GASEOUS WIRE DETECTORS*

J. Va'vra

Stanford Linear Accelerator Center, Stanford University,
Stanford, CA 94309, U.S.A.

Abstract

This article represents a series of three lectures describing topics needed to understand the design of typical gaseous wire detectors used in large high energy physics experiments; including the electrostatic design, drift of electrons in the electric and magnetic field, the avalanche, signal creation, limits on the position accuracy as well as some problems one encounters in practical operations. Reader should also refer to Ref. 1-4.

Chapter 1

1.1. Two-dimensional electrostatic field in a drift cell

This lecture will cover the following topics:

a) numerical solution for wires only,
b) analytical solution for wires and pads,
c) numerical solution for any shape of electrodes,
d) numerical solution for wires, pads and dielectric.

We start the lecture by reminding the **Gauss law in the integral form**:

$$\Phi = \oint_S \vec{E} \cdot d\vec{S} = \frac{1}{\varepsilon_r \varepsilon_0} \sum_i Q_i \qquad (1.1)$$

where $\sum Q_i$ is sum of all charges within the volume defined by surface S, Φ is total flux of electric field $\vec{E}$ through surface S, ε_r is effective relative dielectric constant, $\varepsilon = \varepsilon_r \varepsilon_0$ is the dielectric constant of the medium within the surface S and $\varepsilon_0 = 8.85$ pF/m is permitivity of free space.

1.2. Two-dimensional electrostatic field
(wires only, no dielectric)

---

* Lectures given at ICFA Instrumentation School, Guanajuato, Mexico, June 7-18, 1997.

Example #1 - Find the potential and the electric field of an infinitely long charged line with charge per unit length λ (surface S is defined a cylinder of length L and radius r; $\varepsilon_r = 1$):

$$\Phi = 2\pi r L E = \frac{\lambda L}{\varepsilon_0} \tag{1.2}$$

The potential distribution V(r) is then:

$$V(r) = -\int E(r)\,dr = -\frac{\lambda}{2\pi\varepsilon_0}\int \frac{dr}{r} = -\frac{\lambda}{2\pi\varepsilon_0}\ln r + C \tag{1.3}$$

To determine the linear charge λ, we need a boundary condition. This is done by relating the wire voltage V_0 and the linear charge λ on the surfaces of the conductor, such as a grounded tube surrounding the wire (r_a is anode wire radius, r_c is cathode tube radius):

$$V_0 = \int_{r_a}^{r_c} \vec{E}\cdot d\vec{r} = \frac{\lambda}{2\pi\varepsilon_0}\ln\frac{r_c}{r_a} \tag{1.4}$$

The electric field on the surface of the wire with radius r_a is then:

$$E_a = \frac{1}{2\pi\varepsilon_0}\frac{\lambda}{r_a} \tag{1.5}$$

In case of anode wire, E_a will be used to estimate the wire gain. Typical values of the electric field on the anode wire surface is 200-400 kV/cm; whereas on the cathode wire surface it is less than 20 kV/cm.

Example #2 - Determine the electric gradient on the surface of cathode wires:

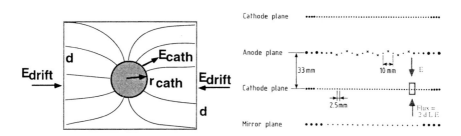

Fig.1 - Geometry of the drift cell; cathode plane is made of 100 μm dia. Cu-Be wires separated by d = 2.5 mm gap; drift field is E = 950 V/cm, wire length L.

The field lines in the middle of drift cell have to end on the cathode wire charges. Therefore field flux $\Phi = 2\,d\,L\,E_{\text{drift field}} = \lambda\,L/\varepsilon_0 = 2\,L\,\pi\,R_{\text{cath}}$. $E_{\text{cathode wire radius}}$ and $E_{\text{cathode wire radius}} = d\,E_{\text{drift field}}/(\pi r_{\text{cath}}) = 0.25 \times 950/(\pi\,50 \times 10^{-4}) \sim 15.1$ kV/cm.

Example #3 - Determine the two-dimensional potential and electric field in the drift cell.

Input variables: V_i - voltages on each wire, $i = 1,...N$
   $\vec{r}_i$ - wire positions, $i = 1,...N$
   a_i - wire radii, $i = 1,...N$
   L - length of the wire

Output variables: λ_i - charge per unit length, $i = 1,...N$

Using the **superposition principle** and the **Gauss law** (see Example #1), the potential at any point within the cell is a result of a summation of potentials resulting from individual linear charges on the wires:

$$V(\vec{r}) = \frac{1}{2\pi\varepsilon_0} \sum_{k=1}^{N} P_{ik}\,\lambda_i\,, \quad \vec{r} \neq \vec{r}_i \qquad (1.8)$$

where P_{ik} are called **potential coefficients**:

$$P_{ik} = \ln(|\vec{r}_i - \vec{r}_k|)\,,\ i \neq k \qquad (1.9)$$
$$P_{ii} = \ln(a_i)$$

This method is quite general - it can be applied to calculate the potential for charges of any type (point, line, ring, etc.). To determine the linear charges λ_k, we need a boundary condition. This is done by relating the wire voltages V_i and the linear charges λ_k on the surfaces of conductors:

$$V_i = \frac{1}{2\pi\varepsilon_0} \sum_{k=1}^{N} P_{ik}\,\lambda_k \qquad (1.10)$$

In addition, one can assume that the total charge of the system is zero $\sum_{i=1}^{N} \lambda_i = 0$.

One can rewrite the equation (1.10) in the matrix form:

$$\begin{pmatrix} 0 \\ V_1 \\ \cdot \\ V_N \end{pmatrix} = \frac{1}{2\pi\varepsilon_0} \begin{pmatrix} 0 & 1 & \cdot & 1 \\ 1 & \cdot & & \cdot \\ \cdot & & P_{ik} & \cdot \\ 1 & \cdot & \cdot & \cdot \end{pmatrix} \begin{pmatrix} \lambda_0 \\ \lambda_1 \\ \cdot \\ \lambda_N \end{pmatrix} \qquad (1.11)$$

Solution is:
$$\lambda_i = 2\pi\varepsilon_0 \sum_{k=1}^{N} C_{ik} V_k \qquad (1.12)$$

where C_{ii} are called coefficients of capacitance and C_{ik}, $i \neq k$ are called coefficients of induction. Once we know the linear charges λ_i on each wire, the potential and electric field in space anywhere between the wires is given by:

$$V(\vec{r}) = \frac{1}{2\pi\varepsilon_0} \sum_{k=1}^{N} \lambda_i \ln(|\vec{r} - \vec{r}_i|), \quad \vec{r} \neq \vec{r}_i \qquad (1.13)$$

$$\vec{E}(\vec{r}) = -\text{grad } V(\vec{r}) = \frac{1}{2\pi\varepsilon_0} \sum_{k=1}^{N} \frac{-\lambda_i}{(|\vec{r} - \vec{r}_i|)^2} (\vec{r} - \vec{r}_i) \qquad (1.14)$$

The electrostatic force per unit length of wire i created by the rest of the system can be calculated as follows (wire i has a length L):

$$\vec{F}(\vec{r}_i) = \frac{1}{2\pi\varepsilon_0} \lambda_i \vec{E}(\vec{r}_i) = (\frac{1}{2\pi\varepsilon_0})^2 L \lambda_i \sum_{k=1}^{N} \frac{\lambda_k}{(|\vec{r}_i - \vec{r}_k|)^2} (\vec{r}_i - \vec{r}_k) \qquad (1.15)$$

1.3. Analytic solution of the Laplace equation for geometries with wires and pads (no dielectric)

The following method is discussed by P.M. Morse and H. Feshbach [4].

1.3.1. The simplest problem is a **single wire and a conducting plane** (pad plane):

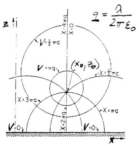

Fig.2 - Single wire and a pad plane.

We assume that the y axis is parallel to the direction of the wires, zero potential of the conducting plane (z = 0), and charge per unit of length λ. The solution of the Laplace equation is a complex potential F (in MKS units):

$$F(Y = x + iz) = V + i\chi = -\frac{\lambda}{2\pi\varepsilon_0} \ln\frac{(Y - Y^0)}{(Y - \overline{Y}^0)} \qquad (1.16)$$

where $Y = x + iz$ is a coordinate of a general point, $Y^0 = x_0 + iz_0$ is the position of the wire, $\overline{Y}^0 = x_0 - iz_0$ is the complex conjugate of Y^0. The real potential:

$$V(x,z) = \mathrm{Re}\, F(Y) = -\frac{\lambda}{2\pi\varepsilon_0}\ell n\left[\frac{(x-x_0)^2+(z-z_0)^2}{(x-x_0)^2+(z+z_0)^2}\right] \quad (1.17)$$

1.3.2. We can expand the problem to a **system of many wires** located above the conducting plane using the superposition principle - see Fig. 3:

$$F(Y = x + iz) = -\frac{\lambda}{2\pi\varepsilon_0}\sum_{k=-\infty}^{k=+\infty}\ell n\frac{(Y-Y_k^0)}{(Y-\overline{Y}_k^0)} \quad (1.18)$$

where Y_k^0 is the coordinate of the k-th wire. If all wires of the grid are spaced with uniform pitch s then one can write:

$$Y_k^0 = x_0 + ks + iz_0, \quad (k = \ldots -2, -3, 0, 1, 2, \ldots) \quad (1.19)$$

where x_0 and z_0 are the wire coordinates for $k = 0$. The complex potential can then be written as a sum, which can be solved:

$$F(Y = x + iz) = -\frac{\lambda}{2\pi\varepsilon_0}\sum_{k=-\infty}^{k=+\infty}\ell n\frac{(Y-Y^0-ks)}{(Y-\overline{Y}^0-ks)} =$$

$$-\frac{\lambda}{2\pi\varepsilon_0}\ell n\frac{\sin[(\pi/s)(Y-Y^0)]}{\sin[(\pi/s)(Y-\overline{Y}^0)]} \quad (1.20)$$

and the corresponding real potential:

$$V(x,z) = \mathrm{Re}\, F(Y) = -\frac{\lambda}{2\pi\varepsilon_0}\ell n\frac{\sin^2[(\pi/s)(x-x_0)]+\sinh^2[(\pi/s)(z-z_0)]}{\sin^2[(\pi/s)(x-x_0)]+\sinh^2[(\pi/s)(z+z_0)]} \quad (1.21)$$

1.3.3. With similar superposition techniques one can then construct the potential distribution of **more complex electrode structures** [3,5] - see Fig. 4:

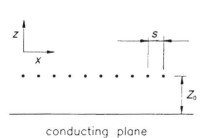

Fig.3 - Wire grid and a pad plane. geometry.

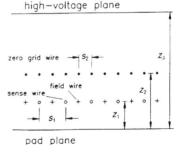

Fig.4 - A typical TPC

1.4. Numerical solution of Laplace equation for any shape of the electrodes (no dielectric)

A good reference is K.J. Binns and P.J. Lawrenson [6]; see also S. Yellin [7]. The numerical solution is based on the following steps: (a) conductor surfaces are specified by linking smooth curves,(b) create equally spaced lattice points with spacing d, (c) determine the topology of a given lattice point relative to the electrode structure; then assign an estimate of the potential per each lattice point, (d) the potential at each interior point of a group of four is average of its nearest neighbors - see Fig.5 for an example of CRID detector, (e) **iterate**; start first with a coarse lattice, then reduce it, etc.

We use the Gauss law in integral form. In absence of charges, the surface integral over the square in Fig. 5 is equal to zero:

$$\Phi = \oint_S \vec{E} \cdot d\vec{S} = \frac{1}{\varepsilon_0} \sum_i Q_i = 0 \qquad (1.22)$$

The electric filed in the n-th iteration is calculated on each surface boundary, for example $E_x^n = (V_{i-1,j}^n - V_{i,j}^n)/d$, etc. - see Fig. 5. By a simple algebra one obtains a simple equation used in the computation:

$$V_{i,j}^{n+1} = \frac{1}{4}[V_{i-1,j}^n + V_{i+1,j}^n + V_{i,j-1}^n + V_{i,j+1}^n] \qquad (1.23)$$

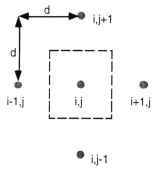

Fig.5 - Grid used to calculate potential distribution.

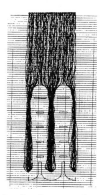

Fig.6 - Barrel CRID single electron detector.

1.5. Numerical solution of the Laplace equation for geometries with wires, pads and dielectric

We can follow the same procedure outlined in the previous problem [6]. However, we have to be more careful to account for the presence of charges within a given grid cubical of Fig.5. Again, we use the Gauss law in integral form:

$$\Phi = \oint_S \vec{E} \cdot d\vec{s} = \frac{1}{\varepsilon_r \varepsilon_0} \sum_i Q_i = \frac{1}{\varepsilon_r \varepsilon_0}(Q_{volume} + Q_{surface} + Q_{leakage} + Q_{ionization}) \quad (1.24)$$

The potential in the center of cubical of Fig.5 is now calculated taking into account the charges within the cubical:

$$V_{i,j}^{n+1} = \frac{1}{4}[V_{i-1,j}^n + V_{i+1,j}^n + V_{i,j-1}^n + V_{i,j+1}^n + \frac{1}{\varepsilon_r \varepsilon_0} d^2 \sum_i Q_i] \quad (1.25)$$

Example of a practical application of this method can be found in Ref. 8. For an alternative method of solution see Ref. 9.

Chapter 2

2. Electrostatic stability of a large system of wires

Equation describing the wire stability can be written as follows:

$$T\frac{d^2 y}{dx^2} + F_{electrostatic} + F_{gravity} = 0 \quad (2.1)$$

where T is mechanical tension on the wire per unit length, x is the coordinate along wire direction, y(x) is the displacement perpendicular to wire, $T d^2 y/dx^2$ is restoring force per unit length, $F_{electrostatic}$ is electrostatic force per unit length and $F_{gravity}$ is gravitational force per unit length.

2.1. Solution A:

$F_{gravity}$ represents a constant force per unit length. This is not generally true for the electrostatic force $F_{electrostatic}$, which generally depends on a value of the displacement y(x). We are going to solve this problem **iteratively** assuming that in each step the electrostatic force is constant force per unit length just like the gravitational force (in the solution B this will not be assumed).

a) **Gravitational force on a wire i:**

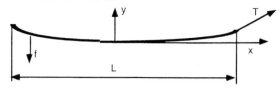

For the gravitational force alone, the solution of equation (2.1) is a parabola:

$$y(x) = \frac{f x^2}{2 T} \tag{2.2}$$

A wire sagitta at $x = L/2$ is: $\quad s_g = y(x = \frac{L}{2}) = \frac{F_{gravity} L^2}{8 T} = \frac{F_{TOT} L}{8 T}, \tag{2.3}$

where L is length of the wire, $F_{gravity}$ is force per unit length, F_{TOT} is a total force ($F_{gravity}$ L) and T is mechanical tension. In a large system of wires, the gravitational force on wire i is (i=1,...N):

$$F_{TOT}(\vec{r}_i) = L\, r_i\, p\, R_i^2\, g, \tag{2.4}$$

where p_i is wire density, R_i is wire radii, L is length of the wire, g is gravitation constant.

Example of gravitational deflections:

Material	$p[g/cm^3]$	$R[\mu m]$	$\pi R^2 p\,[g/cm^3]$	L[cm]	$s_g[\mu m]$	T[g]
W	19.3	10	6.063x10⁻⁵	240	72	60
Cu-Be	8.23	50	6.464x10⁻⁴	240	72	646
Al	2.7	50	2.121x10⁻⁴	240	72	212
C	1.8	50	1.414x10⁻⁴	240	72	141
s.s.	7.5	50	5.890x10⁻⁴	240	72	589

b) **Electrostatic force created by the wire system:** see equation (1.15).

METHOD TO DETERMINE THE WIRE STABILITY [10]:

(a) Solve the 2-dimensional electrostatic problem for λ_i and determine the electrostatic and gravitational forces on each wire; (b) Include the 3-rd dimension by calculating the wire deflections $\vec{d}_i$ using the equation (2.3); (c) Move each wire by $\vec{d}_i$ in the 2-dimensional electrostatic problem and recalculate the electrostatic problem again and determine new λ_i; (d) Iterate in this way ~15 times. If the design is stable in 3-4 iteration, it is safe to build.

Does this simple minded approach work ?

Example #1 - 8-wire prototype for the OPAL central drift chamber) [10]:

Anode wires : T = 101 g, 25 μm dia. W(Re),
Cathode wires : T = 620 g, 100 μm dia. Cu-Be

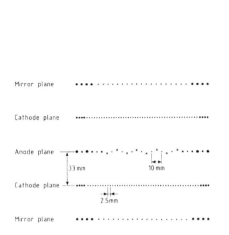

Fig.1 - Geometry of 4.5 m long drift cell; six wires (diameters 300, 175, 300, 175, 100 and 175 μm) terminate the anode plane at each side.

Fig.2 - Additional deflection as obtained in each iteration step of the electrostatic calculation: (A) NiCoTi wire, 35 μm dia., length 4 m, T = 0.9 N, (B) NiCoTi wire, 35 μm dia., length 4 m, T = 1.2 N, (C) NiCoTi wire, 30 μm dia., length 4 m, T=0.9 N, (D) W(Re) wire, 25 μm dia., length 4.5 m, T = 1.0 N.

Without the iterative electrostatic program, this prototype would have never worked. It is very difficult to guess stability of a 4.5 m long chamber like this. As a result of this study, the OPAL chamber was shortened from 4.5 m length to less than 4 m, and the NiCoTi wire was eliminated.

Example #2 - early study for the Mark II vertex chamber at SLC [11]:
Anode (T = 60 g, 20 µm dia. W), **Cathode**(T = 500 g, 150 µm dia. Cu-Be):

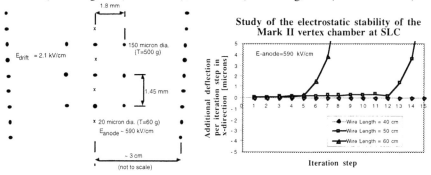

Example #3 - early study for the BaBar drift chamber at PEP II [12]:
Length: 2.5 m & 5 m
Anode: (T = 50 g, 20 µm dia. W), **Cathode:** (T = 25 & 50 g, 55 µm dia. Al):

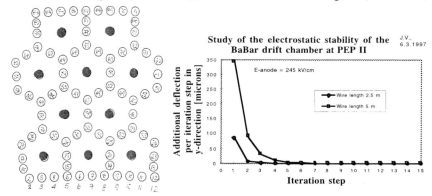

Notes:

(a) Vertex chambers with small wire spacing have similar severe electrostatic instability problems as very long chambers; (b) Ideal drift cells are symmetric such as BaBar drift cell;

(c) A typical drift chambers design, such as the jet chamber, stagger the sense wires by some offset d to solve the left-right ambiguity. This will uniquely define a direction the sense wires will move; (d) One always balances gravitational deflection among different wires in the drift structure; (e) One must stay bellow a certain critical tension for a given choice of wires. This must be tested for each wire choice; (f) The advantage of the numerical method to evaluate the electrostatic stability is that one can find sensitivity to errors in wire tension, wire position, etc.

2.2. Solution B:

We will try to solve the equation (2.1) differently. We will assume that $F_{electrostatic}$ is proportional to the wire displacement $y(x)$:

$$F_{electrostatic} \sim k\, y(x) \qquad (2.5)$$

Combining equations (2.1), (2.3) and (2.5) we obtain:

$$T\frac{d^2 y}{dx^2} + k\, y(x) + \frac{8 T s_g}{L^2} = 0 \qquad (2.6)$$

which has a solution:

$$y(x) = \frac{8 s_g T}{L^2 k}\left(\frac{\cos\sqrt{\frac{k}{T}}(x-\frac{L}{2})}{\cos\sqrt{\frac{k L}{T 2}}} - 1\right) \qquad (2.7)$$

where s_g is the gravitational sagitta of the wire, T is mechanical tension on the wire per unit length, L is length of the wire and k is proportionality constant (do not know yet).

For $x = L/2$ we get:

$$y(\frac{L}{2}) = \frac{8 s_g T}{L^2 k}\left(\frac{1}{\cos\sqrt{\frac{k L}{T 2}}} - 1\right) = \frac{2 s_g}{\chi^2}\left(\frac{1}{\cos\chi} - 1\right) \qquad (2.8)$$

One can see that the wire displacement diverges when $\chi = \sqrt{k/T}\, L/2 \sim \pi/2$, i.e. the chamber becomes unstable. What is the constant k ? It can be shown that:

$$F_{electrostatic} = \frac{V^2}{2}\frac{dC}{dy} = \frac{V^2}{2}\{4\pi\varepsilon_0 \frac{1}{[a\,\ell n(a/r)]^2}\} y(x) = k\, y(x) \qquad (2.9)$$

where dC/dy is change in capacitance per unit length due to displacement in y, V is potential of the wire, r is wire radius, a is typical distance of the wire to the other electrodes. From equation (2.9) we now understand why vertex chambers with small wire spacing have electrostatic problems. As the wire spacing gets small the inter-electrode capacitance gets larger, and therefore we need larger voltages to get the same wire charge. The instability is proportional to voltage square.

The wires may also start vibrating in a presence of large radiation [13], and one has to consider the wire friction of gas in equation (2.1).

Chapter 3:

3.1. Drift of electrons and ions in gases (macroscopic view)

A single electron moving in electric and magnetic fields, $\vec{E}$ and $\vec{B}$, and under the influence of a **frictional force,** can be described by a system of linear differential equations:

$$m\frac{d\vec{v}}{dt} = e\vec{E} + e[\vec{v} \times \vec{B}] - K\vec{v} \tag{3.1}$$

where m - is the mass of the electron, e - is the electric charge of a particle, $\vec{v}$ - is drift velocity vector, $K\vec{v}$ - "Langevin" frictional force, K is a constant, m/K - has a dimension of time (will call it the characteristic time $\tau \equiv m/K$).

We are interested in a steady state solution of equation (1), $d\vec{v}/dt = 0$, which occurs for $t \gg \tau$. From the equation (3.1) we get :

$$\frac{d\vec{v}}{dt} = 0 = \frac{e}{m}\vec{E} + \frac{e}{m}[\vec{v} \times \vec{B}] - \frac{K}{m}\vec{v},$$

$$\frac{e}{m}\vec{E} = \frac{K}{m}\vec{v} - \frac{e}{m}[\vec{v} \times \vec{B}], \quad \frac{e}{m}\vec{E} = \frac{1}{\tau}\vec{v} - \frac{e}{m}[\vec{v} \times \vec{B}] \tag{3.2}$$

Let's define the new variables (ω is the cyclotron frequency):

$$\vec{\omega} = \frac{e}{m}\vec{B}, \quad \vec{\varepsilon} = \frac{e}{m}\vec{E}, \quad \mu = \frac{e}{m}\tau \tag{3.3}$$

Equation (3.2) changes to:
$$\vec{\varepsilon} = \frac{1}{\tau}\vec{v} - [\vec{v} \times \vec{\omega}] \tag{3.4}$$

where
$$\vec{v} \times \vec{\omega} = \begin{pmatrix} i & j & k \\ v_x & v_y & v_z \\ \omega_x & \omega_y & \omega_z \end{pmatrix} \tag{3.5}$$

Expressing equations (3.4) and (3.5) explicitly: (3.6)

$$\varepsilon_x = \frac{1}{\tau}v_x - \omega_z v_y + \omega_y v_z, \quad \varepsilon_y = \omega_z v_x + \frac{1}{\tau}v_y - \omega_x v_z, \quad \varepsilon_z = -\omega_y v_x + \omega_x v_y + \frac{1}{\tau}v_z$$

This can be rewritten in a matrix form: $M\vec{v} = \vec{\varepsilon}$ (3.7)

where
$$M = \begin{pmatrix} \frac{1}{\tau} & -\omega_z & \omega_y \\ \omega_z & \frac{1}{\tau} & -\omega_x \\ -\omega_y & \omega_x & \frac{1}{\tau} \end{pmatrix} \tag{3.8}$$

The solution is obtained by inverting matrix M: (3.10)

$$M^{-1} = \frac{\tau}{1+\omega^2\tau^2} \begin{pmatrix} 1+\omega_x^2\tau^2 & \omega_z\tau+\omega_x\omega_y\tau^2 & -\omega_y\tau+\omega_x\omega_z\tau^2 \\ -\omega_z\tau+\omega_x\omega_y\tau^2 & 1+\omega_y^2\tau^2 & \omega_x\tau+\omega_y\omega_z\tau^2 \\ \omega_y\tau+\omega_x\omega_z\tau^2 & -\omega_x\tau+\omega_y\omega_z\tau^2 & 1+\omega_z^2\tau^2 \end{pmatrix}$$

where
$$\omega^2 = \omega_x^2 + \omega_y^2 + \omega_z^2 = \left(\frac{e}{m}\right)^2 B^2 \quad (3.11)$$

is the square of the cyclotron frequency of the electron.

The final solution can be rewritten after some algebra in a form:

$$\vec{v} = \frac{\mu}{1+(\omega\tau)^2}\left[\vec{E} + \frac{\omega\tau}{B}[\vec{E}\times\vec{B}] + (\omega\tau)^2 \frac{\vec{E}\cdot\vec{B}}{B^2}\vec{B}\right] \quad (3.12)$$

where the drift direction is governed by the dimensionless parameter $\omega\tau$. For $\omega\tau = 0$, $\vec{v}$ is parallel to $\vec{E}$, and equation (3.12) yields: $\vec{v} = \mu\vec{E}$, where μ is the electron mobility, which is proportional to the characteristic time between collisions. From equation (3.11) we obtain $\omega\tau = e/m\, B\,\tau$.

For $\quad \omega\tau = 0 \quad\quad\quad ==> \quad \vec{v} = \mu\vec{E}$, i.e. $\vec{v}$ is aligned with $\vec{E}$,

$\quad\quad\quad \omega\tau$ large $\quad\quad\quad ==> \quad \vec{v}$ tends to be aligned along $\vec{B}$,

$\quad\quad\quad \omega\tau$ large & $\vec{E}\cdot\vec{B} = 0 \quad ==> \quad \vec{v}$ tends to be aligned along $\vec{E}\times\vec{B}$.

In practical chambers we have these conditions typically: $\mu \sim 10^4$ cm^2 V^{-1} s^{-1} for electrons, $\mu \sim 1$ cm^2 V^{-1} s^{-1} for ions, $B \leq 1$ T $= 10^{-4}$ V s cm^{-2}, $\omega\tau = B\mu \approx 10^{-4}$ for ions, $\omega\tau = B\mu \approx 1$ for electrons, $\tau \approx$ 2-5 psec for electrons, $1/\tau \approx$ (2-5) x 10^{11} Hz collision rate for electrons. The effect of typical magnetic fields on ion drift is negligible.

Example #1 - $\vec{E}$ is perpendicular to $\vec{B}$, i.e. $\vec{E}\cdot\vec{B}=0$, $\vec{E} = (E_x,0,0)$, $\vec{B} = (0,0,B_z)$:

From equation (3.12) we obtain:

$$v_x = \frac{\mu}{1+(\omega\tau)^2} E_x \equiv \frac{\mu}{1+(\omega\tau)^2}|\vec{E}|,$$

$$v_y = -\frac{\mu}{1+(\omega\tau)^2}\frac{\omega\tau}{B_z}E_x B_z \equiv -\frac{\mu}{1+(\omega\tau)^2}\omega\tau|\vec{E}|,$$

$$v_z = 0 \quad (3.13)$$

Lorentz angle θ_{xy}: $\quad\quad \tan\theta_{xy} = \frac{v_y}{v_x} = -\omega\tau \quad (3.14)$

By measuring the Lorentz angle we determine $\omega\tau$. The drift velocity magnitude:

$$v(E,B) = \sqrt{v_x^2 + v_y^2} = \frac{\mu}{\sqrt{1+(\omega\tau)^2}} |\vec{E}| = \mu |\vec{E}| \cos\theta_{xy} =$$

$$= v(E, B=0) \cos\theta_{xy} = v(E\cos\theta_{xy}, B=0) \tag{3.15}$$

This is known as **Tonk's theorem** [14]. Experimental verification of the Tonk's theorem for methane c an be found in Ref. 15.

Example #2 - $\vec{E}$ is parallel to $\vec{B}$, i.e. $\vec{E} \times \vec{B} = 0$, $\vec{E} = (0,0,E_z)$, $\vec{B} = (0,0,B_z)$

From equation (3.12) we obtain: (3.16)

$$v_x = 0, \quad v_y = 0, \quad v_z = \frac{\mu}{1+(\omega\tau)^2}[E_z + (\omega\tau)^2 \frac{E_z \cdot B_z}{B^2} B_z] \equiv \mu |\vec{E}|$$

Example #3 - $\vec{E}$ is nearly parallel to $\vec{B}$, i.e. $|\vec{B}| \approx B_z$, $\vec{E} = (0,0,E_z)$, $\vec{B} = (0, B_y, B_z)$, $B_y \ll B_z$:

First we evaluate: $\vec{E} \cdot \vec{B} = E_x B_x + E_y B_y + E_z B_z = E_z B_z$

$$\vec{E} \times \vec{B} = \begin{pmatrix} i & j & k \\ E_x & E_y & E_z \\ B_x & B_y & B_z \end{pmatrix} = \begin{pmatrix} i & j & k \\ 0 & 0 & E_z \\ 0 & B_y & B_z \end{pmatrix} = i\, E_z B_y$$

From equation (3.12) we obtain:

$$v_x = \frac{\mu}{1+(\omega\tau)^2} \frac{\omega\tau}{B} E_z B_y \approx \frac{\omega\tau}{1+(\omega\tau)^2} \frac{B_y}{B_z} v(B=0)$$

$$v_y = \frac{\mu}{1+(\omega\tau)^2} (\omega\tau)^2 \frac{E_z \cdot B_z}{B^2} B_y \approx \frac{(\omega\tau)^2}{1+(\omega\tau)^2} \frac{B_y}{B_z} v(B=0) \tag{3.17}$$

$$v_z = \frac{\mu}{1+(\omega\tau)^2}[E_z + (\omega\tau)^2 \frac{E_z \cdot B_z}{B^2} B_z] \approx \mu E_z = v(B=0)$$

where $v(B=0) = \mu E_z$ is drift velocity for $B = 0$. We can define two Lorentz angles:

$$\tan\theta_{yz} = \frac{v_y}{v_z} = \frac{(\omega\tau)^2}{1+(\omega\tau)^2} \frac{B_y}{B_z}, \quad \tan\theta_{xy} = \frac{v_y}{v_x} = \omega\tau \tag{3.18}$$

SLD CRID example:

(a) $B_z(r,z) = B_z^0 + \frac{1}{2} B_r^0 \frac{r^2 - 2z^2}{r_0 z_0}$, $B_z^0 = 0.6$ T, $B_r(r,z) = B_r^0 \frac{rz}{r_0 z_0} = \kappa z$, $r_0 = 1.2$ m, $z_0 = 1.5$ m, $B_r^0 = 0.0214$ T, (b) assume that B_r is parallel with y axis

($B_r \approx B_y$), (c) electric field E ~ 400 V/cm, (d) drift velocity of 4.3 cm/µs, (e) $\omega\tau$ ~ 0.87 and θ_{xy} ~ 41° for C_2H_6 gas [16], (f) TPC active length is between $z_1 = 0.1$ m and $z_2 = 1.2$ m.

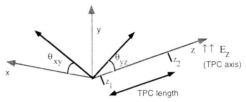

Expected distortion in x-direction:

$$\delta_x = \int_{t_1}^{t_2} v_x \, dt \approx \frac{\omega\tau}{1+(\omega\tau)^2} \frac{v(B=0)}{B_z} \int_{t_1}^{t_2} B_r \, dt =$$

$$= \frac{\omega\tau}{1+(\omega\tau)^2} \frac{1}{B_z} \int_{z_1}^{z_2} B_r \, dz = \frac{\omega\tau}{1+(\omega\tau)^2} \kappa [z_2^2 - z_1^2] \approx 9.9 \text{ mm} \qquad (3.19)$$

where κ is a constant: $\quad \kappa = \frac{1}{2} \frac{B_r^0}{B_z} \frac{r}{r_0 z_0}$.

Similarly for y-direction:

$$\delta_y = \int_{t_1}^{t_2} v_y \, dt \approx \frac{(\omega\tau)^2}{1+(\omega\tau)^2} \frac{v(B=0)}{B_z} \int_{t_1}^{t_2} B_r \, dt =$$

$$= \frac{(\omega\tau)^2}{1+(\omega\tau)^2} \frac{1}{B_z} \int_{z_1}^{z_2} B_r \, dz = \frac{(\omega\tau)^2}{1+(\omega\tau)^2} \kappa [z_2^2 - z_1^2] \approx 8.6 \text{ mm} \qquad (3.20)$$

These calculations were verified by the measurement using the UV fiducial fibers [17].

3.2. The drift of electrons in gases (simple microscopic view)

The simple macroscopic theory, based on a concept of the friction force, cannot predict $\omega\tau$, which has to be obtained by measuring the Lorentz angle. Can we do better by introducing the following details of the electron-molecule collisions?

1) As electron moves in the gas it suffers random collisions with molecules of the gas. We assume that there is no correlation in the direction before and after the collision.

2) Number of collisions n in drift distance x is related to the average drift velocity v as follows:
$$n = (x/v)(1/\tau) \quad (3.21)$$
where τ is an average time between collisions and $1/\tau$ - is average rate of collisions.

3) The average time between collisions τ is related to the electron instantaneous velocity V_{inst}, the collision cross-section σ and the density of the gas N as follows:
$$1/\tau = N \sigma v_{inst} \quad (3.22)$$

3) We will introduce the differential probability P of having the next collision between time t and t + dt:
$$P = 1/\tau \, e^{-t/\tau} \, dt \quad (3.23)$$

We will show that this "new picture" will confirm results obtained using the macroscopic concept of the friction force. However, as we will see, this new picture is still too simplistic.

Example #1 - a uniform electric field E and no magnetic field:

An electron between collisions moves accordingly to the equation of motion:
$$m \frac{dv}{dt} = eE \quad (3.24)$$

Its solution is the electron displacement as a function of time is $x(t) = \frac{1}{2} \frac{e}{m} E t^2$.

The average displacement $<x>$ at average time between collisions τ is obtained by averaging x(t) over time t, using the probability distribution of t, i.e. equation (3.23):

$$<x> = \int_0^\infty \frac{1}{2} \frac{e}{m} E t^2 \, \frac{1}{\tau} e^{-t/\tau} \, dt = \frac{e}{m} E \tau^2 \quad (3.25)$$

The average electron drift velocity is defined as:
$$<v> = \frac{<x>}{\tau} = \frac{e}{m} E \tau = \mu E \quad (3.26)$$

Example #2 - $\vec{E}$ is perpendicular to $\vec{B}$, i.e. $\vec{E}\cdot\vec{B}=0$, $\vec{E} = (E_x,0,0)$, $\vec{B} = (0,0,B_z)$:

An electron between collisions moves accordingly to the equation of motion:

$$m \frac{d\vec{v}}{dt} = e\vec{E} + e[\vec{v} \times \vec{B}] \qquad (3.27)$$

which can be rewritten as a system of differential equations:

$$m \frac{dv_x(t)}{dt} = eE_x + ev_y B_z, \quad m \frac{dv_y(t)}{dt} = -ev_x B_z, \quad m \frac{dv_z(t)}{dt} = 0 \qquad (3.28)$$

We introduce the cyclotron frequency $\omega = e/m\, B_z$ and obtain the solution for the initial conditions $v_x(0) = v_y(0) = v_z(0) = 0$:

$$v_x(t) = \left(\frac{e}{m} E_x \frac{1}{\omega}\right) \sin \omega t, \quad v_y(t) = \left(\frac{e}{m} E_x \frac{1}{\omega}\right)(\cos \omega t - 1), \quad v_z(t) = 0 \qquad (3.29)$$

The drift velocity is then given by the following averages ($\mu = e/m\, \tau$):

$$<v_x(t)> = \frac{e}{m} E_x \frac{1}{\omega} \int_0^\infty \sin \omega t \, \frac{1}{\tau} e^{-t/\tau} dt = \frac{e}{m} \frac{E_x \tau}{1+(\omega\tau)^2} = \frac{\mu}{1+(\omega\tau)^2} E_x$$

$$<v_y(t)> = \frac{e}{m} E_x \frac{1}{\omega} \int_0^\infty (\cos \omega t - 1) \frac{1}{\tau} e^{-t/\tau} dt =$$

$$-\frac{e}{m} \frac{E_x \omega \tau^2}{1+(\omega\tau)^2} = \frac{\mu}{1+(\omega\tau)^2} \frac{\omega \tau}{B_z} E_x B_z \qquad (3.30)$$

$$<v_z(t)> = 0$$

We have obtained the same results as in equations (3.15) which was derived from the friction force model. The Lorentz angle θ_{xy} is:

$$\tan \theta_{xy} = \frac{v_y}{v_x} = -\omega \tau(E,B) \equiv -\frac{e}{m} B \tau = -\mu B = -\frac{v(E,B=0)\, B}{E} \qquad (3.31)$$

However, in practice the equation (3.31) is only approximate because our modeling of electron collisions with the gas is too simple minded. The correct equation is:

$$\tan \theta_{xy} = \psi \frac{v(E,B=0)\, B}{E} \qquad (3.32)$$

where $\psi = \psi(E/N, B/N)$ is the **magnetic deflection factor.**

Measurement of the magnetic deflection factor ψ was done by T. Kunst, B. Goetz and B. Schmidt [15] - see Fig. 1. The fact that it is not really a constant casts a doubt that we are dealing with a real theory so far.

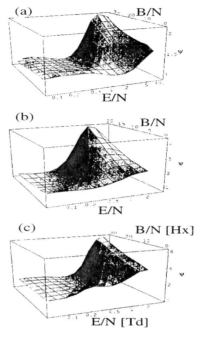

Fig. 1 - Magnetic deflection factor ψ for (a) CH$_4$, (b) 90% Ar+10% CH$_4$, (c) 95% Ar + 5% CH$_4$ gases.

Example #3: - $\vec{E}$ is nearly parallel to $\vec{B}$, i.e. $|\vec{B}| \approx B_z$, $\vec{E} = (0,0,E_z)$, $\vec{B} = (0,B_y,B_z)$, $B_y \ll B_z$:

From equation (3.27) we obtain a system of differential equations: (3.33)

$$m\frac{dv_x(t)}{dt} = e(v_y B_z - v_z B_y), \quad m\frac{dv_y(t)}{dt} = -e\, v_x B_z, \quad m\frac{dv_z(t)}{dt} = e E_z + e\, v_x B_y$$

The solution for the initial conditions $v_x(0) = v_y(0) = v_z(0) = 0$:

$$v_x(t) = \frac{B_y E_z}{B^2}(\cos\omega t - 1), \quad v_y(t) = \frac{B_y B_z E_z}{B^3}(\omega t - \sin\omega t) \qquad (3.34)$$

$$v_z(t) = \frac{E_z}{B^3}(B_z^2\, \omega t + B_y^2 \sin\omega t)$$

The drift velocity is then given by the following averages:

$$<v_x(t)> = \frac{B_y E_z}{B^2} \int_0^\infty (\cos\omega t - 1)\frac{1}{\tau} e^{-t/\tau} dt = -\frac{\mu^2}{1+(\omega\tau)^2} B_y E_z \quad (3.35)$$

$$<v_y(t)> = \frac{B_y B_z E_z}{B^3} \int_0^\infty (\omega t - \sin\omega t)\frac{1}{\tau} e^{-t/\tau} dt = \frac{\mu^3}{1+(\omega\tau)^2} B_y B_z E_z$$

$$<v_z(t)> = \frac{E_z}{B^3} \int_0^\infty (B_z^2 \omega t + B_y^2 \sin\omega t)\frac{1}{\tau} e^{-t/\tau} dt = \frac{1+\mu^2 B_z^2}{1+(\omega\tau)^2} \mu E_z$$

We have obtained the same results as in equations (3.17). This supports our previous intuition that the characteristic time in the friction model is the same as the average time between collisions in the simple microscopic model. However, in all these examples, $\omega\tau$ and μ had to be obtained in the end from the experiment, and -not- from the theory !! Before we mention the real theory, we have to introduce a concept of electron diffusion.

3.3. Electron diffusion

Drifting electrons scatter on the gas molecules. Their motion can be described by their random motion, which is characterized by the mean energy ε and gives rise to diffusion, and by their collective motion, which is characterized by the average drift velocity. The motion follows the continuity equation (the total electron current is given by the sum of the drift current and the diffusion current: $\vec{J} = n\vec{v} - D\vec{\nabla}n$). Its solution has in the simplest case the isotropic distribution, i.e. the point-like cloud of electrons at time $t = 0$ will create a Gaussian density distribution at time t:

$$n = \left(\frac{1}{\sqrt{4\pi D t}}\right)^3 \exp\left(-\frac{r^2}{4Dt}\right), \quad (3.36)$$

where D is the **diffusion coefficient.** From equation (3.37) follows that the diffusion width of an electron cloud σ_x, after starting point-like and traveled time interval t, is:

$$\sigma_x = \sqrt{2Dt} \quad (3.37)$$

One can show that the diffusion coefficient is related to electron energy ε as follows:

$$D = \frac{2}{3}\frac{\varepsilon}{m}\tau = \frac{\varepsilon_k}{m}\tau \tag{3.38}$$

where ε is electron energy, ε_k is so called **"characteristic energy"**, m is mass of the electron and τ is average time between collisions. Recalling the expression for electron mobility, we obtain expression for the characteristic electron energy:

$$\varepsilon_k = \frac{Dm}{\tau} = \frac{Dm}{\frac{\mu m}{e}} = \frac{De}{\mu}. \tag{3.39}$$

The diffusion width σ_x of an electron cloud width, after starting point-like and traveled over a distance x:

$$\sigma_x = \sqrt{2Dt} = \sqrt{\frac{2Dx}{\mu E}} = \sqrt{\frac{2\varepsilon_k x}{eE}}. \tag{3.40}$$

The smallest diffusion corresponds to a thermal energy $\varepsilon_k = kT$, ($\varepsilon = 3/2\, kT \sim 0.04$ eV at 24°C) resulting in the smallest possible diffusion width σ_x of an electron cloud (so called **"thermal limit"**):

$$\sigma_x = \sqrt{2Dt} = \sqrt{\frac{2kTx}{eE}}. \tag{3.41}$$

However the reality is unfortunately more complicated:

1. Electric field alters the diffusion so that it is necessary to introduce two diffusion coefficients D_L and D_T, one for the longitudinal and one transverse direction in respect to electric field. Fig. 2 shows an example in methane [18].

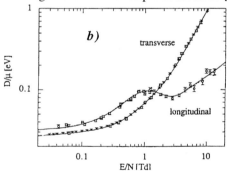

Fig.2 - Diffusion coefficient in methane.

2. Magnetic field alters the diffusion so that the transverse diffusion coefficient $D_T(B)$ in respect to its direction gets smaller:

$$\frac{D_T(B)}{D_T(B=0)} = \frac{1}{1+(\omega\tau)^2}, \qquad (3.42)$$

while the longitudinal diffusion coefficient remains the same $D_L(B) = D_L(0)$.

3.4. Boltzmann equation method.

A full theory of electron transport in gases can get rather complicated. We will follow more simple path of Schultz and Gresser [19]. For more complete description see L. G. H. Huxley and R.W. Crompton [20]. For a theory which includes the ionization and attachment processes see K.F. Ness and R.E. Robson, [21].

Drifting electrons scatter on the gas molecules. This motion follows the Boltzmann transport equation, which expresses the conservation of number of electrons. If f(v,r,t) is the distribution function of electrons at r, v of the phase space at time t, the simplest 1-dimensional form of the Boltzmann equation is:

$$\frac{\partial f}{\partial t} + \frac{\partial f}{\partial r}\frac{\partial r}{\partial t} + \frac{\partial f}{\partial v}\frac{\partial v}{\partial t} - \frac{\partial f}{\partial t}\bigg|_{Coll.} = 0 \qquad (3.43)$$

where $\partial f/\partial t$ represents time evolution of f(v,r,t), $(\partial f/\partial r)(\partial r/\partial t)$ represents loss of electrons in interval dr due to diffusion, $(\partial f/\partial v)(\partial v/\partial t)$ represents loss of electrons in interval dv due to acceleration caused by field E and $\partial f/\partial t|_{Coll.}$ represents loss of electrons in interval dv due to collisions of electrons with molecules of a gas.

To solve the equation (3.47) we introduce the following simplifications:
a) We express the distribution function f using electron energy $\varepsilon = 1/2\,mv^2$, and the mean
   free path $\ell(\varepsilon)$ between two elastic collisions,
b) we assume that the electric field E is parallel with x axis (no magnetic field for now),
c) we assume a stationary case, i.e. no x or t dependence,
d) we assume **no ionization and no attachment processes**,
e) we expand the distribution function f in terms of the Legendre polynomial expansion,
   and use in our case **two terms only**:

$$f = f_0(\varepsilon) + f_1(\varepsilon)\cos\theta + \ldots \qquad (3.44)$$

One gets two coupled equations :

$$eE\frac{\partial f_0}{\partial v} + mv\frac{\partial f}{\partial x} = -\frac{mvf_1}{\ell(\varepsilon)}, \quad \frac{eE}{2v}\frac{\partial}{\partial v}(v^2 f_1) + \frac{1}{2}mv^2\frac{\partial f_1}{\partial x} = \frac{m^2}{M}\frac{3}{2v}\frac{\partial}{\partial v}(\frac{v^4 f_0}{\ell(\varepsilon)}) \quad (3.45)$$

where M is mass of the molecule.

We now assume $\partial f/\partial x = 0$, i.e. uniform distribution along x direction. This yields these two equations:

$$eE\frac{\partial f_0}{\partial v} = -\frac{mvf_1}{\ell(\varepsilon)}, \quad \frac{eE}{2v}\frac{\partial}{\partial v}(v^2 f_1) = \frac{m^2}{M}\frac{3}{2v}\frac{\partial}{\partial v}(\frac{v^4 f_0}{\ell(\varepsilon)}) \quad (3.46)$$

One can now eliminate f_1 and solve for f_0:

$$\frac{2}{3}\frac{(eE)^2}{m}\frac{\partial}{\partial \varepsilon}[\varepsilon \ell(\varepsilon)\frac{\partial(f_0/v)}{\partial \varepsilon}] + \frac{2m}{M}\frac{\partial}{\partial \varepsilon}[\frac{\varepsilon v f_0}{\ell(\varepsilon)}] = 0 \quad (3.47)$$

Equation (3.47) can be easily solved (first, we assume $\ell(\varepsilon) = $ const.):

$$f_0(\varepsilon) = C\sqrt{\varepsilon}\,\exp[-\frac{3m}{M}(\frac{\varepsilon}{eE\ell(\varepsilon)})^2] \quad (3.48)$$

where a constant C is obtained from a normalization: $\int_0^{\varepsilon_{max}} f_0(\varepsilon)d\varepsilon = 1$. A fraction of energy lost by electron scattering elastically from molecule of mass M can be approximated as:

$$\frac{\Delta \varepsilon}{\varepsilon} = \frac{2m}{M}(1-\cos\theta), \quad (3.49)$$

while the mean fraction of energy lost is $\Lambda = 2m/M$. However, the solution (3.48) must be changed if $\ell(\varepsilon)$ is not a constant and if the mean fraction of energy lost in the collision is not equal to $\Lambda = 2m/M$, but it is $\Lambda = \Lambda(\varepsilon)$. In this case equation (3.48) becomes:

$$f_0(\varepsilon) = C\sqrt{\varepsilon}\,\exp[-\int_0^{\varepsilon}\frac{3\Lambda(\varepsilon')\varepsilon'}{(eE\ell(\varepsilon'))^2}d\varepsilon'] \quad (3.50)$$

Unfortunately, one has to introduce several complications:

a) If the energy of electrons is similar as the thermal energy of the molecules ($\varepsilon = kT \sim 0.025$ eV), it is necessary to introduce an additional term in equation (3.47):

$$\frac{2}{3}\frac{(eE)^2}{m}\frac{\partial}{\partial \varepsilon}[\varepsilon \ell(\varepsilon)\frac{\partial(f_0/v)}{\partial \varepsilon}] + \frac{2m}{M}\frac{\partial}{\partial \varepsilon}[\frac{\varepsilon v f_0}{\ell(\varepsilon)}] +$$

$$\frac{\partial}{\partial \varepsilon}[\frac{2\Lambda(\varepsilon)}{m}\frac{\varepsilon^2}{\ell(\varepsilon)}kT\frac{\partial(f_0/v)}{\partial \varepsilon}] = 0 \quad (3.51)$$

This changes solution (3.50) to:

$$f_0(\varepsilon) = C\sqrt{\varepsilon} \, \exp[-\int_0^\varepsilon \frac{3\Lambda(\varepsilon')\varepsilon'}{[(eE\ell(\varepsilon'))^2 + 3\Lambda(\varepsilon')\varepsilon' kT]} d\varepsilon'] \quad (3.52)$$

b) If we wish to add inelastic collisions, we must add to equation (3.51) additional term:

$$\sum_k [\frac{\sqrt{2/m}\sqrt{(\varepsilon+\varepsilon_k)}}{\ell_k(\varepsilon+\varepsilon_k)} f_0(\varepsilon+\varepsilon_k) - \frac{\sqrt{2/m}\sqrt{\varepsilon}}{\ell_k(\varepsilon)} f_0(\varepsilon)], \quad (3.53)$$

where ε_k is the excitation energy of the k-th state and ℓ_k is the mean free path between two collisions which gives rise to the excitation. In practice, the inelastic collisions are vibrational and rotational molecular excitations caused by electrons of sufficient energy. We can approximate equation (3.53) by :

$$\sum_k \varepsilon_k \frac{\partial}{\partial \varepsilon}[\frac{v f_0(\varepsilon)}{\ell_k(\varepsilon)}] \quad (3.54)$$

and the solution (3.50) is still valid provided we use :

$$\Lambda(\varepsilon) = \frac{2m}{M} + \sum_k \frac{\varepsilon_k}{\varepsilon} \frac{\ell_e(\varepsilon)}{\ell_k(\varepsilon)} \quad (3.55)$$

The last equation can be expressed in terms of cross sections :

$$\Lambda(\varepsilon) = \frac{2m}{M} + \sum_k \frac{\varepsilon_k}{\varepsilon} \frac{\sigma_k(\varepsilon)}{\sigma_e(\varepsilon)} \quad (3.56)$$

where $\sigma(\varepsilon) = 1/N\ell(\varepsilon)$ is the cross section, $N = N_0 (p/760)(273/T)$ is the Loschmidt number, N_0 is the Loschmidt number defined at 0°C (2.687 x 10^{25} molecules/m^3), p is pressure in Torr and T is absolute temperature in K.

c) Finally, one includes magnetic field. This results in the following modification of the solution (3.50) ($\vec{E}$ is perpendicular to $\vec{B}$):

$$f_0(\varepsilon) = C\sqrt{\varepsilon} \, \exp[-\int_0^\varepsilon \frac{3\Lambda(\varepsilon')G(B)\varepsilon'}{[(eE\ell(\varepsilon'))^2 + 3\Lambda(\varepsilon')\varepsilon' kTG(B)]} d\varepsilon'] \quad (3.57)$$

where

$$G(B) = 1 + \frac{e^2 B_z^2 \ell_e(\varepsilon)^2}{2m\varepsilon} \quad (3.58)$$

d) Gas mixtures are calculated as follows:

$$\sigma = \sum_i \delta_i \sigma^i, \quad \sigma\Lambda = \sum_i \delta_i \sigma^i \Lambda_i \quad (3.59)$$

Once we obtain the function $f_0(\varepsilon)$, the electron transport coefficients are calculated as follows ($\vec{E}$ is perpendicular to $\vec{B}$):

1. **Drift velocity :**

$$v_x = -\frac{2}{3}\frac{eE}{m}\int_0^{\varepsilon_{max}} \frac{\varepsilon \ell_e(\varepsilon)}{G(B)} \frac{\partial(f_0(\varepsilon)/v(\varepsilon))}{\partial \varepsilon} d\varepsilon,$$

$$v_y = \frac{eEeB}{3m}\int_0^{\varepsilon_{max}} \frac{\ell_e^2(\varepsilon)v(\varepsilon)}{G(B)} \frac{\partial(f_0(\varepsilon)/v(\varepsilon))}{\partial \varepsilon} d\varepsilon$$

$$v = \sqrt{v_x^2 + v_y^2} \tag{3.60}$$

2. **Lorentz angle :** $\quad \theta = \tan^{-1}(\frac{v_y}{v_x})$ (3.61)

3. **The diffusion coefficient :** $\quad D_T = \frac{1}{3}\int_0^{\varepsilon_{max}} \frac{\ell_e(\varepsilon)v(\varepsilon)}{G(B)} f_0(\varepsilon) d\varepsilon$ (3.62)

From the Lorentz angle we determine the $\omega\tau$ term needed in many equation in the earlier part of the lecture. The theory works with a 5-10% accuracy. The following three examples were calculated by the author using a code written by P. Coyle following the Schultz and Gresser theory [22]:

Example #1 ($f_0(\varepsilon)$ function in methane) :

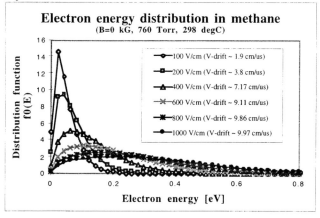

Example #2 ($f_0(\varepsilon)$ function in various gases) :

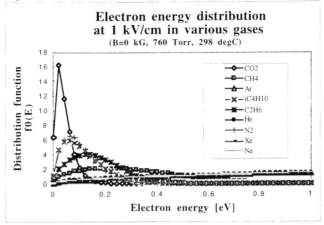

Example #3 (Calculated Lorentz angles):

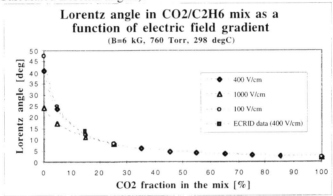

We can now predict the Lorentz angle, and therefore $\omega\tau$. However, the presented theory is not sufficiently precise in some cases because of the following arguments: (a) the presented version of the theory in this lecture does not calculate the longitudinal diffusion coefficient D_L, and (b) if we want 1% precision, the two-term approximation is not sufficient in gases with strong inelastic processes (hydrocarbons, CF_4, etc.). B. Schmidt [18] points out that to explain the data in methane it is necessary to introduce (a) six terms, (b) an anisotropy in the elastic scattering and (c) introduce higher order of vibration cross-sections - see his results on Figs 3 and 4. However, Schmidt's theory does not include the ionization and attachment processes. This particular problem was done, for example, by S. Biagi [23].

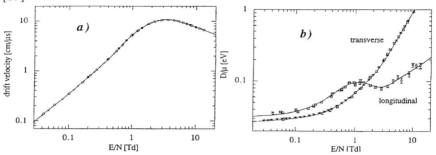

Fig.3 - Drift velocity in methane.   **Fig.4** - Diffusion coefficient in methane.

3.5. Monte Carlo method.

I will mention a recent attempt by H. Pruchova and B. Franek [24]. A computer program follows the electrons and ions in small steps (fraction of a psec) and evaluates electrostatic forces, position and velocities of each electron and ion, and probability of various physics processes using the electron-molecule scattering cross-sections: (a) elastic scattering (electron does not lose energy), (b) inelastic scattering (electron loses energy), (c) ionization scattering (a new electron is produced), (d) attachment scattering (electron is absorbed by a molecule), etc. - see Figs. 5-9 (one should also mention that similar attempts were made earlier, for example, by J. Groh [25], and by M. Matobe et al.[26]).

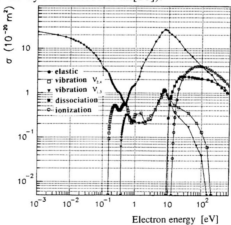

Fig.5 - Electron molecule cross-sections in methane.

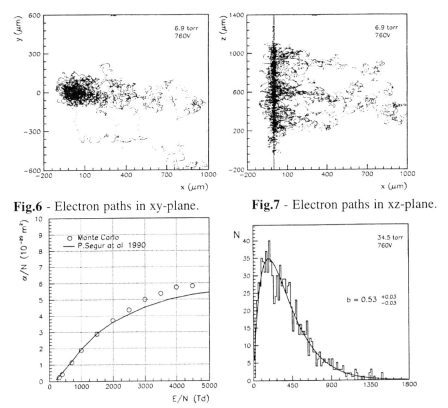

Fig.6 - Electron paths in xy-plane.

Fig.7 - Electron paths in xz-plane.

Fig.8 - Prediction of the 1-st Townsend ionization coefficient using the Monte Carlo and the Boltzmann equation methods.

Fig.9 - Gas gain.

Difficulties of the Monte Carlo method:

1. Very demanding on the CPU time.
2. It is difficult to obtain the "correct" cross-sections. Methane is about the best studied gas of all. Some cross-sections in literature are altered to obtain the best match between theory and data using the Boltzmann equation procedure. There is some risk that the incorrectly "tuned" cross-sections would yield inconsistencies if used in the Monte Carlo method.
3. Still some difficulties to predict the practical quantities such as drift velocity, diffusion, etc., even in methane. Nevertheless, the results are still impressive.

More work is needed in this area !

Chapter 4:

4.1. Gain - phenomenological parametrization

The multiplication of ionization is described by the 1-st Townsend ionization coefficient α, which is defined as the mean number of secondary electrons produced by a free electron per centimeter of its path. The increase of the number of electrons dN per path dr is given by:

$$dN = N \alpha \, dr \quad (4.1)$$

No simple general expression for α exists; it has to be measured for every gas mixture. It can be shown experimentally that α is proportional to gas density ρ, provided that we keep E/ρ fixed:

$$\alpha = f(E/\rho) \, \rho \quad (4.2)$$

The amplification gain on the wire is given by the integration of equation (4.1) between the point r_c where the field is sufficient to start the avalanche and the anode radius r_a:

$$G = \frac{N}{N_0} = \exp \int_{r_c}^{r_a} \alpha(r) \, dr = \exp \int_{E(r_c)}^{E(r_a)} \alpha(E) \frac{\partial r}{\partial E} \, dE \quad (4.3)$$

where N & N_0 is final and initial number of electrons in the avalanche, $\partial E/\partial r$ is the electric field gradient, $E(r_a)$ is the electric field on the surface of the anode wire and $E(r_c)$ is the electric field at critical radius beyond which the field is too low to support charge multiplication. Electric field near an anode wire whose radius is small compared to the inter-electrode distances is determined from equation (1.9). Inserting equation (1.5) into (4.3):

$$G = \frac{N}{N_0} = \exp \int_{E(r_c)}^{E(r_a)} \alpha(E) \frac{\lambda}{2\pi\varepsilon_0 E^2} \, dE \quad (4.4)$$

In a homogeneous electric field, such in the parallel plate chamber with a gap L, the equation (4.2) gives a simple expression $G = N/N_0 = \exp(\alpha L)$.

4.2. Parametrization of α

4.2.1. The Diethorn parametrization [27]:

He approximates α as follows:

$$\alpha(E) = \beta E \quad (4.5)$$

This approximation is valid for noble gases for electric field between 10^2-10^3 [V/cm Torr], a typical range of fields near the thin anode wires. Inserting equation (4.5) into (4.4):

$$\ln G = \int_{E(r_c)}^{E(r_a)} \alpha(E) \frac{\lambda}{2\pi\varepsilon_0 E^2} dE = \frac{\lambda}{2\pi\varepsilon_0} \int_{E(r_c)}^{E(r_a)} \frac{\beta}{E} dE = \frac{\beta\lambda}{2\pi\varepsilon_0} \ln \frac{\lambda}{2\pi\varepsilon_0 r_a E(r_c)} \quad (4.6)$$

Potential difference between $r = r_a$ and $r = r_c$:

$$\phi(r_a) - \phi(r_c) = \int_{r_a}^{r_c} E(r) dr = \frac{\lambda}{2\pi\varepsilon_0} \ln \frac{r_c}{r_a} = \frac{\lambda}{2\pi\varepsilon_0} \ln \frac{\lambda}{2\pi\varepsilon_0 r_a E(r_c)} \quad (4.7)$$

Assuming that it takes energy $e \Delta V$ in average to produce one more electron, the potential difference $\phi(r_a) - \phi(r_c)$ gives rise to Z generations:

$$G = 2^Z, \quad Z = \frac{\phi(r_a) - \phi(r_c)}{\Delta V} \quad (4.8)$$

This results in:
$$\ln G = \frac{\ln 2}{\Delta V} \frac{\lambda}{2\pi\varepsilon_0} \ln \frac{\lambda}{2\pi\varepsilon_0 r_a E(r_c)} \quad (4.9)$$

Therefore $\beta = \ln 2 / \Delta V$. Using equations (4.2), (4.4) and (4.9) we get **the final Diethorn formula:**

$$\ln G = \frac{\ln 2}{\Delta V} \frac{V}{\ln(r_a/r_a)} \ln \frac{V}{\ln(r_c/r_a) r_a E(r_c, \rho_0)(\rho/\rho_0)} \quad (4.10)$$

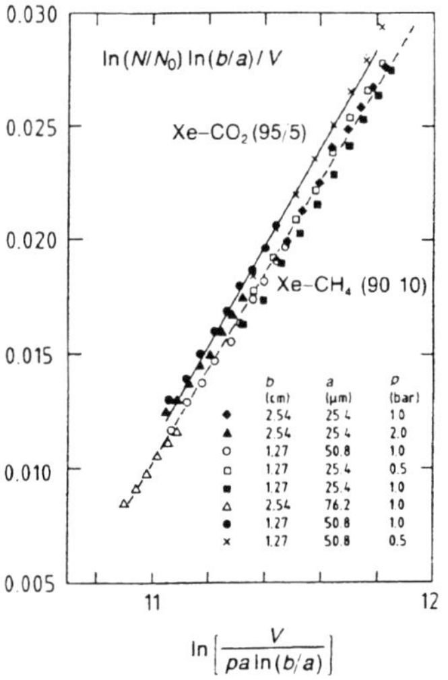

Fig.1 - Diethorn plot ($p = \rho / \rho_0$); different wire diameters fall on the same curve for a given gas - this allows scaling from one design to another.

1) Experimentally we vary $p = \rho/\rho_0$, a and $V/\ln(r_c/r_a)$, and measure G.

2) A plot of $\frac{1}{V}\ln G \ln(r_c/r_a)$ versus $\ln[V/\ln(r_c/r_a) r_a (\rho/\rho_0)]$ must be linear and yields two constants ΔV and $E(r_c, \rho_0)$ - see Fig.1.

Table#1 - Examples of measured Diethorn parameters for several gases:

Gas mixture	$E(r_c)$ [kV/cm]	ΔV [Volts]
90% Ar + 10% CH4	48 ± 3	23.6 ± 5.4
95% Ar + 5% CH4	45 ± 1	21.8 ± 4.4
92.1% Ar + 7.9% CH4	47.5	30.2
23.5% Ar + 76.5% CH4	196	36.2
9.7% Ar + 90.3% CH4	21.8	28.3

Note: ΔV and $E(r_c)$ can be considered fundamental gas constants

Application of the Diethorn formula:

a) Dependence of gain on gas density:

This is especially important for chambers operating at atmospheric pressure (the gas density is proportional to it). From equation (4.10) we obtain:

$$\frac{dG}{G} = -\frac{\lambda \ln 2}{\Delta V \, 2\pi \varepsilon_0} \frac{d\rho}{\rho} \qquad (4.11)$$

The factor that multiplies $d\rho/\rho$ ranges typically between 5 and 8.

b) Dependence of gain on geometry, voltage and space charge:

All these effects change the local charge density of the wire λ. This in turn changes the wire gas gain as follows:

$$\frac{dG}{G} = (\ln G + \frac{\lambda \ln 2}{\Delta V \, 2\pi \varepsilon_0})\frac{d\lambda}{\lambda} \qquad (4.12)$$

The factor that multiplies $d\lambda/\lambda$ ranges typically between 10 and 20.

4.2.2. The Zastawny parametrization [28]:

This parametrization is valid over larger range of s, i.e. it is more general compared to Diethorn's parametrization. First, **introduce a new variable** $S(r) \equiv E(r)/\rho$, where E is electric field intensity and ρ is gas density. Values S_a and S_c correspond to $r = r_a$ (anode surface) and $r = r_c$ (beginning of amplification, i.e. $\alpha(r) = 0$ for $r > r_c$). In the cylindrical geometry:

$$S(r) \equiv \frac{E(r)}{\rho} = \frac{1}{\rho}\frac{E_a \, r_a}{r} = \frac{S_a \, r_a}{r} \quad (4.13)$$

Equation (4.3) can then be rewritten:

$$\ln G = \int_{r_c}^{r_a} \alpha(r)\,dr = \int_{S_c}^{S_a} \alpha \frac{\partial r}{\partial S}\,dS = \rho\, r_a\, S_a \int_{S_c}^{S_a} \frac{\alpha}{\rho}\frac{1}{S^2}\,dS \quad (4.14)$$

Based on equation (4.2), α/ρ is a function of S only. We introduce so called **"reduced gas gain"** Ψ:

$$\Psi \equiv \frac{\ln G}{\rho\, E_a\, r_a} = \int_{S_c}^{S_a} \frac{\alpha}{\rho}\frac{1}{S^2}\,dS = f(S_a) \quad (4.15)$$

The reduced gas gain Ψ is only a function of variable S_a - see Fig. 2.

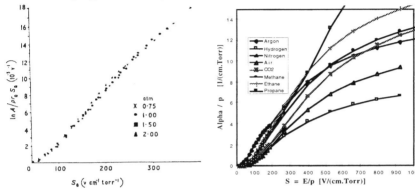

Fig.2 - Reduced gas gain $\Psi = f(S_a)$ in CO_2 gas.

Fig.3 - The first Townsend coefficient $\alpha/\rho = f(S)$.

This is a significant result ! All data are following one single curve for a given gas choice, independent of gas pressure, voltage or anode radius. This is used to predict the gas gain in a new drift chamber geometry: (a) we parametrize the "reduced gas gain" $\psi = f(S_a)$ in a known geometry such as a tube wire chamber, (b) then we scale it to any other geometry, provided that we can calculate S_a in this geometry. This allows to know the gas gain when we simulate the drift chamber geometry in the electrostatic program.

Zastawny parametrized α as follows (in fact this is an old Korff parametrization [29]):

$$\frac{\alpha}{\rho} = A\exp(-\frac{B}{S}) \quad (4.16)$$

147

where for a given gas choice, A and B are constants only in certain E/p intervals (typically 2-3 intervals). In this parametrization the **"reduced gas gain"** ψ has the following simple form:

$$\Psi \equiv \frac{\ln G}{\rho E_a r_a} = \int_{S_c}^{S_a} \frac{\alpha}{\rho} \frac{1}{S^2} dS = f(S_a) = A \exp\left(-\frac{B}{S_a}\right) \qquad (4.17)$$

Fig. 3 shows an example of author's calculations of α using Zastawny's fits to his data.

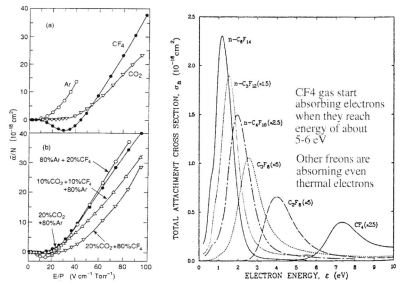

Fig.5 - The effective ionization coefficient $\bar{\alpha}/p$ or CF_4 gas is [30].

Fig.6 - The attachment cross-sections in Freons.

4.3. Gain in the presence of electron attachment

In this case one introduces **the attachment coefficient η,** and the gain equation (4.2) is modified as follows:

$$G = \frac{N}{N_0} = \exp \int_{r_c}^{a} (\alpha(r) - \eta(r)) \, dr \qquad (4.18)$$

where $\bar{\alpha} \equiv \alpha(r) - \eta(r)$ **is the effective ionization coefficient.** From Fig. 5 one can see that we need to add some gas to CF_4, for example iC_4H_{10}, to eliminate its absorption of electrons for: 10 kV/cm < E/p < 35 kV/cm. The explanation of this effect can be traced to the attachment cross-section in fluorocarbons - see Fig. 6. Is this significant for the practical applications ? Yes, it can affect the "near wire resolution" in detectors such as tube wire chambers [31] - see Fig. 7.

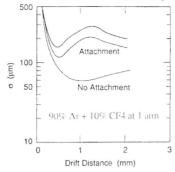

Fig.7 - Monte Carlo simulation of the resolution in 4 mm straw tube.

4.4. Statistical fluctuation of the gas gain

The total charge developed during a pulse is proportional to:

$$Q = n_0 \, e \, M \qquad (4.19)$$

where $n_0 = E/W$ is the number of individual avalanches, E is the energy deposited by the incident radiation, W is the energy required to form one ion pair, M is the average multiplication factor and e is the electron charge. The average multiplication factor from all the avalanches which contribute to a given pulse is:

$$M = \frac{1}{n_0} \sum_{i=1}^{n_0} A_i \equiv \overline{A} \qquad (4.20)$$

where eA_i is the charge contributed to the i-th avalanche. The pulse amplitude, which is proportional to Q, is subject to fluctuations because of fluctuations in n_0 and M :

$$\left(\frac{\sigma_Q}{Q}\right)^2 = \left(\frac{\sigma_{n_0}}{n_0}\right)^2 + \left(\frac{\sigma_M}{M}\right)^2 \qquad (4.21)$$

where σ_M can be rewritten as:

$$\sigma_M^2 = \left(\frac{1}{n_0}\right)^2 \sum_{i=1}^{n_0} \sigma_A^2 = \frac{1}{n_0} \sigma_A^2 \qquad (4.22)$$

Combining equations (4.21) and (4.22):

$$\left(\frac{\sigma_Q}{Q}\right)^2 = \left(\frac{\sigma_{n_0}}{n_0}\right)^2 + \frac{1}{n_0}\left(\frac{\sigma_{\overline{A}}}{\overline{A}}\right)^2 \qquad (4.23)$$

1. **Variation in the number of ion pairs:**

$$\left(\frac{\sigma_{n_0}}{n_0}\right)^2 = \frac{F}{n_0} \qquad (4.24)$$

where F is so called **Fano factor** (typically 0.05-0.20).

2. **Variation in single electron avalanches:**

The distribution follows the **Polya distribution** (proposed by Byrne [32]):

$$P(A) = \left[\frac{A(1+\theta)}{\bar{A}}\right]^\theta \exp\left[-\frac{A(1+\theta)}{\bar{A}}\right] \qquad (4.25)$$

which has a variance $(\sigma_{\bar{A}}/\bar{A})^2 = 1/\bar{A} + b$, $b \equiv (1+\theta)^{-1}$, b is typically 0.5 and the parameter θ has a value between 0 and 1 - see Fig.3. At small E/p the parameter θ approaches zero, i.e. the distribution is the exponential (so called **Furry distribution**, which has a variance $(\sigma_{\bar{A}}/\bar{A})^2 = 1$):

$$P(A) = \frac{1}{\bar{A}} \exp\left(-\frac{A}{\bar{A}}\right) \qquad (4.26)$$

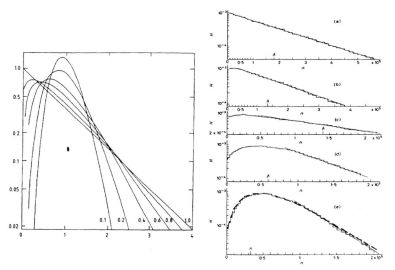

Fig.3 - Polya distributions as a function of $A/\bar{A}$ for various values of the parameter $b \equiv (1+\theta)^{-1}$; ($b = 1$ corresponds to $\theta = 0$).

Fig.4 - Single electron pulse height spectra measured by Schlumbohm for various values of χ: (a) $\chi=26$, (b) $\chi=22.6$, (c) $\chi=10.5$, (d) $\chi=5.3$ and (e) $\chi=4.1$.

Raether [33] suggested that the critical quantity to decide the shape of the Polya distribution is:

$$\chi = \frac{eE/\alpha}{eV_{ion}} \qquad (4.27)$$

where α is the 1-st Townsend coefficient, $1/\alpha$ is the mean free path, eE/α is the energy gained between two subsequent collisions and eV_{ion} is the ionization energy. For $\chi > 25$ the pulse height spectrum tends to exponential ($\theta \sim 0$), for $\chi < 20$ the pulse height spectrum exhibits a turnover ($\theta > 0$). Fig.3 shows the pulse height measurement in avalanches started by single electrons as measured by Schlumbohm [34], in methylal, in parallel-plate geometry - see Fig.4.

From the last chapter about the detector problems we will see that this is not the whole story. The parameter θ can be negative if the detector has problem with quenching, i.e. if the an avalanche has a tendency to breed the secondary avalanches.

The amplification on the wire **is non-linear** at large gains - see Fig. 5. At a gain of about 1.5×10^8 the limited streamer regime starts where the output pulse height does not depend any more whether the initial charge is only one electron or 220 electrons. One can also see the non-linearity of the output earlier.

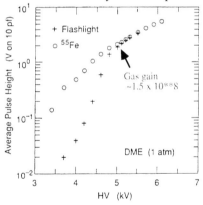

Fig.5 - Gain in a wire tube chamber as a function cathode voltage for a) Fe^{55} source (~220 el. deposited), and b) single electron source [36].

The non-linearity of charge amplification on the wire is good because it limits the gas gain and prevents the development of a spark (see last chapter). This is not so in the parallel plate chamber which tends to be more linear in this respect.

Examples of gas gain for typical anode surface gradients (E_a):

1. The gas gain as a function of the electric field on the anode surface:

Gas (Fe^{55} source)	Gas pressure	Total wire gain	E_a [kV/cm]	Wire dia. [μm]
90% Ar+10% CH_4	1 atm	~10^4	~200	20
90% Ar+10% CH_4	1 atm	~10^5	~240	20
90% Ar+10% CH_4	4 atm	~4×10^4	~320	25
90% Ar+8% CH_4+2% C_4H_{10}	4 atm	~4×10^4	~360	20
90% Ar+8% CH_4+2% C_4H_{10}	4 atm	~4×10^4	~309	25
90% Ar+10% C_4H_{10}	6.1 atm	~10^5	~760	7.8
50% Ar+50% C_2H_6	1 atm	~5×10^4	~275	20

2. The gas gain at 1 atm as a function of the electric field on the anode surface of a 20 μm anode wire [35]:

Gas (single electron source)	Parameter θ	Visible wire gain	E_a [kV/cm]
50% Ar+50% C_2H_6	0.252±0.026	~5×10^5	~311
95% CF_4+5% DME	0.272±0.044	~9.2×10^5	~377
80% CF_4+20% C_4H_{10}	0.624±0.043	~3.4×10^5	~349
90% CF_4+10% CH_4	0.222±0.042	~6.3×10^5	~396
50% He+50% C_2H_6	0.287±0.019	~2.9×10^5	~302
78% He+15% CO_2+7% C_4H_{10}	-0.031±0.019	~5.8×10^5	~283
50% Ar+50% C_2H_6	0.252±0.026	~5×10^5	~311
96.4%He+3.6%DME	-0.897±0.050	1.9×10^4	~188
80.5%He+19.5%DME	0.321±0.058	4.1×10^5	~264
100% DME	1.768±0.079	2.6×10^4	~188

Chapter 5:

5.1. Creation of the electrical signal

Moving charges create electrical signals on nearby electrodes [2].
Avalanche electrons have to travel only few tens of microns, positive ions travel much larger distance toward the cathode. Energy ε of the electrostatic field of the capacitor is:

$$\varepsilon = 1/2\, Q_0 V, \qquad (5.1)$$

where V is potential of the capacitor and Q_0 is its charge. A small charge q that travels between two points 1 and 2 under the influence of the field E changes the electric energy ε of the capacitor by the amount:

$$\Delta\varepsilon = \int_1^2 q\,\vec{E}.d\vec{r} = q(V_1 - V_2) = q\,\Delta V, \qquad (5.2)$$

This change in energy Δε of the capacitor is the source of the anode signal. The signal shape depends on the nature of the electrical configuration - see Fig. 1.

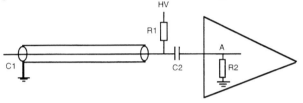

Fig.1 - A simple equivalent circuit of a tube wire detector.

There are two limiting cases:
a) The potential of the wire is re-established during the development of the pulse, which requires that charges flow quickly enough into the detector, which acts as a current source. In this case it is required that time constant R2.C1 is small compared to the pulse rise-time. The signal is the current I(t) that flows through R2. Using equation (5.1), the change in energy $\Delta\varepsilon = (1/2)\Delta Q_0(t) V$ of C1 capacitor causes a current flow through R2 resistor:

$$I(t) = d/dt[\Delta Q_0(t)] = (2/V)d/dt[\Delta\varepsilon] \qquad (5.3)$$

b) The potential of the wire is not re-established quickly because charges are not flowing quickly enough into the detector. This results in a drop of wire potential of the detector, which acts as a voltage source. In this case time constant R2.C1 is large compared with the pulse rise-time. We have a voltage pulse on R2 resistor:

$$\Delta V = 2/Q_0\, \Delta\varepsilon \qquad (5.4)$$

In practice, we are typically closer to the case (a). I give two examples:

1) CRID detector:

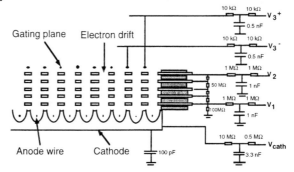

2) OPAL drift chamber:

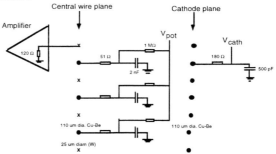

Let's consider an example of the cylindrical wire tube. I will show that the contribution of the charge motion of the electrons can be neglected [1]. To obtain a signal from the motion of a charge q over a distance dr, we consider a case (b) discussed above, and equations (5.4) and (1.9):

$$dV = \frac{2}{Q_0} d\varepsilon = 2 \frac{q E(r) dr}{Q_0} = \frac{2q}{2\pi\varepsilon_0 L} \frac{1}{r} dr \qquad (5.5)$$

Assuming that all ion pairs are created at a distance d from anode, we obtain a contribution from electron and ion motions separately by integrating equation (5.5):

electrons: $\qquad \Delta V^- = \frac{2q}{2\pi\varepsilon_0 L} \int_{r_a}^{r_a+d} \frac{1}{r} dr = \frac{2q}{2\pi\varepsilon_0 L} \ln \frac{r_a+d}{r_a} \qquad (5.6)$

ions: $\qquad \Delta V^+ = \frac{2q}{2\pi\varepsilon_0 L} \int_{r_a+d}^{r_c} \frac{1}{r} dr = \frac{2q}{2\pi\varepsilon_0 L} \ln \frac{r_c}{r_a+d} \qquad (5.7)$

Total induced signal on anode is then:

$$\Delta V = \Delta V^- + \Delta V^+ = \frac{2q}{2\pi\varepsilon_0 L} \ln\frac{r_c}{r_a} = \frac{2q}{C} \qquad (5.8)$$

where C is the total capacitance of this detector is:

$$C = \frac{Q_0}{V_0} = 2\pi\varepsilon_0 L \frac{1}{\ln r_c/r_a} \qquad (5.9)$$

From here it follows that to get as large a signal as possible, one wants to keep the detector capacitance C1 in Fig. 1 as low as possible. Because d is only few tens of microns, the signal due to electron motion is much smaller than the signal due to positive ions. Therefore, the electron signal is neglected in our considerations. The time dependence of the motion of positive ions is estimated as follows:

$$v_{\text{ion drift}}(t) = \frac{dr(t)}{dt} = \mu^+ \frac{E(r)}{p} = \frac{\mu^+ C V_0}{2\pi\varepsilon_0 L p} \frac{1}{r} \qquad (5.10)$$

where μ^+ is the ion mobility and p is the gas pressure. By integrating equation (12), and assuming that $\mu^+ \sim$ conts., we obtain:

$$r(t) = r_a \sqrt{1 + \frac{t}{t_0}}, \qquad (5.11)$$

where t_0 is the **characteristic time** of the chamber:

$$t_0 = \frac{\pi\varepsilon_0 L p r_a^2}{\mu^+ C V_0} = \frac{r_a}{2\mu^+ E(r_a)} \qquad (5.12)$$

The characteristic time, typically 0.1-2 ns, is expressed in terms of the wire radius, ion mobility and electric field on the surface of the wire. Using equations (5.2) and (1.5), we calculate the change in energy $\Delta\varepsilon$ as a function time from the motion of charge q representing the positive ions starting at anode radius and ending at radius r(t):

$$\Delta\varepsilon = \int_{r_a}^{r(t)} q\vec{E}\cdot d\vec{r} = \int_{r_a}^{r(t)} q\frac{\lambda}{2\pi\varepsilon_0}\frac{1}{r}\cdot d\vec{r} = \frac{q\lambda}{2\pi\varepsilon_0}\ln\frac{r(t)}{r_a} = \frac{q\lambda}{4\pi\varepsilon_0}\ln(1+\frac{t}{t_0}) \qquad (5.13)$$

For normalization purposes, it is useful to calculate the total energy change (using equations (1.8) and (5.13)):

$$\Delta\varepsilon|_{tot} = \int_{r_a}^{r_c} q\vec{E}\cdot d\vec{r} = \frac{q\lambda}{2\pi\varepsilon_0}\ln\frac{r_c}{r_a} = q V_0 \qquad (5.14)$$

This allows us to re-write equation (5.13) as follows:

$$\Delta \varepsilon = \frac{q\lambda}{4\pi\varepsilon_0} \ln(1+\frac{t}{t_0}) = q\, V_0 \ln(1+\frac{t}{t_0})[1/2 \ln\frac{r_c}{r_a}] = q\, V_0\, F(t) \qquad (5.15)$$

Going back to our earlier discussion related to Fig.1, and using equations (5.3) and (5.4):

Case (a) - **current pulse:**

$$I(t) = \frac{d}{dt}\Delta Q(t) = \frac{2}{V}\frac{d}{dt}\Delta\varepsilon = \frac{2}{V}\frac{q\lambda}{4\pi\varepsilon_0}\frac{1}{t+t_0} = [q\big/\ln\frac{r_c}{r_a}]\frac{1}{t+t_0} \qquad (5.16)$$

Case (b) - **voltage pulse:** (5.17)

$$\Delta V = \frac{2}{Q_0}\Delta\varepsilon = \frac{2}{Q_0} q\, V_0 \ln(1+\frac{t}{t_0})\frac{1}{2\ln\frac{r_c}{r_a}} = \frac{2q}{C}\ln(1+\frac{t}{t_0})\frac{1}{2\ln\frac{r_c}{r_a}} = \frac{2q}{C}F(t)$$

5.2. Pulse shape prediction

5.2.1. - CRID photon detector with an amplifier [37] - see Figs. 3 and 4.

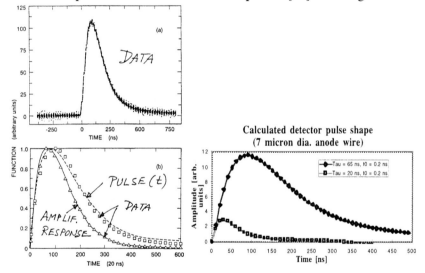

Fig. 3 - (a) Measured chamber single electron pulses, (b) measured and calculated amplifier response to an impulse charge and the chamber pulse.

Fig. 4 - The effect of the shaping time on pulse height.

In this case the avalanche is initiated by a single electron created by a photo-ionization of a photo-sensitive molecule such as TMAE or TEA. The pulse is the result of a convolution of the amplifier response to the impulse charge and the positive ion response:

a) amplifier response: $t\, e^{(-t/\tau)}$, where $\tau \sim 65$ ns (shaping time),

b) positive ion response: $I(t) = \left(q \middle/ \ell n \dfrac{r_c}{r_a}\right) \dfrac{1}{t+t_0} = \dfrac{A}{t+t_0}$

c) Additional constants needed for the problem (for CH_4 gas): $r_c = 0.146$ cm, $r_a = 3.5 \times 10^{-4}$ cm, $\mu^+ (CH_4) \sim 2.26$ cm^{2}V^{-1}sec^{-1}, $E(r_a) = 810$ kV/cm, $t_0 = 0.1$ ns (unusually small !!; normal drift chambers have thicker wires, $t_o \sim 1-2$ ns and $E(r_a) \sim 300-400$ kV/cm).

The convolution of two functions is the following integral:

$$\text{Pulse}(t) = A \int_0^t \dfrac{t'\, e^{(-t'/\tau)}}{t - t' + t_0} dt' = A \exp\left(-\dfrac{t+t_0}{\tau}\right)(t+t_0). \qquad (5.18)$$

$$\cdot \left\{ \ell n\left(1 + \dfrac{t}{t_0}\right) + \sum_{n=1}^{\infty} \left(\dfrac{t_0}{\tau}\right) \dfrac{1}{n \cdot n!}\left[\left(1+\dfrac{t}{t_0}\right)^n - 1\right]\right\} - \tau\left[1 - \exp\left(-\dfrac{t}{\tau}\right)\right]$$

The comparison between this simple theory and the measurement was done in the CRID detector operating with the CH_4 gas, the early version of the CRID amplifier was coupled to the LeCroy waveform digitizer. It is interesting to determine the influence of the shaping time on the pulse height. If CRID detector would use 20 ns shaping time, the pulse height would have been 4-5 times smaller, resulting in the necessity to increase operating cathode voltage by ~150 Volts - see Fig. 4. Running with shorter shaping time would mean a necessity to run higher voltage of about 150 V on cathode (the final CRID gas is actually C_2H_6, which gives $t_o \sim 0.2$ns for $E(r_a) = 725$ kV/cm and $\mu^+(C_2H_6) \sim 1.4$ cm^{2}V^{-1}sec^{-1}).

5.2.2. - Drift chamber pulses [38].

In this case the avalanche is initiated by a number of electrons arriving in clusters at slightly different times and each subject to a different avalanche fluctuation, which will create an additional randomness compared to the photon detector response. To treat this problem correctly, one has to use the Monte Carlo method. The drift chamber pulse is a result of a numerical convolution of three basic responses:

$$\text{Pulse}(t) = I_{\text{drift}}(t) \times I_{\text{avalanche}}(t) \times I_{\text{electronics}}(t), \qquad (5.19)$$

where

1. $I_{\text{drift}}(t)$ is generated by:

    a) working with a correct electrostatic field,

    b) creating the primary ionization,

c) drifting each electron within each cluster independently,

c) using the correct drift velocity in each step,

d) including the effect of the diffusion,

e) including the effect of the magnetic field.

2. $I_{avalanche}(t)$ is generated by:

a) including the effect of the motion of the positive ions,

b) including the effect of the avalanche fluctuations.

3. $I_{electronics}(t)$ is generated by:

a) using the measured or calculated response of the amplifier,

b) using the effect of filters, cables, noise, etc.

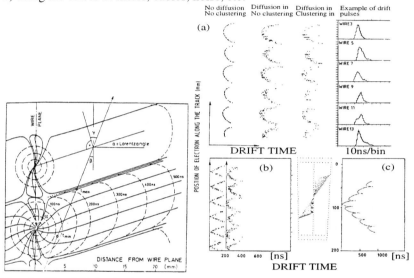

Fig. 5 - The drift in the jet chamber. **Fig. 6** - The non-isochronic charge collection in the jet chamber operating with 90%Ar+10%C_4H_{10} at 1 atm and 10 kG: (a) the effect of diffusion and clustering on the drift chamber pulse randomness (avalanche fluctuations and electronics included), (b) the effect of impact parameter (0 and 6 mm), (c) the effect of angle of the track (B = 0 kG).

Why all this trouble ? My hope at that time, i.e. ~15 years ago, was to discover some new timing strategy which would allow to create a extremely high resolution drift chamber. Let's apply this simulation model to the jet chamber - see Fig. 5. The electrons do not arrive to the anode wire at the same time. This is a major factor limiting the high resolution capability <u>for short drifts</u> - see Fig. 6:

Can we improve the non-isochronic behavior of the jet chamber by a choice of the gas operating point ? Yes, but the magnetic field will spoil it - see Fig. 7:

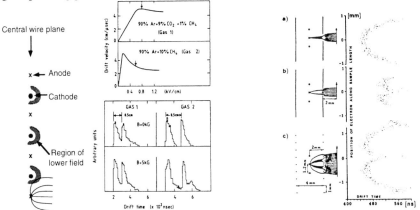

Fig. 7 - The effect of choice of operating point in the drift velocity curve on the isochrony of the jet chamber.

Fig. 8 - The effect of the drift geometry on the isochrony of the charge collection.

Can we improve the non-isochronic behavior by a choice of geometry ? Yes, but the magnetic field and angle of tracks will spoil it again - see Fig. 8.

5.2.3. - Propagation of signals along the wires.

First, I will show a practical example of the ~4.5 m long 8-wire prototype built during the R&D stage for the OPAL central drift chamber development [10] - see Fig. 9 and 10.

For both terminations we clearly observe reflections. The chamber impedance is of the order ~300 Ω; a 50 Ω termination at the end is an obvious mismatch which will cause a reflection. The charge division needs a low impedance amplifier. Therefore we will have inherently the reflection in the problem, and therefore the non-linearity in the z-coordinate.

The theory in the early part of this chapter did not include one significant ingredient: the

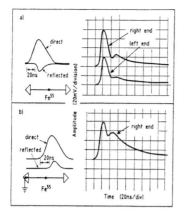
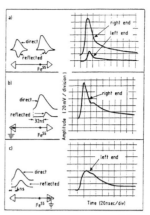

Fig. 9 - Signals of a Fe^{55} source painted on a potential wire in the middle of the chamber. (a) Low impedance at both ends (~50 Ω), (b) high impedance at left end (~MΩ).

Fig. 10 - Signals of a Fe^{55} source painted on a potential wire 50 cm from the wire end. (a) Low impedance at both ends, (b) high impedance at left end, (c) high impedance at right end.

propagation of signals along the wires. Such propagation follows the **Telegrapher** equation for the system of wires:

$$\frac{\partial I_i}{\partial z} = -\sum_k C_{ik} \frac{\partial V_k}{\partial t}, \quad \frac{\partial V_i}{\partial z} = -R_i I_i - \sum_k L_{ik} \frac{\partial I_k}{\partial t} \quad (5.23)$$

where C_{ik}, L_{ik} are wire to wire capacitance and inductance per unit length. The solution of this problem allows to design a proper drift chamber termination by minimizing the signal reflections, at least in principle.

Let's discuss two possible wire terminations of the 4 m long OPAL central drift chamber [39].

a) Simple minded termination of the OPAL full size prototype, which was "tuned" by the 8-wire prototype [10] - see Figs. 9 & 10.

b) Termination of the final OPAL chamber after the Bock's analysis [39]:

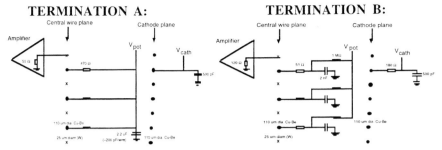

Model for the input pulse shape in the analysis (4 free parameters):

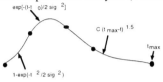

Fig. 11 shows the result of the P.Bock's calculation of the pulse shapes.

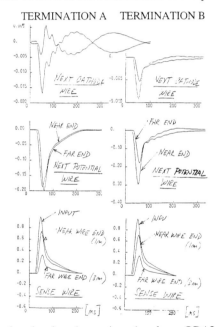

Fig. 11 - Simulated pulse shapes in a 4 m long OPAL drift chamber.

However, this termination did not remove the basic reflection problem discussed earlier, which is inherent in the problem of charge division. All it really does is to remove the 2-nd order ripple effects.

5.2.4. - currents induced on nearby electrodes.

Ramo's theorem allows us to predict the induced signal on any electrode [40]. It says: the instantaneous current I_i flowing into one particular electrode i due to a motion of charge q at position $\vec{r}$ with a velocity $\vec{v}(\vec{r})$ can be calculated using equation:

$$I_i = -q \frac{\vec{v}(\vec{r}) \cdot \vec{E}(\vec{r})}{V_i} \qquad (5.24)$$

where $\vec{E}(\vec{r})$ is field created by rising the electrode i to potential V_i and grounding all other electrodes, in absence of charge. A consequence of this theorem is the well known fact that the signal induced on the wire, which has an avalanche, has a sign opposite to that of the signal on the neighboring wires, which have the cross-talk.

Example:

The radioactive Fe^{55} source painted on the potential wires of the 8-wire prototype allowed also to study the wire to wire cross-talk. The cross-talk between nearest neighbors amounts to ~7%, it has the same time structure as the prompt signal and has the opposite polarity [10] - see Fig. 12.

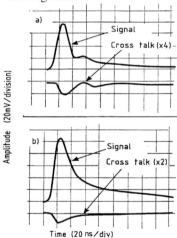

Fig. 12 - Cross-talk from one wire to the next. (a) Middle position of Fe^{55} source, (b) source positioned 50 cm from one wire end.

Coefficients of induction C_{ik} can be used to estimate the cross-talk from a signal on neighboring wires as follows [41].

5.2.5. - charge division.

The propagation of signals along the wire is governed by the equation:

$$\frac{\partial^2 V(z,t)}{\partial z^2} = \frac{L_w C_w}{\ell^2} \frac{\partial^2 V(z,t)}{\partial t^2} + \frac{R_w C_w}{\ell^2} \frac{\partial V(z,t)}{\partial t} \qquad (5.24)$$

where V(z,t) is voltage, Z is position along the wire (z = 0 at center), t is time, ℓ is length of the wire, R_w, L_w, C_w is wire resistance, inductance and capacitance, $L_w C_w/\ell$. The $\partial^2 V(z,t)/\partial t^2$ term represents wave propagation, $R_w C_w/\ell^2 \, \partial V(z,t)/\partial t$ represents diffusive propagation. Assuming the boundary condition $V(-\ell/2,t) = V(+\ell/2,t) = 0$, the equation (5.24) can be solved by the Fourier series method. The general solution has a form:

$$V(z,t) = \sum_{n=1}^{\infty} A_n T_n(t) \sin\left[\frac{n \pi z}{\ell}\right] \qquad (5.25)$$

Time equation then becomes

$$\frac{L_w C_w}{\ell^2} \frac{\partial^2 T_n(t)}{\partial t^2} + \frac{R_w C_w}{\ell^2} \frac{\partial T_n(t)}{\partial t} + \left(\frac{n \pi}{\ell}\right)^2 T_n(t) = 0 \qquad (5.26)$$

To get some insight into the integration time needed to obtain the linear charge division, we simplify this problem by assuming [42]: (a) neglect the wave propagation by assuming that the $L_w C_w$ is small, i.e. the equation (23) becomes the diffusion equation; (b) assume that the boundary condition assumes the grounded ends, i.e. $V(-\ell/2,t) = V(+\ell/2,t) = 0$ (z = 0 at center); (c) we inject a δ-function charge Q at position x along the wire; (d) we detect the charge using the integrating charge amplifier.

The current flowing into charge integrated amplifier is:

$$I(z,t) = \frac{Q}{R_w C_w} \left\{ \sum_{n=1}^{\infty} 2 n \pi \exp\left[-\frac{t n^2 \pi^2}{R_w C_w}\right] \sin\left[\frac{n \pi z}{\ell}\right] \right\} \qquad (5.27)$$

The integrated charge $q_1(z,t)$ at one wire end after time t is: (5.28)

$$q_1(z,t) = \int I(z,t) \, dt = = Q \left\{ \frac{1}{2} + \frac{z}{\ell} + \frac{2}{\pi} \sum_{n=1}^{\infty} \frac{(-1)^n}{n} \exp\left[-\frac{t n^2 \pi^2}{R_w C_w}\right] \sin\left[n \pi \left(\frac{1}{2} + \frac{z}{\ell}\right)\right] \right\}$$

Clearly, the solution has transients represented by the sum. The longest lasting transient term (n = 1) decays with a time constant $R_w C_w / \pi^2$. For linear relation between $q_1(z,t)$ and the position along the wire, the sum in equation (25) should be negligible. The time required for the position non-linearity to be less than 0.2 % is t $\geq R_w C_w / 2$. Examples of two detectors using the charge division:

a) Opal 8-wire prototype: $R_w = 850\ \Omega$, $C_w = 40\text{-}50$ pF, => $R_w C_w \sim 35$ns

b) CRID detector: $R_w = 40\ k\Omega$, $C_w = 2\text{-}3$ pF, => $R_w C_w \sim 100$ns

We will now discuss the position sensitivity and resolution of the charge division:

a) Without coupling capacitors between the wire and the amplifier (include the contact resistance between the wire and PC board).

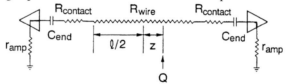

Let q_1 and q_2 be the charge reaching the preamps at each end of the wire of resistance R_w. Let $R_{contact}$ and r_{amp} are the contact resistance and amplifier input impedance. If $Q=q_1+q_2$ is the charge injected, then: (5.29)

$$\frac{q_1}{q_1+q_2} = \frac{\frac{1}{2}(R_w + 2r_{amp} + 2R_{contact}) + \frac{zR_w}{\ell}}{R_w + 2r_{amp} + 2R_{contact}} = \frac{1}{2} + \frac{z}{\ell}\frac{R_w}{R_w + 2r_{amp} + 2R_{contact}}$$

b) With coupling capacitors between the wire and the amplifier.

Let's assume that the charge Q divides at t = 0 such that the initial charges deposited on the capacitors are:

$$\frac{q_1(t=0)}{q_1+q_2} = \frac{1}{2} + \frac{z}{\ell}\frac{R_w}{R_w + 2r_{amp} + 2R_{contact}} \qquad (5.30)$$

Usually we measure charge over a certain time period corresponding to gate T. There will be a relaxation of charge between the capacitors through the resistive wire. What is measured is a charge $q_1(T)$ and not $q_1(t=0)$. To estimate $q_1(T)$ consider the equivalent circuit below, where the charges $q_1(t=0)$ and $q_2(t=0)$ are placed on the capacitors at t = 0.

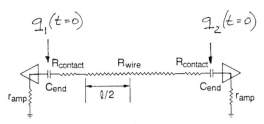

Solving the differential equation, we find:

$$\frac{q_1(T)}{q_1+q_2} = \frac{1}{2} + \frac{z}{\ell}\left(\frac{\beta R_w}{R_w + 2r_{amp} + 2R_{contact}}\right) \quad (5.31)$$

where

$$\beta = \exp\left[-\frac{T}{(R_w + 2r_{amp} + 2R_{contact})C_{end}}\right] \quad (5.32)$$

If β is small, sensitivity is lost. It is therefore mandatory that $R_w C_{end} \gg T$. We must also require that $R_w \gg r_{amp} + R_{contact}$. In typical chamber designs $\beta \sim 1$.

To get the position resolution as a function of Q and z, we use equation (5.31) and solve for z:

$$\frac{z}{\ell} = \frac{R_w + 2r_{amp} + 2R_{contact}}{2\beta R_w} \frac{q_1 - q_2}{q_1 + q_2} \quad (5.33)$$

Therefore

$$\left(\frac{\sigma_z}{\ell}\right)^2 = \left(\frac{\partial \frac{z}{\ell}}{\partial q_1}\right)^2 (\sigma_{q_1})^2 + \left(\frac{\partial \frac{z}{\ell}}{\partial q_2}\right)^2 (\sigma_{q_2})^2 + \left(\frac{\partial \frac{z}{\ell}}{\partial \beta}\right)^2 (\sigma_\beta)^2 \quad (5.34)$$

After some rearranging we obtain the following expression:

$$\left(\frac{\sigma_z}{\ell}\right)^2 = \left(\frac{\sigma_{PED}}{\sqrt{2}}\right)^2 \frac{\left[\left(\frac{R_w + 2r_{amp} + 2R_{contact}}{\beta R_w}\right)^2 + \left(\frac{2z}{\ell}\right)^2\right]}{(q_1+q_2)^2} + \left(\frac{\sigma_\beta}{\beta}\right)^2 \left(\frac{z}{\ell}\right)^2 \quad (5.35)$$

where σ_{PED} is the pedestal arising from amplifier noise, cable pick-up noise, etc. We assume that $\sigma_{PED} = \sigma_{q_1} + \sigma_{q_2}$. The resistance of the wire adds noise also, so called Johnson noise, which is a position independent noise. If we add it in quadrature, we get the final equation:

$$\left(\frac{\sigma_z}{\ell}\right) \cong \sqrt{\left(\frac{\sigma_J}{q_1+q_2}\right)^2 + \left(\frac{\sigma_{amp}/\sqrt{2}}{q_1+q_2}\right)^2 \left[\left(\frac{R_w + 2R_{contact} + 2r_{amp}}{R_w}\right)^2 + \left(\frac{2z}{\ell}\right)^2\right]} \quad (5.36)$$

I will now discuss specifically the **CRID detector**. The Johnson noise is estimated using the following formula:

$$\sigma_J = 2.718\sqrt{\frac{kT\tau}{2(R_w + 2R_{contact} + 2r_{amp})}} \quad (5.37)$$

where k is the Boltzmann constant; T is absolute temperature; τ is the amplifier shaping time (65 ns). The CRID amplifier noise with RC-CR shaping and with the FET input can be calculated as follows [43]:

$$\sigma_{amp} \cong 2.718 \sqrt{\frac{k T R_{eq}(C_{in}+C_{ch})^2}{2\tau}} \qquad (5.38)$$

where R_{eq} is the equivalent noise resistance of the FET (~50Ω), C_{in} is the amplifier input capacitance (~10 pF), C_{ch} is the detector capacitance (~15 pF), ℓ = 103.5±0.5 mm, q_1+q_2 is the visible charge (1-2 x 10^5 el.), R_{wire} = 41.3±2.64kΩ, $R_{contact}$ = 94.6±116Ω, r_{amp} = 680±50 Ω. Equations (5.34) and (5.35) yield σ_J ~ 960 and σ_{amp} ~ 530 electrons. We can now use the equation (5.33) to estimate the charge division resolution. Fig.13 compares the calculation with the data [44].

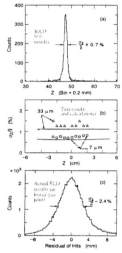

Fig. 13 - Charge division resolution.

The final z-coordinate has to take into account the gains of the amplifiers resulting in the final expression:

$$z = C \frac{g_1 q_1 - g_2 q_2}{g_1 q_1 + g_2 q_2} \frac{R_w + 2(r_{amp} + R_{contact})}{R_w} \frac{\ell}{2} \qquad (5.39)$$

where q_1, q_2 is measured charges at both ends of the wire, g_1, g_2 are calibration factors reflecting the amplifier gain variations (obtained from the special amplifier calibration runs), R_w is carbon wire resistance, r_{amp} is the amplifier impedance, $R_{conract}$ is contact resistance between the carbon wire and the PC board trace, ℓ is wire length, C is calibration factor reflecting errors in resistors, capacitors, etc.

Chapter 6

Limit of accuracy for the high resolution drift chambers - can they compete with the silicon ?

In this chapter I will discuss a limit of highest possible resolution in the most precise drift chambers. There are two methods to estimate the high precision drift chamber resolution:

6.1. Simple estimate:

Generally, there are several major contributions to the resolution.

$$\sigma^2(x) = \sigma^2_{diffusion}(x) + \sigma^2_{ionization}(x) + \sigma^2_{track} + \sigma^2_{electronics} \quad (6.1)$$

where $\sigma_{electronics}$ is an offset caused by the electronics noise, σ_{track} is the finite size of ionization trail left by a track, $\sigma_{ionization}(x)$ are fluctuations in primary ionization statistics, $\sigma^0_{diffusion\ near\ wire}$ is the constant term describing the diffusion near the wire in a presence of a very large field, $\sigma_{diffusion}(x)$ is the x-dependent term describing the diffusion in the middle of drift cell where the field is low. We neglect other effects, which are important for the TPC resolution, such as angle of a track in respect to wire, pad resolution, etc.; see Ref.3 for description of these effects.

(a) σ_{track} - This term is typically neglected. The primary ionization is contained within less than 2 µm of the original track direction.

Table 1 - Physical width of the track in Argon gas

%	Atomic shell	Average energy	Electron range 9.9×10^{-6} (E / keV) g cm^{-2}
~92	M-shell	≤ 30 eV	< 2 µm
~8	L-shell	≤ 400 eV	~ 20 µm
~0.1	K-shell	≤ 4 keV	~ 200 µm

(b) $\sigma_{ionization}(x)$ - This term is important near the anode wire where the number of electrons available is small and subject to ionization fluctuations.

(c) $\sigma^0_{diffusion\ near\ wire}$ - This term has not been even considered in any analysis I know so far. It has been suggested by F. Villa [45], who pointed out that the electron energy

increases as a function of electric field at least quadratically, i.e. one could have significant contribution near the anode wire where the electric field is typically 200-400 kV/cm. I have decided to explore this question experimentally [35], and concluded that such term indeed could play a role in the highest resolution applications - see Fig.1. One can see that the $\sigma^0_{\text{diffusion near wire}}$(1 electron) term appears to be dominant. to explain the offset at $\sigma(x = 0)$. In the final application more than one electron contribute, i.e. we expect $\sigma^0_{\text{diffusion near wire}}$(N electron) = $\sigma^0_{\text{diffusion near wire}}$(1 electron)$/\sqrt{N}$. For example in case of CF_4 gas, we see in Fig.1, $\sigma^0_{\text{diffusion near wire}}$(1 electron) ~ 170 μm. In an application where we would have N~100 electrons, we expect $\sigma^0_{\text{diffusion near wire}}$(N electron)~ $170/\sqrt{100} = 17\,\mu m$. This is certainly not negligible, and I should add that 100 electrons is rather unusual; typically we end up with 10-20 electrons. The measurement of $\sigma^0_{\text{diffusion near wire}}$(1 electron) in cool gases such as DME or CO_2 does not exist at present.

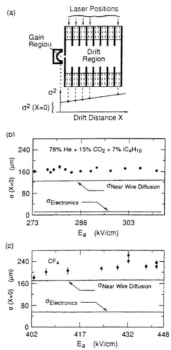

Fig. 1 - Diffusion near the wire in a presence of a very large field dominates $\sigma(x = 0)$.

(d) $\sigma_{diffusion}(x)$ - As we discussed earlier, one electron time dispersion of an original point-like charge distribution is:

$$\sigma_x(1\ electron) = \sqrt{2Dt} = \sqrt{\frac{2Dx}{\mu E}} = \sqrt{\frac{2\varepsilon_k x}{eE}} = \frac{1}{\sqrt{p}}\sqrt{\frac{2\varepsilon_k x}{e\left(\frac{E}{p}\right)}} \quad (6.2)$$

From here we obtain four practical dependencies:

a) If $\varepsilon_k \sim$ const. and $x \sim$ const.: $\quad \sigma_x(1\ electron) \sim \dfrac{1}{\sqrt{E}} \quad (6.3)$

b) If in addition $E/p \sim$ const.: $\quad \sigma_x(1\ electron) \sim \dfrac{1}{\sqrt{p}} \quad (6.4)$

c) For <u>cool gases</u> ($\varepsilon_k \sim kT$): $\quad \sigma_x(1\ electron) = \sqrt{2Dt} = \sqrt{\dfrac{2kTx}{eE}} \quad (6.5)$

d) Because of the diffusion dependence on the drift distance x the resolution data are usually fitted by:

$$\sigma^2 = \sigma_0^2 + \delta x \quad (6.6)$$

The resolution data are usually presented in this form to verify that other effects, such as the electron attachment, do not affect this dependence.

CO_2 and DME gases are frequently used in high precision drift chambers, because they approach the cool gas limit - see Fig. 2.

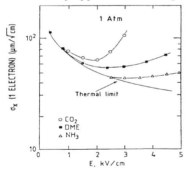

Fig. 2 - Single electron diffusion for 1 cm drift and 1 atm pressure as a function of electric field [46].

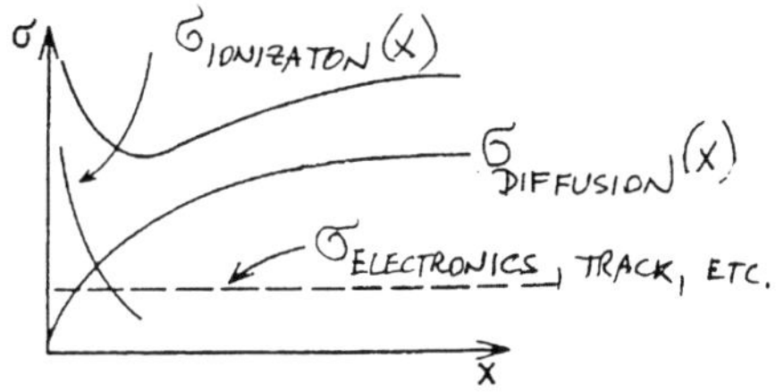

Fig. 3 - Typical contributions to the resolution as a function of drift distance.

We can see that the cool gases can reach $\sigma_x(1\ electron) \sim 50\text{-}80\,\mu m$ at 1 atm and 1 cm drift, and the transverse and longitudinal diffusions are close to each other (we assume the same). As we discussed earlier, in hot gases $\sigma_L(1\ electron)$ could be as low as half of σ_x(1 electron). Typically, in hot gases σ_L (1 electron) ~ 120-200 μm at 1 atm and 1 cm drift. It is the longitudinal diffusion which influences the arrival time distribution.

Individual contributions in typical high precision chambers can be qualitatively described in Fig.3. The relative size of terms in equation (1) depends on the choice of gas, pressure, electronics, method of charge collection and method of analysis. For example, the $\sigma_{ionization}(x)$ term can be greatly suppressed by increasing the gas pressure.

Table 1 - Cluster size distribution in percent

k (cluster size)	CH_4	Ar	He	CO_2
1 e⁻	78.6 %	65.6	76.6	72.5
2	12	15.0	12.5	14.0
3	3.4	6.4	4.6	4.2
4	1.6	3.5	2.0	2.2
5	0.95	2.25	1.2	1.4
6	0.6	1.55	0.75	1.0
7	0.44	1.05	0.50	0.75
8	0.34	0.81	0.36	0.55
9	0.27	0.61	0.25	0.46
10	0.21	0.49	0.19	0.38
11	0.17	0.39	0.14	0.34
12	0.13	0.30	0.10	0.28
13	0.10	0.25	0.08	0.24
14	0.08	0.20	0.06	0.20
15	0.06	0.16	0.048	0.16
16	0.050	0.12	0.043	0.12
17	0.042	0.095	0.038	0.09
18	0.037	0.075	0.034	0.064
> 20	$(11.9/k^2)$	$(21.6/k^2)$	$(10.9/k^2)$	$(14.9/k^2)$

Note: 20-30% of the time a cluster will have more than one electron.

6.1.1. The first electron timing:

Cramer [47] derived a formula describing time accuracy based on arrival time of the M-th electron in a given sample of N electrons. It is valid for large N, and M << N:

$$\sigma_{diffusion}(M-\text{th arriving electron}) = \frac{\sigma_L(1\text{ electron})}{\sqrt{2 \ln N}} \sum_{i=M}^{N} \frac{1}{i^2}, \quad (6.7)$$

which becomes for M = 1:

$$\sigma_{diffusion}(1-\text{st arriving electron}) = \frac{\sigma_L(1\text{ electron})}{\sqrt{2 \ln N}} \sum_{i=1}^{N} \frac{1}{i^2} \approx$$

$$\approx \frac{\sigma_L(1\text{ electron})}{\sqrt{2 \ln N}} \sqrt{\frac{\pi^2}{6}} = \frac{0.91\, \sigma_L(1\text{ electron})}{\sqrt{\ln N}} \quad (6.8)$$

However, as we discussed in the previous chapter, the electrostatic field of the drift cell causes non-isochronic charge collection. This creates a non-Gaussian tail in the electron arrival distribution, i.e. not all electrons can contribute to the 1-st electron timing because they arrive too late.

In addition, the electrons are produced in clusters, which cause additional variability in the ionization arrival. Table 1 shows the measurement of the cluster size distribution in percent for several gases [48]. Table 2 shows the ionization available per 1 cm of track length at 1 atm of pressure:

Table 2 - Primary and total ionization per cm at 1 atm (compilation of numbers from A.V. Zarubin [49], except the last three - taken from A. Pansky et al. [50]).

Gas	Primary ionization (clusters) [cm^{-1}]	Total ionization (electrons) [cm^{-1}]	Average energy needed to create one ion pair w [eV]	dE/dx (min) [keV/cm]
He	3.3	7.6	42.3	0.322
N$_2$	20.8	60.5	34.7	2.097
O$_2$	23.2	76.5	30.8	2.360
Ne	10.9	39.9	36.4	1.452
Ar	24.8	96.6	26.3	2.541
Kr	33	197.5	24.05	4.750
Xe	44.8	313.3	21.9	6.862

CO$_2$	33.6	100.0	32.8	3.280
CH$_4$	24.8	59.3	27.1	1.608
C$_2$H$_6$	40.5	117.7	24.4	2.870
iC$_4$H$_{10}$	83.6	232.8	23.2	5.402
DME	62	120 (?)	-	-
TEA	144	-	-	-
TMAE	281	-	-	-

Note: If the sample available for detection is only few mm long, we are dealing with rather small number of electrons available at 1 atm.

By including a correction for non-isochronous charge collection, gas pressure, and by assuming that E/p is constant while changing the pressure p, we can modify the equation (6.8):

$$\sigma_{diffusion}(1-\text{st arriving electron}) = \frac{0.91}{\sqrt{\ln(\eta(x)Np)}} \frac{\sigma_L(1 \text{ electron})}{\sqrt{p}} \qquad (6.9)$$

where $\eta(x)$ is a correction describing the fraction of electrons, which can contribute to the first electron timing, x is the drift distance, p is the gas pressure, N is the total number of electrons per sample. Fig. 4 shows an example of the choice of gas pressure to control the drift chamber tracking resolution [51].

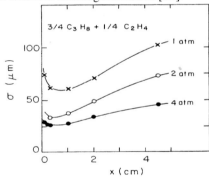

Fig. 4 - Chamber resolution as a function of pressure.

It is not straightforward to guess the $\eta(x)$ factor. One needs some knowledge of the charge collection, i.e. one needs a drift simulation program. In typical examples of high accuracy drift chambers the factor $0.91/\sqrt{\ln(\eta(x)Np)}$ is close to 0.4-0.6 [38].

As an example, for the DME gas at 1 atm pressure, 1 cm sample size and 1 cm drift (<u>consider the diffusion contribution only</u>): in this case one has N ~ 120 electrons / cm, σ_L(1 electron)~ 55 μm /√cm, η(x) ~ 0.2 (assume) and $0.91/\sqrt{\ell n\,(\eta(x)\,N\,p)}$ ~ 0.5; therefore one could achieve resolution of about ~30 μm.

6.1.2. Center of gravity timing:

Let's assume that we have electronics capable of measuring every arriving electron. If we assume that the electron cloud is Gaussian then:

$$\sigma_{diffusion}(\text{center of gravity timing}) = \frac{1}{\sqrt{\eta(x)\,N\,p}} \frac{\sigma_L(1\ electron)}{\sqrt{p}}, \qquad (6.10)$$

where η(x) is the correction describing the fraction of electrons which are used for centroid timing; it is influenced by variables such as the geometry, speed of digitizing clock, method of analyzing data, etc. One can see, that now it would make more sense to invest into "being clever", i.e. to design the isochronous charge collection geometry.

The same example for the DME gas at 1 atm pressure, 1 cm sample size and 1 cm drift (<u>consider the diffusion contribution only</u>): in this case one has η(x) ~ 1.0 (assume an ideal case); therefore one could achieve resolution of about ~5 μm !!!

The Flash ADC digitizers came on the market ~15 years ago. Great improvements in resolution of high precision drift chambers were expected at that time. However, the hopes for an "ultra-high ~5μm accuracy" <u>short drift</u> wire chambers did not materialize because of (a) non-isochronous charge collection, (b) fluctuation in the ionization statistics and (c) η(x) factor is always less than one.

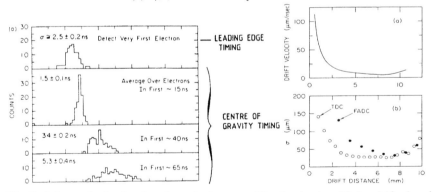

Fig.5 - Monte Carlo simulated resolution with a hypothetical infinitely fast electronics capable of digitizing every arriving

Fig.6 - Aachen Univ. drift chamber test; (a) drift velocity near the anode wire, (b) chamber resolution using the

electron in the jet chamber with 4 mm wire spacing, 90% Ar + 10% C_4H_{10} at 1 atm, B = 10 kG and 7.5 mm drift distance. leading edge timing and center of gravity FADC timing.

Fig. 5 shows that the center of gravity timing is better than the 1-st electron timing only if we average over the electrons located in the near-isochronous central part of the sample. As we average over larger part of the sample the center of gravity is disturbed by the fluctuations in the ionization statistics coupled with the nonisochrony of the charge collection ($\eta(x) < 1$). In practice, of course, we do not have infinitely fast electronics. Closest practical approximation is the combination of a slow gas and the time expansion chamber, which uses extremely slow drift velocity. Fig. 6 illustrates that the centroid timing is actually worse than the leading edge timing if we are very close to the anode wire where the charge collection is very non-isochronous [52].

To average over the near isochronous limit means to use only the leading edge portion of the drift pulse, i.e. to use the first FADC bins of the pulse.

One should also remark that for the long drift distances (>10 mm) the FADC timing method gives finally better results than the leading edge timing method because the diffusion starts washing out the non-isochronous collections and the clustering effects.

6.3. Monte Carlo estimate
Example #1 - Mark II vertex chamber at SLC [53].
1. Assume the following drift chamber operating conditions:

92% CO_2 + 8% iC_4H_{10} gas at 4 atm pressure, tracks are either parallel to anode plane or 10° inclined, anode surface gradient $E_a \sim$ 450 kV/cm at 2 atm and 590 kV/cm at 4 atm on the surface of 20 µm wire, the average drift field E ~ 2.1 kV/cm at 4 atm, giving an average drift velocity of about 4 µm/ns (slow gas) and magnetic field was off for this particular study.

2. Calculate the drift chamber pulses:

Creating first the weighted drift time distribution by:

a) Starting with a track segment assuming 3.4 clusters/mm of track (CO_2 gas). The probability distribution of electrons within each cluster is taken from argon data [53].

b) Stepping each electron in electrostatic field using the drift velocity and diffusion is determined at each point of the drift according to E/p at that point.

c) Assigning a weight x to each electron according to the Furry exponential distribution $A(x) \approx x \exp(-1.5x)$, to simulate the avalanche fluctuations.

3. The weighted drift time distribution is then convoluted with:

a) Response of the amplifier which is assumed to be a simple triangle with a 5 ns rise time and a 15 ns fall time.

b) Include the response of positive ions. This response, as we discussed earlier, has the form: $1/(1+t/p t_0)$, where p is pressure, t is time and t_0 is the characteristic time. To improve the multiple hit capability, the design of the electronics minimized the pulse tail using so called zero-pole filter technique [55]. The zero-pole filters look as follows:

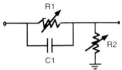

the design of these filters goes as follows. First, one has to express the "1 / t" tail in terms of three exponential curves:

$$\frac{1}{1+\frac{t}{p t_0}} = A\exp(-\frac{t}{\alpha p t_0}) + B\exp(-\frac{t}{\beta p t_0}) + C\exp(-\frac{t}{\gamma p t_0}) \quad (6.11)$$

Notice that the first term is the "fast" term followed by two "slow" terms (A = 0.79, α = 1.6, B = 0.185, β = 13.5, C = 0.024, γ = 113.0, t_0 = 0.76 ns, p = 4 atm, $\mu(CO_2)$ = 1.09 cm^2 V sec^{-1}). One adjusts the components of the filter to cancel the middle term in equation (6.11). The resulting response is:

$$V(t) = 0.99\exp(-\frac{t}{\alpha p t_0}) + 0.01\exp(-\frac{t}{\gamma p t_0}) \quad (6.12)$$

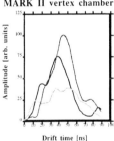

Fig.7 - Monte Carlo simulation of drift pulses in Mark II drift vertex chamber at SLC; 4 atm, 92% CO_2+8%iC_4H_{10} gas, v_{drift} = 4 μm/ns, 12 mm drift distance; 0° angle in respect to wire plane.

3. Timing strategy with pulses of Fig.7:

a) The timing strategies with infinitely fast electronics which detects individual arrival times of each electron. This method can determine the limit of how well we can do.

b) The leading edge timing with the "realistic" pulses was obtained by fitting the first 5 points of the pulse waveform with the 3-rd order polynomial, we choose a threshold to be 2-3% of the average pulse peak, and then find the crossing point with the fitted curve.

c) The FADC centroid timing with the "realistic" pulses was done by assuming the 100 MHz clock, which determined the corresponding values on the waveform (every 10 ns). With these points one determines a simple center of gravity.

d) Other timing methods were tried: multiple threshold, the reference pulse timing for the FADC algorithm, the parabola fit, etc. None of them made an improvement in the above mentioned methods.

Table 1 - Simulated resolution in Mark II vertex chamber (4 atm in 92% CO_2 + 8% iC_4H_{10} gas, 12 mm drift) [53].

Timing method	Design (a) 0 degrees	Design (a) 10 degrees	Design (d) 0 degrees	Design (d) 10 degrees
First electron timing with infinitely fast electronics	20 ± 3 µm	34 ± 3	25 ± 2	32 ± 3
Centroid timing with infinitely fast electronics	7 ± 1	14 ± 2	12 ± 1	13 ± 1
Leading edge timing with realistic pulses (threshold ~2-3% of the average ampl.)	20 ± 3	48 ± 4	29 ± 2	36 ± 3
Centroid timing with the realistic pulses and 100 MHz digitizer	20 ± 1	27 ± 2	36 ± 3	43 ± 3

Table 2 - Simulated resolution in Mark II vertex chamber
(2 atm in 92%CO_2+ 8% iC_4H_{10} gas, 12 mm drift)

Timing method	Design (a) 0 degrees	Design (d) 0 degrees
First electron timing with infinitely fast electronics	41 ± 4 μm	46 ± 4
Centroid timing with infinitely fast electronics	17 ± 2	21 ± 2
Leading edge timing with realistic pulses (threshold ~2-3% of the average ampl.)	41 ± 5	42 ± 5
Centroid timing with the realistic pulses and 100 MHz digitizer (use a simple centroid timing)	41 ± 3	42 ± 5

Experimental results from the vertex chamber of the Mark II experiment at SLC [56]:

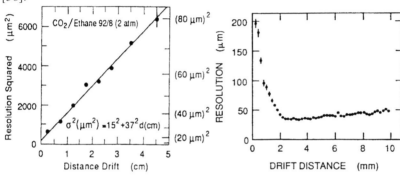

Fig.8 - Measured tracking resolution in Mark II vertex chamber in the middle of drift cell at 2 atm.

Fig.9 - Measured tracking resolution in Mark II vertex chamber near the anode wire at 2 atm.

The resolution in the middle of the drift cell follows the well known diffusion law: $\sigma(\mu m) = \sqrt{15^2 + 37^2 d(cm)}$, i.e. 46 ± 4 μm for 12 mm drift distance and 2 atm pressure, which is close to the Monte Carlo prediction (41 ± 5 μm) [53] - see Figs. 8 and 9.

Studies with various timing "tricks" to improve the accuracy were considered studying the micro-jet chamber prototype for possible use at HRS [57]. However, in this particular case, where one deals with short drift distances, the first electron timing was the best result on could achieve.

Radial drift chamber

All previously mentioned methods either used only a small portion of the ionization available or were sensitive to clustering effects.

D. Nygren suggested a technique to utilize all available ionization with equal weight and be insensitive to the clustering effect [58] - see Fig. 13. One should note that A. H. Walenta proposed similar concept called "induction" chamber [59].

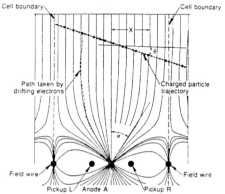

Fig.13 - Principle of the radial drift chamber.

The drift is radial using the low diffusion DME gas. Each electron's original position is reconstructed by measuring both <u>drift time</u> and <u>angle</u> α using two avalanche charge pick-up electrodes L&R - see Fig. 13. One expects $\eta(x) \sim 1$ in equation (6.10), i.e. $\sigma_{diffusion}$ (center of gravity timing) $\sim 55/\sqrt{120} \sim 5\,\mu m$ for DME gas.

D. Nygren's analysis of a practical chamber design operating with DME gas at 1 atm (total ionization in DME is N= 120/cm, $\sigma_{diffusion} \sim 55\,\mu m/\sqrt{cm}$), $N_S=10$ is the number of samples per cm, $<N_e>= 12$ is the average number of electrons per sample, $\sigma_{el.} \sim 20\,\mu m$ is the electronics noise, $\sigma_{aval.} \sim 88\,\mu m$ is the r-ϕ avalanche noise and radial distance of 3 cm with 30 samples. This appears to be the best one can do with a gaseous detector.

$$\sigma_{r-\phi} = \sigma_{el.} \oplus \frac{\sigma_{diffusion}}{\sqrt{<N_e>}} \oplus \frac{\sigma_{aval.}}{\sqrt{<N_e>}} = 20 \oplus \frac{55}{\sqrt{12}} \oplus \frac{88}{\sqrt{12}} \sim 36\,\mu m \text{ per sample}$$

$$\sigma_z = \sigma_{el.} \oplus \frac{\sigma_{diffusion}}{\sqrt{<N_e>}} = 20 \oplus \frac{55}{\sqrt{12}} \sim 25\,\mu m \text{ per sample}$$

Chapter 7

7.1. Wire ageing

The avalanche creates a plasma condition which induces the polymerization process. To get a introduction to the problems involved, I recommend to read two reviews about this subject by J. Va'vra [60] and J. Kadyk [61].

Physicist's view of the avalanche: **Chemist's view of the avalanche:**

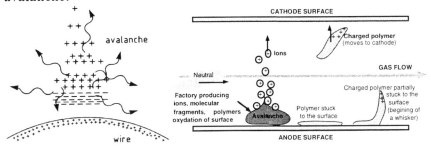

7.1.1. Anode related problems.

Fig.1 Whisker formation [61].   **Fig. 2** Film / liquid droplet formation [62,63]

Reprinted with kind permission of Elsevier Science.

© 1987, 1988 IEEE, Reprinted with permission.

For example, droplets were observed on anode wires in CH_4 + TMAE after obtaining a charge dose ~5-10 mC/cm, and then exposing the chamber to air - see Fig. 2.

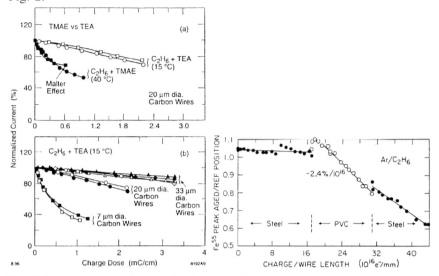

Fig. 3 Dependence on the anode wire diameter and some specific molecule [64].

Fig. 4 Dependence on the gas tubing material [65].

Reprinted with kind permission of Elsevier Science.

General comments about wire ageing:

1. It is a very complicated <u>chemical process</u>, which is not well understood quantitatively, except in few isolated cases.

2. However, what is already very clear is that one should:

    a) build the wire chambers <u>as cleanly as possible</u>,

    b) avoid soft glues or outgasing materials,

    c) test all materials used in the construction under the most representative condition - see Fig. 4.

3. The ageing rate depends on:

    a) wire radius (smaller radius larger gain drop) - see Fig. 3,

    b) wire alloy (oxidation of surfaces),

    c) gas type -see Fig. 3,

    d) gas additive (water, alcohol, etc.),

    e) gas flow,

    f) gas pressure, etc.

4. What to do about the wire ageing problem:

    a) wire replacement,

    b) wire washing in alcohol (TMAE ageing),

    c) wire heating (TMAE ageing),

    d) <u>run the gas gain as low as possible</u> (preventive),

    e) limit the charge doses by making the detectors thin (preventive).

7.2. Cathode related problems

7.2.1. Malter effect [66,67] - see Fig. 5.

If a radiation source is present, there is going to be a charge build up on the surface of the insulator, and one can reach a very high gradient across the thin insulator film. This can cause the electron emission which can lead to a positive feedback mechanism. Discharging time constant of the film is $RC \sim \varepsilon_r \varepsilon_0 \rho_{film}$, where $\varepsilon_r \sim 4$, $\varepsilon_0 \sim 8.85$ pF/m, $\rho_{film} \sim 10^{12} - 10^{15}$ $\Omega \bullet$cm => For $\rho_{film} \sim 10^{15}$ $\Omega \bullet$cm the time constant is $RC \sim 15$ min. The problem is defined by a relative rate of charging and discharging of the insulator film.

a) The Malter effect shows up as a continuous current present even when the source is removed.

b) or only occasional sporadic bursts [68].

7.2.2. Photo-cathode damage - see Fig.6:

The damage can be caused by [69]: a) sparking, b) environmental damage, c) electrolytic currents within the photo-cathode material, d) light exposure, e) gas gain. Example of a damage by an operation with gas gain [70] can be seen in Fig. 6. All photosensitive materials and their ageing by-products are good insulators. Therefore they could suffer from the Malter effect.

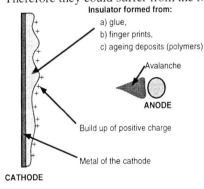
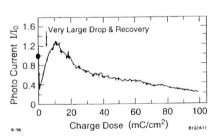

Fig. 5 Malter effect origin.   **Fig. 6** Aging of CsI [70].

Reprinted with kind permission of Elsevier Science.

7.3. Quenching problems

Avalanches produce photons which have to be absorbed by the hydrocarbon molecules, which have many vibrational modes.

Example of extremely bad quenching is CF_4 + TMAE gas. The avalanche excited CF_4 molecule emits photons at 170 nm, which are capable of photoionizing the TMAE molecule. This results in an extremely unstable operation where an avalanche breeds secondary avalanches [71] - see Fig. 7.

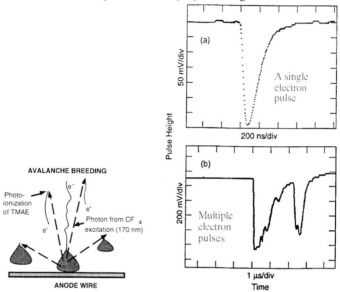

Fig. 7 Avalanche breeding.

One can also photo-ionize the nearby surfaces, which may have been coated by the polymerization products.

The avalanche breeding can be recognized by observing an excessive tail in the single electron pulse height distribution resulting in the negative θ parameter in the Polya function. To fix the problem of avalanche breeding, one has to add a C_4H_{10} molecule which absorbs the 170 nm photons [71]. - see Fig. 8.

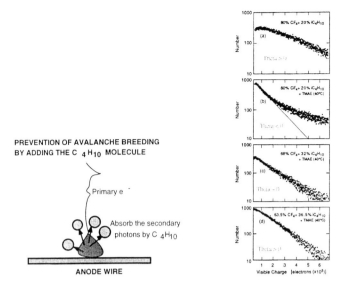

Fig. 8 Avalanche breeding can be stopped by an addition of C_4H_{10}.

To be able to detect the single electrons, the chambers used a relatively high gas gain (2-3×10^5); people built "barricades" (see Fig. 9) around the anode wires to limit the photon feedback in a presence of large dE/dx track deposits [72]. A real fix is to run lower gas gain and make the detectors thin.

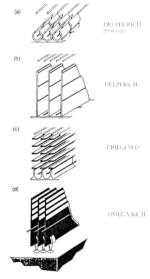

Fig. 9 An attempt to shield avalanche photons.

7.4. High voltage problems

a) **Sparking condition** [73]:

$$N_{\text{primary ionization deposit}} * G > 10^8$$

where $N_{\text{primary ionization deposit}}$ is the total primary ionization deposit, G is the average gas gain. A consequence of this law is that the α-particles may cause sparking sooner than the minimum ionizing particles for the same gas gain. This explains why sometime a chamber works in the lab and not in a real background environment next to the beam line !!

b) **High voltage insulation using several insulating layers.**

A bad HV design is a combination of G-10 sheet and Kapton printed circuit - see Fig. 10. There will be always some pin hole in Kapton causing a spark !!

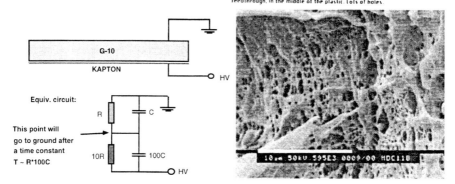

Picture taken by R. Malchow, Col. St. Univ.

Fig. 10 Example of bad HV design.   **Fig. 11** Voids in Delrin coasing a current leakage.

c) **Delrin pins in large drift BES chamber.**

As a result of not following the DuPont Co. molding procedure, the feedthrough developed the high voltage problem after several months of running - see Fig. 11.

7.5. Sensitivity to various drifts :

1. Gas mixture changes:	Change	$\frac{\Delta PH}{PH}[\%]$
90% Ar + 10% CH_4	$\Delta Ar = 1\%$	11
90% Ar + 10% CO_2	$\Delta Ar = 1\%$	10
50% Ar + 50% C_3H_8	$\Delta Ar = 1\%$	5

2. **Barometric pressure changes** (EPI):		
95% Ar + 5% CH$_4$	$\Delta P = 1\%$	7
3. **Leaks** (MAC at PEP I):		
86% Ar + 14% CH$_4$	add 0.6% of N$_2$	10
86% Ar + 14% CH$_4$	add 0.15% of O$_2$	10
4. **Voltage drifts** (JADE):		
90% Ar + 10% CH$_4$	$\Delta V_{gain} = 1\%$	20
90% Ar + 10% CH$_4$	$\Delta V_{drift} = 5\%$	25

References:

1. F. Sauli, CERN Yellow Report 77-09, 1977.
2. D.H. Wilkinson, "Ionization chambers and counters," 1950.
3. W. Blum and L. Ronaldi, "Particle detection with Drift Chambers," Spring-Verlag, 1994.
4. P.M. Morse and H. Feshbach, Methods of theoretical physics, McGraw Hill, New York, 1953.
5. R. Venhof, Drift Chamber Simulation Program GARFIELD, CERN/DD Garfield Manual, 1984.
6. K.J. Binns and P.J. Lawrenson, "Analysis and computation of electric and magnetic field problems," Pergamon Press, 1973, p.241.
7. S. Yellin, CRID note #40, SLD, SLAC, April 4, 1988.
8. S. Parker, Nuclear Instr.&Meth., A275(1989)494.
9. R. Bellazzini and M.A. Spezziga, Rivista Del Nuovo Cimento, Vol.17, N.12, 1994.
10. H.J. Burckhart et al., Nuclear Instr.&Meth., A244(1986)416.
11. J. Va'vra, Mark II Vertex chamber internal note, SLAC, February 2, 1985.
12. J. Va'vra, BaBar Internal DC note, SLAC, April 19, 1993.
13. I.R. Boyko et al., Nuclear Instr.&Meth., A367 (1995) 321.
14. L. Tonks, Phys. Rev. 97(1955)1443.
15. T. Kunst, B. Goetz and B. Schmidt, Nuclear Instr.&Meth., A324(1993)127.
16. P. Coyle et al., SLAC-PUB-4403.
17. K. Abe et al., Nuclear Instr.&Meth., A343 (1994) 74.

18. B. Schmidt, Dissertation, Univ. of Heidelberg, 1986.
19. G. Schultz & Gresser, NIM 151(1978)413, and G. Schultz, Thesis, 1976.
20. L. G. H. Huxley & R.W. Crompton, "The diffusion and drift of electrons in gases."
21. K.F. Ness and R.E. Robson, Phys. Rev A, 34(1986)2185.
22. P.Coyle, Lorentz program.
23. S. Biagi, Nuclear Instr.&Meth., A273(1988)533 (available as MAGBOLTZ program).
24. H. Pruchova and B. Franek, Nucl. Instr&Meth., A366(1995)385; Issue of ICFA Bullctin, SLAC-PUB-7376, 1997; and H. Pruchova's Ph.D. thesis, Prague Tech. Univ.
25. J. Groh, Interner Bericht DESY FH1T-89-03 May 1989.
26. M. Matobe et al., IEEE Trans. Nucl. Sci. NS-32,541(1985).
27. Diethorn, USAEC Report NY06628 (1956).
28. Zastawny, J. Sci. Instrum., 1966, Vol. 43, p.179.
29. S.A. Korff, "Electrons and Nuclear counters," 1955.
30. P.G. Datskos, J.G. Carter, L.G. Christophorou, J. Appl. Phys. 71(1992)15.
31. S. Biagi, Nucl. Instr. & Meth.. A310 (1991) 133.
32. Byrne, Proc. R. Soc. Edinburgh 66A(1962)33.
33. Raether, "Electron Avalanches and Breakdowns in Gases," Butterworths, London, 1964.
34. Schlumbohm, Z. Phys. 151 (1958)563.
35. J.Va'vra et al., Nuclear Instr.&Meth., A324(1883)113.
36. Y. X. Wang, G. Godgrey, Nuclear Instr.&Meth., A320 (1992) 238.
37. J. Va'vra, CRID note #50, SLD, SLAC, 1987.
38. J. Va'vra, Nuclear Instr.&Meth., A244(1986)391 and Nuclear Instr.&Meth., 225(1984)445.
39. P. Bock, OPAL, Heidelberg Univ., 1984, unpublished.
40. S. Ramo, Proc. I.R.E. 27(1939)584.
41. J. Va'vra, SLAC-PUB-2635, October 1980.
42. R.B. Owen and M.L Awcock, IEEE Trans. Nucl. Sci, NS-15 (1968) 290.
43. C.F.G. Delaney, Electronics for the Physicist, Ellis Horwood, Chichester, West Sussex, UK, 1980.
44. J. Va'vra, SLD, CRID note #75, Nov. 24, 1992, D. Aston et al., Nucl. Instr.

& Meth. A283 (1989) 590, and K. Abe et al., Nuclear Instr.&Meth., A343(1994)74.
45. F. Villa , Nuclear Instr.&Meth., A217(1983)273.
46. S. Bobkov et al., CERN-EP/83-81.
47. Cramer, "Mathematical Methods of Statistics," Princeton Univ. Press 1951.
48. Fischle et al., Nuclear Instr.&Meth., A301 (1991) 202.
49. A.V. Zarubin, Nuclear Instr.&Meth., A283 (1989) 409.
50. A. Pansky et al., Nuclear Instr.&Meth., A323 (1992) 294.
51. W. Farr et al, NIM 154 (1978) 175.
52. V. Commichau et al., Aachen Univ. preprint, PITHA 84-34.
53. J. Va'vra, SLAC, Mark II vertex chamber internal notes Oct. 12, 1984 and Oct. 31, 1984; and summarized in Nuclear Instr.&Meth., A244 (1986) 391.
54. F. Piuz, Nuclear Instr.&Meth., 175 (1980) 297.
55. Boie et al., Nuclear Instr.&Meth., 192(1982)365.
56. D. Durrett et al., SLAC-PUB-5259, May 1990.
57. J.Va'vra, Nuclear Instr.&Meth., 217(1983)322 and Nuclear Instr.&Meth., 225(1984)445.
58. J. Huth and D. Nygren, TPC-LBL-85-7.
59. A. H. Walenta, SLAC-PUB-259, 1982.
60. J. Va'vra, Nuclear Instr.&Meth., A252(1986)547.
61. J. Kadyk, Nuclear Instr.&Meth., A300(1991)436.
62. J. Va'vra, IEEE Trans.Nucl.Sci. NS-35, 1(1987)487.
63. C. Woody, IEEE Trans. Nucl. Sci. NS-35, 1(1988)493.
64. J. Va'vra, Nuclear Instr.&Meth., A387(1997)183.
65. Kothaus, LBL workshop, Berkeley, 1986.
66. L. Malter, Phys. Rev. 50(1936).
67. Guentherschultze, Z. Phys. 86(1933)778.
68. J. Va'vra, Nuclear Instr.&Meth., A367(1955)353.
69. P. Krizan et al., Nuclear Instr.&Meth., A387(1997)146.
70. A. Breskin, Nuclear Instr.&Meth., A371 (1996)116.
71. J.Va'vra et al., Nuclear Instr.&Meth., A370 (1996) 352.
72. J. Va'vra, Nuclear Instr.&Meth., A371 (1996) 33 and Nuclear Instr.&Meth., A387(1997)137.
73. H. Raether, Z. Phys. 112(1939)464.

REVIEW TALKS

Visible Light Photon Counters (VLPCs) for High Rate Tracking Medical Imaging and Particle Astrophysics

Muzaffer Atac

Fermi National Accelerator Laboratory
Batavia, IL 60510, U.S.A.
University of California at Los Angeles
Los Angeles, CA 90024, U.S.A.

Abstract. This paper is on the operation principles of the Visible Light Photon Counters (VLPCs), application to high luminosity-high multiplicity tracking for High Energy Charged Particle Physics, and application to Medical Imaging and Particle Astrophysics. The VLPCs as Solid State Photomultipliers (SSPMs) with high quantum efficiencycan detect down to single photons very efficiently with excellent time resolution and high avalanche gains.

INTRODUCTION

High Energy Particle Physics experiments designed to run at high luminosity and high multiplicities require fine granularity, fast, good time resolution and good spatial resolution tracking. Scintillating fibers with the VLPC readout fulfill all these requirements. Scintillating fibers with a diameter less than 1mm can provide good multitrack resolution and high speed. The VLPCs with quantum efficiency around 80%, avalanche gain around 30,000, excellent time resolution (about 1nsec), and surface area of 1mm in diameter can provide an efficient tracking for High Energy Particle Physics (1) (2) (3). The VLPCs were developed by Rockwell International Science Center (now part of Boeing Co.) jointly with M. Atac, originally for UCLA under DOE contracts. The author has pioneered the development in 1987-88 under the contract for developing scintillating fiber tracking for High Energy Particle Physics, working together with Rockwell

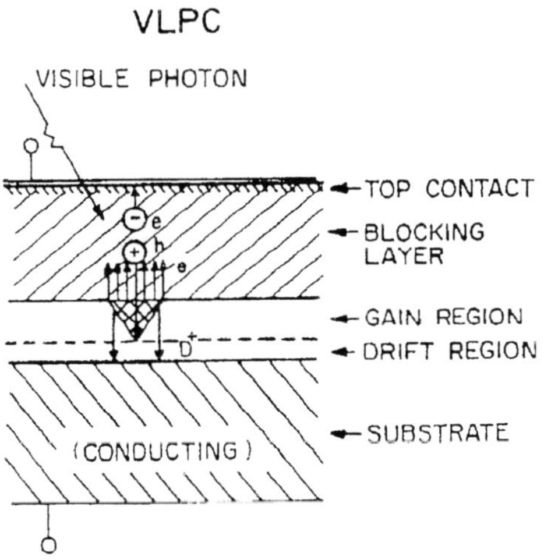

VLPC

FIGURE 1. Schematic of the operational principles of the VLPC.

International Science Center, Anaheim, California. We called this research and development High Intensity Scintillating-fiber Tracking Experiment (HISTE) program. During the last few years the VLPCs have been purchased by E835 and D0 experiments at Fermilab. Fermilab physicists and engineers have been developing large systems for D0. 100,000 pixel VLPC system is needed for D0. E835 has done an experiment at Fermilab using a fiber tracker with a modest number of VLPCs (about 1200 channels). COSMOS (E803) collaboration has decided to put together a scintillating fiber tracking system with 40,000 fibers and VPLCs.

In the following we will also talk about application of VLPCs to Medical Imaging and Particle Astrophysics.

Before the VLPCs became practical in using for fiber tracking, UA2 Experiment used image intensifier readout with vacuum photocathode (4). This resulted in an inefficient tracking. Most image intensifier tubes are rather sensitive to magnetic fields. Tests done with the VLPCs showed no effect up to 12kG field.

OPERATING PRINCIPLES OF THE VLPCS

The operating principles of the VLPCs are given in Reference 2, therefore we will discussed them briefly here. The VLPCs are Impurity Band Conduction (IBC) devices that are minimized in quantum efficiency in the Infrared (IR) region while maximized in quantum efficiency for the wavelengths around 550nm relative to the original device, Solid State Photo Multiplier (SSPM), which was discovered

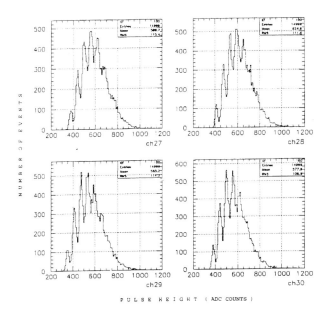

FIGURE 2. Multiple photoelectron peaks resulted when the VLPCs were illuminated with a pulsed LED. It shows uniform responses in quantum efficiency and gain from different channels obtainable under the same bias voltage.

by Rockwell International Science Center (5). The VLPCs are silicon based devices with some level of donor and acceptor concentrations in silicon formed by molecular epitaxy technique.

A schematic diagram of the VLPC is shown in Figure 1. In a VLPC, a neutral donor is a substitutional ion with an electron bound to it in a hydrogen-like orbit with an ionization potential of about 0.05eV. Because of this very small energy gap, the devices need to run at cryogenic temperatures. Nominally they run at a temperature between 6 and 7K. When the concentration of impurities is sufficiently high, they form an energy band separated from the conduction band by the ionization potential. When the applied electric field is sufficiently high, about 2×10^3-10^4V/cm, each initial electron starts an avalanche of free electron-hole pairs within 1ns. The avalanche size could reach up to 5×10^4 when applied voltage reaches −7 volts. The avalanche may occupy about 10 micron diameter area for about few microseconds while the rest of the area is continuously available for detecting photons. The gain and the quantum efficiency (QE) of the devices taken from same wafer are very uniform at a common voltage and temperature as seen in Figure 2 (6). For this, photons from an LED illuminated the

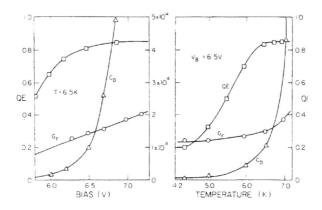

FIGURE 3. Quantum efficiency (QE) at λ=560nm of an AR coated VLPC as functions of bias voltage and temperature is shown. Dark count pulse rate (C_D) and fast gain (G_F) are also indicated.

top of a cassette which housed the VLPCs. The photons then transmitted through optical fibers illuminating the VLPCs. The sensitive area of the devices was 1mm in diameter. They run in a space charge saturated avalanche mode. Due to the small gain dispersion, as seen in the figure, multi photoelectron peaks arenicely separated. Due to their capability of counting photons, we called them Visible Light Photon Counters (VLPCs).

Because the devices are impurity semiconductors and the VLPC with arsenic impurity, that has a bandgap around 50 millielectron volt, it has to be operated at cryogenic temperatures, otherwise dark pulse (thermal electron pulse) rate would be extremely high (saturating the device). Figure 3 shows the quantum efficiency, the fast avalanche gain and dark pulse rate as functions of temperature and bias voltage. Optimum operating temperature is around 6.5K and optimum bias voltage is about 6.5V. The devices were coated with 560nm antireflective (AR) coating to improve the quantum efficiency (QE) from 70% to 85%. Operating the devices at 6.5K, controlling and monitoring the temperature within 0.1K are relatively easy.

Main characteristics of the VLPCs are summarized in the following table:

Table I.

Quantum efficiency optimized for 530nm	80%
Avalanche gain	$3-5\times10^4$
Thermal electron pulse rate at 6.5K	$\sim5\times10^3$/sec.mm
Saturation pulse rate	5×10^7/sec.mm
Pulse risetime	<3nsec
Average power per pixel	1.5 microwatt
Optimum operating voltage	~6.5V
Optimum operating temperature	6-7K
Dynamic range (linear)	3000 photoelectrons
Effect by magnetic field	no effect up to 12kG

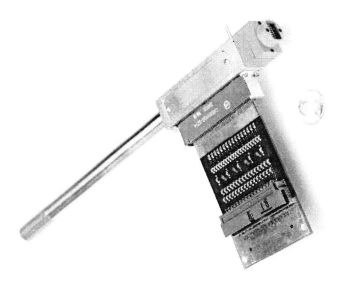

FIGURE 4. A photograph of a 32 channel VLPC cassette. The penny in the picture shows the compactness of the unit.

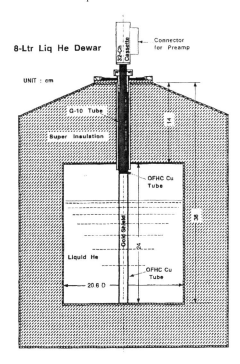

FIGURE 5. The structure of the 8-liter cryostat..

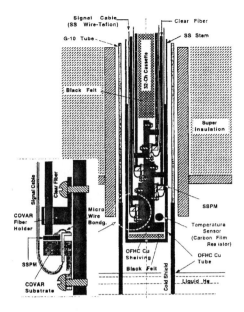

FIGURE 6. Enlarged view of the cassette.

PROTOTYPE CRYOGENIC CASSETTE DESIGN

A 32 channel VLPC cassette was designed by the author and constructed to carry out scintillating fiber tests for determining photoelectron yield from a variety of fibers, attenuation length of the photons in the fibers, and timing and rate capabilities of the VLPCs. Figure 4 shows one of the cassettes with a penny next to it. The penny is there to show how small a cryogenic system is used. At the top of the cassette a 32 channel of a QPAO2 amplifier is shown (7). The unit is designed to pass a small amount (about 50cc) of boil off liquid Helium (He) when inserted into a liquid He dewar. Details of the cassette-dewar assembly are shown in Figure 5. With the help of the Oxygen free High Conductivity Copper (OFHC) cold shield tubing, liquid He temperature is brought to the level of the OFHC housing in which the VLPCs are kept. With this arrangement the required temperature is reached and kept fairly constant over several days, using only 2 liters of liquid He per day. This small amount of usage is due to full usage of enthalpy of boil off He going through the cassette and cooling all the components in the thin wall stainless steel 304 tubing. The details of the arrangement in the cassette are shown in Figure 6 (8).

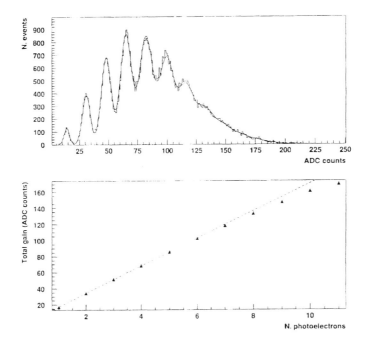

FIGURE 7. Multi-photoelectron peaks and a calibration plot. As we see in the figure, multi-photon peaks are well resolved. For this reason we call the devices "Visible Light Photon Counters" (VLPC).

SCINTILLATING FIBER TRACKING TESTS

Some scintillating fiber tracking tests were carried out using 500 micron overall diameter multiclad scintillating fibers (manufactured by Kuraray Co., Japan) and the above described cassette and cryogenic arrangement. We detected 6 photoelectrons per minimum ionizing track in the average from the middle of 280cm length of scintillating fibers (were mirrored at the end) which were coupled to 500cm length of multiclad clear optical fibers (6). An ADC counts versus number of photoelectron (pe) calibration was done before every measurement as shown in Figure 7. For this an LED was used, illuminating the optical fibers in the cassette. As seen in the figure, after the sixth photoelectron peak a small saturation appears. This is due to the amplifier saturation. The calibration was obtained by making a cut at N>2 and fitting to a convolution of Poisson and Gaussian functions:

$$f(x) = N \sum_n \frac{1}{\sigma_n \sqrt{2\pi}} \exp\left[-\frac{1}{2}\left(\frac{x-n}{\sigma_n}\right)^2\right] \frac{\exp(-n_{pe})}{n!} \quad (1)$$

where the free parameters were normalization factor (N) and the mean value of the Poisson distribution (n_{pe}). Sigma of each gaussian (σ_n) was fixed by determining the peaks with the LED runs for the bias voltage and temperature.

$$\sigma_n^2 = \sigma_0^2 + \sigma_c^2 \cdot n, \quad (2)$$

where σ_c is a sigma of the n-th peak and σ_0 is a sigma value of the pedestal.

Attenuation of photons in various scintillating fibers are shown in Figure 8. The results indicate that attenuation lengths are between four and five meters and there was no appreciable change in this number for fibers of 500 microns to 1mm. Attenuation length, λ, of photons from 3HF (1500 ppm 3-Hydroxyflavone +1% P-therphenyl) scintillating fiber were measured through various diameters of multiclad clear optical fibers. The 3HF emits photons around the peak wavelength of 530nm (manufactured by Kuraray Co.). The results, as seen in the figure, indicated that the attenuation lengths of photons from the 3HF within the experimental error does not depend on fiber diameter between 0.5mm to 1mm.

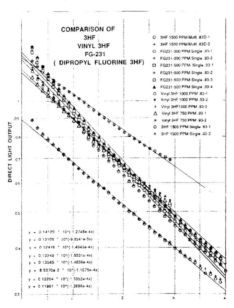

FIGURE 8. Attenuation length plots for various scintillators. There is about 80% more yield from the multi-clad fibers relative to single clad.

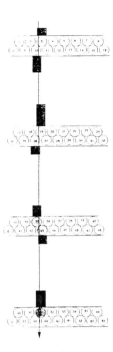

FIGURE 9. A minimizing ionizing track. The number of detected photons is indicated in the bars.

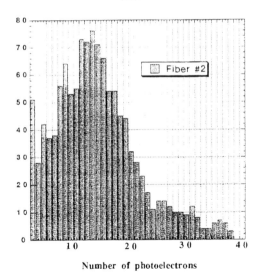

FIGURE 10. Pulse height distribution from VLPC fiber #2 when fiber is illuminated by minimum ionization particles (MIP).

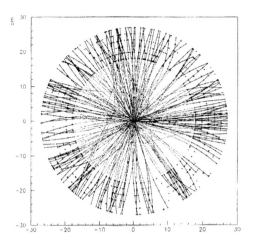

FIGURE 11A. A typical top + 6 MB event in the R-φ view. Crosses are axial hits, solid lines connect hits used, and dotted curves are extrapolation inward of the final parameters.

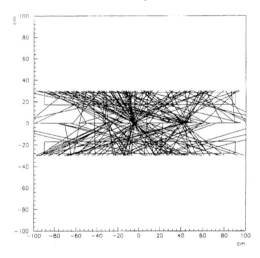

FIGURE 11B. R-Z view of a top + 6 MB event. Crosses show locations of associated stereo-hit/axial-segment points. Lines show extrapolation of fitted segment to beam line.

Some beam tests were carried out at the Meson 6 West test beam using 835 micron (scintillating core of 725 micron) multiclad 3HF fibers using author's designed cassettes (9). Four staggered doublets were used with the tests. Figure 9 shows a 15GeV hadron (most likely a pion) track with the number of photoelectrons indicated in the pulse height histograms. Figure 10 shows the number of photoelectron distribution for a fiber. Weighted average photoelectrons is more than 15. For a staggered doublet the tracking efficiency is better than 99.7%.

The Collider Detector Facility at Fermilab (CDF) has considered using scintillating fiber tracker and some track reconstructions were done using top quark events. Figures 11-a and 11-b show how efficiently fiber tracker could find tracks that were found by the Central Tracker (CTC). Presently Fermilab physicists and engineers are working hard on a large system design for the D0, Short Baseline Neutrino Experiment (COSMOS), and CP Violation Experiment at the Main Injector (KAMI). The number of VLPCs to be needed will be around 240,000.

USE OF VLPCS FOR MEDICAL APPLICATIONS

Single Fiber Tracker for Stereotactic Biopsy and Intraoperative Lumpectomy of Breast Cancer

Breast cancer in women of child bearing age is the second leading cause of death in the USA (10). Early detection has allowed for less extensive surgical procedures and/or decreased need for chemotherapy since a substantial majority of questionable lesions detected by mammography are benign. There is a growing interest among the health care professionals and patients in finding alternatives to surgical biopsy for diagnosing these lesions. State-of-the-art stereotactic breast biopsy is comparable in sensitivity to surgical biopsy, and the procedure is quicker, cheaper, and easier than the standard practice of preoperative, mammographically guided localization followed by surgical biopsy.

The problems mentioned above can be ameliorated by a nuclear medicine procedure using a beta detector on the end of a 0.8mm diameter scintillator and fiber optic cable (11). By positioning the detector within a few millimeters of the suspected area, small lesions, usually not detectable using gamma radiation detectors, can be identified and quantified for activity. The fiber optic cable with a small scintillating plastic fiber attached (fused) to the tip can either be inserted into a core biopsy, or can be used during ductogram to identify the duct system containing microcalcified clusters. When inserted into a surgical wand, it could be used to ensure that all residual tumor was removed during lumpectomy. This diagnostics alone is very much needed to prevent recurrence and spread of malignant tissues.

We have developed a prototype suitable probe that uses a rather small diameter biopsy needle (in the current study an 18 gauge needle with an external diameter of 1.25mm) containing a 0.83mm diameter and 3mm length 3HF (above mentioned)

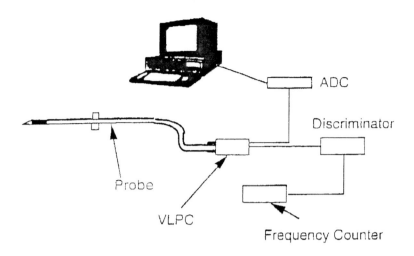

FIGURE 12. A schematic view of the biopsy needle probe together with a simple data acquisition system.

multiclad scintillating fiber, which is fused to the same diameter multiclad clear optical fiber of 200cm length.

Photons emitted from the scintillating fiber by the passage of betas or positrons are transmitted through the optical fiber and are detected by a VLPC. For the set up, a cassette and the cryogenics mentioned in the tracking section above were used. The probe assembly and the rather inexpensive data acquisition system are shown in Figure 12.

The signals from the VLPC were amplified by a transimpedance amplifier (TIA), fed into a discriminator and counted by a commercial scaler. The VLPC produces an avalanche gain around 30,000 per photoelectron. We obtained less than two pulses per minute as background counts when the threshold was set above three photoelectrons. We measured experimentally that the average number of photoelectrons produced in the VLPC was more than 40, by the passage of betas through the thin scintillator. Pulse height spectrum obtained using a Bi^{207} beta source is shown in Figure 13. Only a small fraction of the 1MeV beta particle energy is left in the thin scintillator, giving rise to the pulse height spectrum.

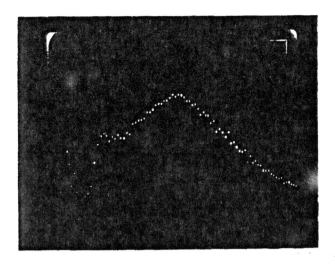

FIGURE 13. Pulse height spectrum obtained using Bi^{207} beta source. The average energy released in the scintillator fiber is about 60KeV. The peak value corresponds to 40 photoelectrons detected by the VLPC.

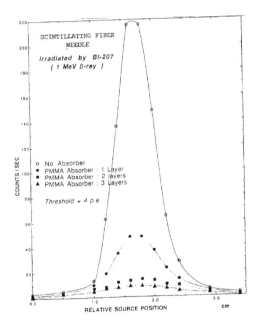

FIGURE 14. Results from the Bi^{207} source test. The curves clearly indicate that the 1MeV betas are rapidly absorbed by the 1.5mm thick Lucite sheets, and there are not many counts from gamma conversions in the scintillator although 90% of the decays from the source are gamma rays in this case.

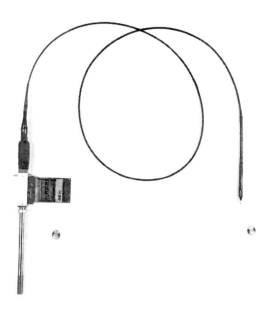

FIGURE 15. Photograph of the probe with the 2 meter long optical fiber between the biopsy needle and the VLPC unit.

FIGURE 16. Test done with a rat having an R3230 AC in the hind leg. The rat was administered 432 microcurie FDG i.v.

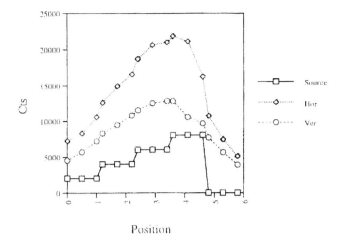

FIGURE 17. Two dimensional scan, even from outside of the skin, indicates where the radionucleide concentration is.

EXPERIMENTAL RESULTS

In order to determine point spread function, we moved the probe linearly relative to 1 microcurie Bi^{207} source without and with 1.5mm thick Lucite sheets (mimicking tissue equivalent density) in between the source and the probe, and recorded the counts per second. The source diameter was approximately 4mm and it was not collimated. The results plotted, in Figure 14 show that the 1MeV betas from the source were very much attenuated after one sheet of Lucite, but we can find the source position after 4.5mm thickness. We expect that the intrinsic resolution of the probe be better than 1mm. The curves also show that the probe is sensitive to betas and not to gammas, although only 8% of the decays produce betas and the rest being the gamma activity. This feature is important due to the fact that the probe will not be sensitive to 511KeV gammas from positron annihilation when a positron source is traced. For a probe like this in a clinical condition, the cryogenic part can be cooled by liquid Helium vapor for safety. A photograph of the probe is shown in Figure 15.

As a first experiment, a preliminary test was done using a rat bearing R3230 adenocarcinoma. The experimental arrangement is shown in Figure 16. As shown in Figure 17, the biopsy needle was moved in an x,y matrix points and the counts were recorded. Even from outside of the skin the probe indicates where the radionucleide concentration is.

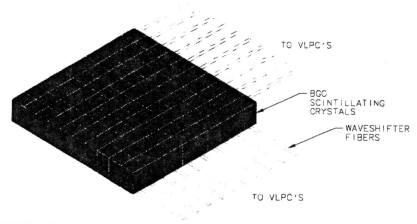

FIGURE 18. The principal scheme for detecting gamma rays in a two dimensional readout. More layers can be added depending on the energy of the gammas to be detected

USE OF VLPCS FOR PARTICLE ASTROPHYSICS

A possible way of using VLPCs as photodetectors for detecting relatively low energy gammas is shown in Figure 18. In this case wavelength shifting fibers are attached to a matrix array of scintillating crystals in an x,y arrays. The crystal size can be sufficiently large to detect multi MeV gammas from outer space. Scintillating crystals like BGO can be used in this case. Rubrene doped polystyrene can be used as wavelength shifter. Optical fiber that is coupled to wavelength shifter carries the photons to the VLPCs. This idea was proposed for medical imaging by M. Petroff, but it did not work so well due to low energy gammas, 511KeV, from positron annihilation. I am convinced that it will work here due to the detection of multi MeV gammas.

REFERENCES

1. Petroff, M.D. and Atac, M. , *IEEE Trans. on Nucl. Sci.* NS-36 (1989) 163.
2. Atac, M., Park, J., Cline, D., Chrisman, D., Petroff, M. and Anderson;E., Nucl. Instr. and Meth. A314 (1992) 56.
3. Atac, M., et al.; *Nucl. Instr .and Meth.* A320 (1992) 155.
4. Alitti, J., et al.; *Nucl. Instr. and Meth.* A279 (1989) 364.
5. Petroff, M.D., Stapelbroek, M.G., and Kleinhans, W.A., *Appl. Phys. Lett.* Vol.51 No.6 (1987) 406.
6. Atac, M., Mishina, M., Takano, T., Valls, J., Yasuokka, K. and Yoshida, T., CDF/ANAL/Tracking/Public/3569, April 11, 1996.
7. Zimmerman, T., *IEEE Trans. Nuc. Sci.* NS-37(2), (1992) 439.

8. Gubinelli, M., Tonet, O., Sorel, M., Atac, M., Mishina, M., and Valls; J., CDF/Pub/Public/4154, July, 1997.
9. Atac, M., Cline, D., Pischalnikov, Y., and Rhoades, J.; Presented at the SciFi Conference at Notre Dame.
10. Titcomb, C.L., *Hawaii Medical Journal*, Vol. 49 (1990) 18.
11. Atac, M., Nalcioglu, O., and Roeck, W., unpublished report.

PERSPECTIVES IN HIGH ENERGY PHYSICS[1]

Arturo Fernandez[1] and Arnulfo Zepeda[2]

[1] *Facultad de Ciencias Físico Matemáticas*
Benemérita Universidad Autónoma de Puebla
P.O. Box 1364, Puebla, Pue. México

[2] *Centro de Investigación y de Estudios Avanzados del IPN*
P.O.Box 14-740, 07000 México D.F. México

Particle physics, including high- and low-energy physics, nonaccelerator physics and astroparticle physics, is nowadays a very exiting and complex enterprise. Its objective is one and only one: to understand the deepest roots of the laws that govern the behavior of the fundamental constituents of matter.

The relevant questions that are today within reach demand a wide variety of new and inventive experiments and imaginative theoretical ideas. Some speculations, in the form of internally consistent proposals, show, on the other hand, the extension of the variety of open possibilities for the New Physics beyond the presently confirmed model, the Standard Model of strong and electroweak interactions. Speculations and sound theoretical calculations serve as a base or as a framework where to develop the understanding of the world. There is however no substitute for the real test of our understanding of nature, the experiment. In other words, measurement, detection and discovery are the keys to real progress toward our objective.

In order to talk about perspectives we should first answer this question: Where are we today? Certainly much more ahead than the discovery of the nucleus by Rutherford and collaborators, certainly much more ahead than the discovery of scaling in deep inelastic scattering. What are the objectives and the real possibilities to test the world today? To answer this question we would like to know

- The description of operating, approved and planed accelerator facilities.

---

[1] Presented by A. Zepeda

- The description of working and planned detectors and their experimental capabilities.

- The description of nonaccelerator experiments related to particle physics.

- The present status of confirmed facts in the physics of quarks and leptons.

- The description of plausible theoretical ideas and their observable signatures.

Some of these items are described in these proceedings or were discussed during the School, in particular you may take a look to the contributions from

- Lectures on new experimental techniques and detector systems for future high energy physics experiments (A. Savoy–Navarro).

- Review talks on neutrino detectors (Nakamura), photodetection (K. Arisaka), cryogenic solid state detectors for dark matter (D. McCammon), and on air shower detectors (C. Escobar).

Therefore we shall limit our review to only a few items.

I OPERATING AND APPROVED ACCELERATOR FACILITIES

High-energy physics has progressed twelve orders of magnitude in energy over the last one-hundred years (from 1eV to 1TeV). Modern high-energy accelerators are time-like machines allowing us to probe fundamental physics at distances of the order of 10^{-16}cm and hence understand the relevant phenomena at times as close as 10^{-10}sec after the Big Bang. The further advancement of elementary-particle physics demands, on the other hand, the increase of energy and luminosity in planned accelerators by at least one or two orders of magnitude. This would allow high-energy physicists to make experiments leading to expected and unexpected discoveries and put constraints on the possibilities for new physics. At the present time any improvement in energy or luminosity will require a crucial innovations in accelerator technologies.

Nowadays there are several accelerator machines working around the world. These machines produce beams of accelerated particles ($e^{\pm}$, p, $\bar{p}$) with he characteristics displayed in Table 1.
Here we briefly describe some of the most important accelerators and its upgrades.

SLAC (Stanford Linear Accelerator Center)

SLAC [1] has played a leading role in the field of liner colliders and its goal now is the construction of a TeV range electron-positron collider, generally referred to as the Next Linear Collider (NLC). SLAC operates at the present time the following major experimental facilities:

- The linac, a three-kilometer (or two-mile) long linear accelerator, capable of producing electron and positron beams with energies up to 50 GeV.

- End Station A (ESA) for fixed target experiments.

- SPEAR, a storage ring 80 meters in diameter now used as the synchrotron radiation source for the Stanford Synchrotron Radiation Laboratory (SSRL).

- PEP, a 30 GeV colliding-beam storage ring, 800 meters in diameter, now being upgraded to PEP-II, which will serve as a B meson factory, colliding 9 GeV electrons with 3.1 GeV positrons.

- SLC, a 100 GeV electron-positron linear collider, and SLD, the collider detector.

- Final Focus Test Beam, a facility for research on future accelerator design.

HERA (Hadron-Electron Ring Accelerator facility)

HERA [2] is the only facility in the world in which electrons or positrons and protons collide. At HERA physicists are able to observe electron/positron proton collisions at center-of-mass energies which are one order of magnitude above energies previously available. In this type of collisions structures inside the proton can be studied down to one part in a thousand of the size of the proton itself.

Inside HERA the two particle beams circulate in opposite directions in separate storage rings and are brought to head-on collision at two interaction points. These are equipped with the detectors H1 and ZEUS, which have both been taking data since the beginning of the HERA research program in 1992. Besides collision experiments HERA accommodates two beam-target experiments: HERMES, in operation since 1995, uses the polarized electron beam to investigate the origin of nucleon spin; HERA-B, scheduled to start in

TABLE 1. Accelerator Facilities

Accelerator	Name	Beam	CM Energy (Gev)
CERN-LEP	CERN Large electron-positron Collider	e^+e^-	180
CERN-SPS	CERN Super Proton Synchrotron	pp	540
DESY-HERA	e-p collisions DESY	ep	314
FNAL	Fermi Nac. Acc. Laboratory	pp	2000
SLAC	Stanford Linear Accelerator Laboratory	e^--e^+	100

1998, will use the proton beam for the study of CP violation in the B-meson system.

The underground storage ring facility HERA has a circumference of 6,336 meters. It was constructed from 1984 to the end of 1990, when the operation began. The first particle collisions were observed in October 1991; the research program started in June 1992. Since then the integrated HERA luminosity continuously increased from 0.05 inverse pb^{-1} (1992), to 1 pb^{-1} (1993), to 6.2 pb^{-1} (1994), to 12.3 pb^{-1} (1995), to 17.2 pb^{-1} (1996). For the 1997 HERA run an integrated luminosity of 25 pb^{-1} is expected.

Fermilab (Fermi National Accelerator Laboratory)

The Fermilab Collider [3] is at the energy frontier and will provide important results and great discovery opportunities until the LHC era. The highest priority for the Laboratory in the pre-LHC era should be the accumulation of the maximum integrated luminosity in collider detectors that can work properly at high luminosity beams. That is one of the challenges for new design of both, accelerators and particle detectors. Fermilab is a six accelerators complex. The fifth, called the Main Injector, is now under construction, but the others are working since the past decade. The game starts accelerating a bunch of negative hydrogen ions (positive protons) with a Cockcroft-Walton accelerator and ends with protons and anti-protons flying around the Tevatron with an energy of approximately 1 Tev, the highest energy ever riched in this kind of machines.

The construction of the Main Injector accelerator at Fermilab started in 1993 with the objective of improving the performance of the Tevatron by a factor of five. The Main Injector is being built tangent to the Tevatron tunnel and will be completed in 1998 using preexisting components from the Main Ring accelerator. Roughly half the size of the Main Ring, the Main Injector will give the Tevatron additional discovery power by increasing the number of protons and antiprotons that collide at high energy. A remarkable aspect of the Main Injector project is that the expected luminosity is now four times the original design. This is due to the improvements to the accelerator complex during Run I and to the Recycler concept. The integrated luminosity goal for each collider detector is now 2 nb^{-1}, greatly expanding the physics possibilities.

CERN

CERN [4] has essentially two main accelerator programs: The Super Proton Synchrotron (SPS) and the Large Electron Positron (LEP) accelerators. The latter machine is the world's largest particle accelerator (26.7 km in circumference some 100 meters underground).

The 400 GeV Super Proton Synchrotron (SPS) is the latest and largest of the CERN accelerators. A first design was put forward to Council in 1964 and, in a considerably modified form, the project was finally approved in February 1971. During the last years, the design and construction of this giant machine has progressed very well, and in the spring of 1995 the commissioning stage was reached.

LHC is an accelerator which will bring protons into head-on collision at energies higher (14 TeV) than ever achieved before. It will allow scientists to penetrate still further into the structure of matter and recreate the conditions prevailing in the Universe just 10^{-12} seconds after the "Big Bang" when the temperature was 10^{16} degrees. The accelerator will produce not only higher energy but also a higher luminosity than has been achieved before and it will reveal the behavior of fundamental particles of matter which has never been studied.

II WORKING AND PLANNED DETECTORS.

Particle spectrometers are made essentially of tracking detectors, calorimeters and particle identification devices. As we saw in the last section, particle detectors have to be able to identify the energy, other kinematical parameters, and the nature of particles produced by very energetic beams which generate a plethora of particles in the final state. The main trend in R&D is therefore towards high multiplicity tracking, fine grain and radiation hardness. For calorimeters, in addition to the above requirements, there are the critical ones of homogeneity of response and hermeticity. The development of particle identification detectors is receiving an strong impulse from preparations for CP violation studies, for which these issues are of critical importance. Also important is to note the parallel progress in readout electronics which is of equivalent relevance and difficulty as the development of particle detectors, due in particular to the high rate of final particles, multiplicity and radiation levels.

For tracking, progress has been reported in solid state detectors, such as silicon microstrips used in CDF [5], gaseous tracking detectors, such as drift chambers which are being used in fixed target experiments (E791) [6], or, more recently, the microstrip gas chambers which will be used in the ATLAS spectrometer [7]. Among other tracking detector systems we may mention the scintillator fiber trackers which will be used in D0 [8], to be placed behind their silicon tracker within the D0 upgrade program.

R&D effort towards future hadron colliders has produced a new generation of calorimeters characterized in particular for a high speed of response, homogeneity of response and resistance to radiation. The use, for instance, of the liquid argon calorimeter of ATLAS with "accordion" structure of the lead absorbers and readout electrodes is justified by the requirements of the LHC physics related to the mentioned characteristics. The CMS [9] experiment proposes to construct a scintillator based on a high resolution homogeneous calorimeter using crystals made of sophisticated chemical formulas ($PbWO_4$). This choice is based on the high density, the fast luminescence, the reasonable light yield and the radiation resistance of the crystal.

Particle identification is essential for both the good selection of decay channels and for flavor tagging. For this purpose there are multicell threshold Čerenkov counters, Time of Flight systems (TOS), RICH detectors and so on. BELLE, [10] a detector for studying CP violation at the KEK B Factory, will use Aerogel Čerenkov counters consisting of 900 (barrel) plus 224 (forward end-cap) modules with typical dimensions of $12\,cm^3$. Their TOF system is build of 128 plastic scintillation counters 4 cm thick and readout from both sides. ALICE [11] is the only experiment designed to exploit the physics of heavy ion collisions at LHC. In the proposed experimental layout a fundamental role is played by the particle identification system (PID). Two arrays are presently under development: a TOF barrel optimized for identification of hadrons with momenta smaller than 2 GeV/c on a event by event basis and a single arm system devoted to inclusive high momentum particle identification. A proximity focusing RICH system based on CsI (solid photocathodes) as photoconverter seems well suited in particle identification at high particle densities when a pad readout is implemented for a two-dimensional determination of the ionizing particles impact points and of the Čerenkov photons conversion points.

III ACHIEVEMENTS AND SEARCHS FOR NEW PHYSICS BY THE PRESENT ACCELERATORS

The main objective of the accelerator machines described above can be summarized as follows.

- Perform precision tests of the Standard Model (SM) at the level of a few *per mille* accuracy. The SM has been established to a high degree of accuracy.

- Count neutrinos. Beyond doubts, it is now established that the number of light neutrinos ($m_\nu < M_Z/2$) is 3.

- Search for the top quark. As is well known now, the top quark has been detected.

- Search for the Higgs boson. The result of all the searches has been negative and a lower bound of $m_H > 65$ GeV has been reached.

- Search for new particles (negative results).

Presicion tests of the standard model

LEP (LPE1) was terminated in 1995 and it produced 16 million of Z's which were analysed by he four LEP experiments. LEP and SLC were tuned to an energy region around $E_{CM} = M_Z = 91.2$ GeV. As a result a number of parameters of the Standard model were measured with great precision [12]:

$$M_Z = 91186.7(2.0) MeV$$
$$\sin^2 \theta_{eff} = 0.23152(32)$$
$$\alpha_s(M_Z) = 0.119(4)$$
$$m_t = 175.6(5.5) GeV. \quad (1)$$
$$(2)$$

with the last quantity obtained at the Tevatron.

The present status of the electroweak data are presented in Table 2 together with their SM values. These correspond to a fit in terms of m_t, m_H and $\alpha_s(m_Z)$ of all the available available data including the CDF/DO value of m_t.

TABLE 2

Quantity	Data(August '97)	Standard Model	Pull
m_Z(GeV)	91.1867(20)	91.1866	0.0
Γ_Z(GeV)	2.4948(25)	2.4966	-0.5
σ_h(nb)	41.486(53)	41.467	0.4
R_h	20.775(27)	20.756	0.7
R_b	0.2170(9)	0.2158	1.4
R_c	0.1734(48)	0.1723	-0.1
A^l_{FB}	0.0171(10)	0.0162	0.9
A_τ	0.1411(64)	0.1470	-0.9
A_e	0.1399(73)	0.1470	-1.0
A^b_{FB}	0.0983(24)	0.1031	-2.0
A^c_{FB}	0.0739(48)	0.0736	0.0
A_b (SLD direct)	0.900(50)	0.935	-1.7
A_c (SLD direct)	0.650(58)	0.668	-0.3
$\sin^2 \theta_{eff}$(LEP-combined)	0.23199(28)	0.23152	1.7
$A_{LR} \to \sin^2 \theta eff$	0.23055(41)	0.23152	-2.4
m_W(GeV)(LEP2+$p\bar{p}$)	80.43(8)	80.375	0.7
$1-\frac{m_W^2}{m_Z^2}(\nu N)$	0.2254(37)	0.2231	0.6
Q_W(Atomic PV in Cs)	-72.11(93)	-73.20	1.2
m_t(GeV)	175.6(5.5)	173.1	0.4

Until a year ago there were some discrepancies between the experimental and expected SM values for R_b and R_c. These discrepancies have disappeared.

Search for new physics

The CDF collaboration at Fermilab has made [13] a search for leptoquarks, Higgs bosons, supersymmetric particles, and heavy stable charged particles. All with negative results. The CDF collaboration collected data from 1992 to 1995 corresponding to about 110 pb^{-1}. This large sample of data led to the discovery of the top quark [14], precision electroweak measurements, studies of b quark mesons, studies of quantum chromodynamics, and the search for new particles and phenomena beyond the standard model.

The two HERA experiments, H1 and ZEUS, observed an excess of events, in high energy collisions of electrons and photons, at high x (or M = $\sqrt{(xs)}$), y, and Q^2 [15], and whose possible explanation is the production of leptoquarks. For Q^2 > 15,000 GeV^2, the joint distribution has a probability of less than one per-cent to come from Standard Model NC DIS processes. Within this model, the predictions are known with confidence. Similar probabilities occur for masses M larger than 175 GeV. The events at high Q-squared and large M are particularly interesting because they occur in a previously unexplored region for deep inelastic scattering. Increased luminosities from the forthcoming data taking period, March-October, 1997, will clarify whether the observed effect is a statistical fluctuation or a signal of new physics.

Motivated by these results the CDF Collaboration has performed a search for events which could correspond to the production of a pair of leptoquarks. These events would have an electron, a positron and two jets in the final state. The signal obtained (12 events) is compatible with background and a leptoquark of mass less than 210 GeV is excluded.

At LHC two detectors, ATLAS and CMS, which will record the interactions created by colliding proton beams at an energy of up to 14 TeV, are already at an advanced stage of development. However the LHC will not be limited to the study of proton-proton collisions, the LHC can also collide heavy ions, such as lead, to produce a total energy of 1148 TeV. A large energy density can be obtained over a wide enough region in the collisions to cause phase transition of nuclear matter into quark-gluon plasma. Studies of such a state of matter are expected to yield important new results. Proton-proton collisions at the LHC will be a copious source of B-mesons. The study of the decay of these mesons will allow physicists a deeper examination of CP-violation and a tailor-made B-physics detector is under development for LHC. At a later stage, proton beams from LHC can also be made to collide with electron beams from LEP opening up another field of research. This wide range of physics possibilities will enable LHC to retain its unique place on the frontiers of physics research well into the next century.

The CDF Collaboration has also searched for the production of neutral Higgs bosons ins association with W or Z bosons. No signal was found for $M_H < 230$ GeV. A search for the production of charged Higgs bosons was also performed and a the negative result excludes a region roughly corresponding to $M_H < 150$ GeV and $\tan\beta < 1$ or $100 < \tan\beta$, where $\tan\beta$ is the ratio of the vacuum expectation values of two higgs doubles in the two Higgs models.

In relation with supersymmetry, interest arouse in one event observed by CDF with two electrons and two photons in the final state and over 50 GeV of missing energy. The supersymmetric explanation of this rare event in the standard model invokes the production of a pair of selectrons decaying each into the next-lightest neutralino and a photon. This explanation calls however for other type of events, such as those with two photons and missing E_T, which do not seem to be present.

The net result of the precision measurements and the search for new physics is that there is no significant evidence for departures from the SM and no convincing hint of new physics (also including the first results from LEP2) [16]

IV ELECTROWEAK SYMMETRY BREAKING

As we have seen the SM has been tested to the level of 10^{-3} and it works extremely well. However the underlying mechanism responsible for the breaking of the electroweak gauge symmetry, $SU(2)_L \otimes U(1)_Y$, remains unknown. That is why there is consensus in that the most pressing question nowadays may be posed as " Who is responsible for electroweak symmetry breaking?". The possible candidates are

- Elementary Higgs, either standard or supersymmetric. As we have seen, however, the search for it has been unfruitful.

- Technicolor. In this type of models, the Higgs boson is not elementary but a condensate. Here we include models where the condensate is made up of top quarks. The condensate is formed due to new strong intercations of the QCD type. These forces imply the formation of new hadrons which would be detectable at LHC. The main problem of this line of though is that it is rather difficult to construct a realistic model.

Technicolor
For pedagogical reasons we elaborate here briefly about the technicolor idea, which has its roots in the fact QCD itself breaks the electroweak symmetry, although its contribution to the W and Z masses is only of about 30 MeV (which is small compared with the present uncertainty in the experimental value of m_W, 80 MeV).

QCD breaks **dynamically** the electroweak symmetry because the colour $SU(3)_c$ forces produce condensates of the form $\langle 0|\bar{u}u|o\rangle \neq 0$, $\langle 0|\bar{d}d|0\rangle \neq 0$. Thus, it is the same mechanism that gives rise to the spontaneous breaking of the **global** $SU(2)_L \otimes SU(2)_R$ symmetry that the QCD lagrangian possesses in the limit of zero u and d quark masses and in the absence of electroweak interactions. The global symmetry in question is actually $U(2) \otimes U(2)$, the Abelian part being a product of an axial $U(1)$ and a baryon number symmetry, $U(1)_A \otimes U(1)_B$.

When the electroweak interactions are turned on two things happen. First the whole global symmetry is reduced to the product of a **local** $SU(2)_L \otimes U(1)_Y$ symmetry and a global $U(1)_A \otimes U(1)_B$ one, the rest is lost by the noncommutativity of T_{3R} with T_{1R} and T_{2R}, where T_{iR}, $1 = 1, 2, 3$, are the generators of $SU(2)_R$. That is, of the three Abelian symmetries generated by T_{3R}, B and A, the part generated by $Y = T_{3R}+B$ is converted into local and the rest, $U(1)_A \otimes U(1)_B$, remains global. Thus in this first step the former global $U(2) \otimes U(2)$ symetry is broken, by the act of gauging the electroweak part, explicitly to $SU(2)_L \otimes U(1)_Y \otimes U(1)_A \otimes U(1)_B$.

The second thing that happens is that the QCD condensates break spontaneously the electroweak symmetry down to the electromagnetic one, $SU(2)_L \otimes U(1)_Y \longrightarrow U(1)_Q$. Since the formation of the condensates is of dynamic origin, we say that the breaking is dynamic, besides being spontaneous.

That the condensates break the symmetry as stated above can been seen from the following reasoning. For two flavors, u and d, with $m_u = m_d = 0$, the QCD Lagrangian is invariant under the transformations

$$M_L \in SU(2)_L \Rightarrow \begin{pmatrix} u \\ d \end{pmatrix}_L \to M_L \begin{pmatrix} u \\ d \end{pmatrix}_L$$

$$M_R \in SU(2)_R \Rightarrow \begin{pmatrix} u \\ d \end{pmatrix}_R \to M_R \begin{pmatrix} u \\ d \end{pmatrix}_R.$$

However there is no $SU(2)_L \otimes SU(2)_R$ multiplet structure in the hadronic spectrum, not even approximate (which would be the case for realistic m_u, $m_d \ll \Lambda_c$, where Λ_c is the QCD scale). Instead there is an approximate $SU(2)_V$ multiplet structure, where

$$SU(2)_V \subset SU(2)_L \otimes SU(2)_R$$

$$T_V = T_L + T_R$$

and besides, there are three pseudoscalar bosons with mass ~ 0. This situation points towards the implication of the Goldstone theorem for the case that the axial $SU(2)_A$ is spontaneously broken,

$$T_i A|0\rangle \neq 0, \ i = 1, 2, 3, \qquad T_A = T_L - T_R$$

Color forces are then expected to produce

$$\langle 0|(\bar{u}, \bar{d}) \begin{pmatrix} u \\ d \end{pmatrix} |0\rangle \neq 0 \tag{3}$$

which is the same as

$$\langle 0|(\bar{u}, \bar{d})_L \begin{pmatrix} u \\ d \end{pmatrix}_R + (\bar{u}, \bar{d})_R \begin{pmatrix} u \\ d \end{pmatrix}_L |0\rangle \neq 0 \tag{4}$$

implying the breaking of both $U(2)_L$ and $U(2)_R$, but conserving $U(2)_V$. In particular it then breaks $SU(2)_L$. This $SU(2)_L$ is a global symmetry of QCD, but it is the same $SU(2)_L$ which when promoted to <u>local</u> is responsible for part of the electroweak dynamics. Therefore QCD is responsible for the breaking of $SU(2)_L \otimes U(1)_Y$ electroweak and thus it contributes to m_w.

The contribution of QCD to m_W is easy to estimate from the electroweak coupling of quarks and W's, g, and from the coupling, f_π, of the axial-vector current of quarks to the Goldstone bosons of the spontaneously broken symmetry, the pions (π). The result, taking into account that $f_\pi \simeq 100$ MeV, is

$$m_W \simeq \frac{gf_\pi}{2} \simeq 30 MeV.$$

The technicolor idea starts by postulating additional elementary fermions, the techniquarks, which in the simplest model come in two techniflavors, U and D. Thechniquarks are subject to a new QCD-like interaction, Technicolor based on a gauge group $SU(N)_C$. The left handed parts of U and D form a double of ordinary electroweak $SU(2)_L$ and the technicolor force produces technicondensates $\langle 0|\bar{U}U|0\rangle$ and $\langle 0|\bar{D}D|0\rangle$. The mechanism of dynamically broken $SU(2)_L \otimes U(1)_Y$ is repeated here and a technipion, Π, is gives mass to the W, $m_W = gF_\Pi/2$. The observed value of m_W is obtained by forcing the technipion decay constant, F_Π to have the desired value, 250 GeV, and no necessity of introducing an elementary Higgs boson, whose vacuum expectation value breaks the electroweak symmetry, arises.

Technicolor by itself is not, however, enough to replace the elementary Higgs since its dynamics is unable to provide the masses of ordinary fermions. To achieve this step it is necessary to extend technicolor to a second level introducing a new gauge interaction between the ordinary quark and the techniquarks. These new theory is called Extended Technicolor and its dynamics is able to give rise to a loop of an extended technigauge field, of mass M_{ETC}, and a techniquark, F, which contributes to the propagator of the ordinary-quark f producing in this way, through the condensate in the technifermion line, a non-zero ordinary-quark mass,

$$m_f = g_{ETC}^2 \frac{<0|\bar{F}F|0>_{ETC}}{M_{ETC}^2}.$$

While the scale of technicolor, Λ_{TC}, is expected to be of the order of 1 TeV, the magnitude of M_{ETC} should satisfy the relation

$$\frac{M_{ETC}}{g_{ETC}} \lesssim 30 TeV \qquad (5)$$

in order that m_f achieves a value of a few GEV. For bigger values of m_f the bound becomes lower. Extended technicolor interactions, on the other hand, induce flavor changing processes in the ordinary quark sector, such as $K^o - \bar{K}^o$ and $B_d^o - \bar{B}_d^o$ mixing, at a rate that is in clash with experimental lower bounds for them if M_{ETC} satisfies relation 5. This reasoning assumes that the size of the technicondensate does not varies too much from the scale of technicolor to the scale of extended technicolor

$$<\bar{F}F>_{ETC} \simeq <\bar{F}F>_{TC}$$

which is true if α_{TC} decreases rapidly above Λ_{TC}. The solution to this problem seems to be the introduction of more technifermions so that the evolution of α_{TC} slows down. This idea is known as walking technicolor.

Signatures of technicolor include the production of technihadrons, additional Z-s [17] which mix with the Z of the Standard Model, departures in the values of the parameters measured in the precision tests of the SM, and flavor changing neutral currents.

V LEPTOQUARKS

Unification of leptons and quarks in some common framework leads necessarily to the introduction of leptoquarks, massive particles of spin = 0 or 1 which decay into a lepton and a quark. They are therefore color triplets with either nonzero or unfixed baryon and lepton number. In Grand Unified Models (GUMs) where leptoquarks have unfixed baryon and lepton number, such as SU(5), leptoquarks induce proton decay and since this process has an stringent lower bound, leptoquarks are expected to be very massive. There are however models where leptoquarks have fixed baryon and letpon number and where other conditions for proton stability are given. In these models the leptoquark masses are not constrained by the experimental bound on the proton lifetime. Here we illustrate this possibility with an specific model, $[SU(6)]^3 \times Z_3$ [18].

$SU(6)_L \otimes SU(6)_C \otimes SU(6)_R \times Z_3$

THE MODEL. The gauge group of the model contains three SU(6) groups. The first, $SU(6)_L$ can be better understood from its decomposition into the product of the SM $SU(2)_L$ group and a "horizontal" $SU(3)_H$ one that acts in the family space,

$$SU(6)_L \supset SU(2)_L \otimes SU(3)_H,$$

in such a way that the fundamental representation of $SU(6)_L$ decomposes as $6 = (2,3)$. The sector of ordinary left-handed quarks is then in a 6 of $SU(6)_L$

$$\psi = \begin{pmatrix} u & c & t \\ d & s & b \end{pmatrix}_L$$

while the right handed quarks are singlets of $SU(6)_L$,

$$q_{IL}^{c(-2/3)}, \quad q_{IL}^{c(1/3)}, \quad I = 1, 2, 3.$$

Before we explain the role of the other factors in the gauge group of the model let us point out that the model leads naturally to a heavy top quark because here it is the only one that can acquire mass at tree level. To see this consider the Higgs field, ϕ, whose vacuum expectation value should break $SU(2)_L$ at some step in the symmetry breaking chain. ϕ should transform as a doublet of $SU(2)_L$ and therefore the simplest possibility is that it is 6 of $SU(6)_L$. ϕ leads then to just one type Yukawa coupling for the quarks:

$$\mathcal{L}_Y = \sum_I \gamma_I q_{IL}^{c(-2/3)T} C \psi_\alpha \phi^\alpha,$$

where γ_I are Yukawa coupling constants which we assume to be of order one and where C is the charge conjugation operator and a sum over the index α from 1 to 6 is understood. After introducing the vacuum expectation values $\langle 0|\phi_\alpha|0\rangle = v_\alpha$, for the neutral components of ϕ, $\alpha = 1, 3, 5$, the following mass term for quark arises:

$$\mathcal{L}_Y \to (\gamma_u u^c + \gamma_c c^c + \gamma_t t^c)_L^T C (v_1 u + v_2 c + v_3 t),$$

where we have done some renaming. Therefore the mass matrix of the charge 2/3 sector is

$$M^{(2/3)} = \begin{pmatrix} \gamma_u v_1 & \gamma_u v_2 & \gamma_u v_3 \\ \gamma_c v_1 & \gamma_c v_2 & \gamma_c v_3 \\ \gamma_t v_1 & \gamma_t v_2 & \gamma_t v_3 \end{pmatrix}$$

and it has obviously only one non-zero eingenvalue γv, where

$$\gamma^2 = \gamma_u^2 + \gamma_c^2 + \gamma_t^2, \quad v^2 = v_1^2 + v_2^2 + v_3^2.$$

Therefore
$$m_t = \gamma v \gtrsim M_W.$$

Since SU(6) is not pseudoreal like SU(2), ϕ^* is not a 6 and therefore there is no other Yukawa coupling and
$$M^{(-1/3)} = 0.$$

Now, the SU(6)$_R$ part of the gauge group arises from the requirement of making it left-right symmetric. In this way SU(6)$_L$ is extended to
$$SU(6)_L \times SU(6)_R \times U(1)_{B-L}.$$

Up to this point the group does not contain color. SU(3)$_c$ is added to the gauge group of the model in a way that there is only one coupling constant above the unifying scale. For this reason SU(3)$_c \otimes$U(1)$_{B-L}$ is extended to $SU(6)_C$ and an operator P that interchanges the three SU(6) factors is introduced. The gauge group is then
$$G \equiv SU(6)_L \otimes SU(6)_C \otimes SU(6)_R \times Z_3,$$

where
$$Z_3 = \{1, P, P^2\}, \quad P(A, B, C) = (B, C, A),$$

A being a representation of SU(6)$_L$, B of SU(6)$_C$ and C of SU(6)$_R$. Therefore
$$Z_3(A, B, C) = (A, B, C) + (B, C, A) + (C, A, B).$$

Notice that G contains the well known left-right symmetric group,
$$G \supset G_{LRS} \equiv SU(3)_c \otimes SU(2)_L \otimes SU(2)_R \otimes U(1)_{B-L}$$

The generalized color group, SU(6)$_C$, contains, besides the familiar three quark colors, three leptonic colors,
$$(6)_C = \underbrace{r, y, b}_{c}, \quad \underbrace{\ell_1, \ell_2, \ell_3}_{\text{leptonic color}},$$

so that a true quark lepton symmetry is achieved.

The matter content of the model is accommodated into a single irreducible representation of G,
$$\psi(108)_L = Z_3 \psi(6, 1, \bar{6})_L$$
$$= \psi(6, 1, \bar{6})_L + \psi(1, \bar{6}, 6)_L + \psi(\bar{6}, 6, 1)_L,$$

where ordinary left-handed fermions are in $\psi(\bar{6},6,1)_L$:

$$\psi(\bar{6},6,1)_L = \begin{pmatrix} d_r & d_y & d_b & E_1^- & L_1^o & T_1^- \\ u_r & u_y & u_b & E_1^o & L_1^+ & T_1^o \\ s_r & s_y & s_b & E_2^- & L_2^o & T_2^- \\ c_r & c_y & c_b & E_2^o & L_2^+ & T_2^o \\ b_r & b_y & b_b & E_3^- & L_3^o & T_3^- \\ t_r & t_y & t_b & E_3^o & L_3^+ & T_3^o \end{pmatrix}_L ,$$

while ordinary right-handed fermions are in $\psi(1,\bar{6},6)_L$. $\psi(6,1,\bar{6})_L$ represents 36 exotic Weyl leptons, 9 with positive electric charges, 9 with negative (the charge conjugated to the positive ones) and 18 are neutrals. The model is easily shown to be anomaly free. In terms of $SU(3)_c \otimes SU(2)_L \otimes U(1)_Y$ we have

$$\psi(\bar{6},6,1)_L = 3(3,2,1/3) + 3(1,2,-1) + 3(1,2,1) + (1,2,-1),$$
$$\psi(1,\bar{6},6)_L = 3(\bar{3},1,-4/3) + 3(\bar{3},1,2/3) + 3(1,1,2)$$
$$+ 9(1,1,0) + 3(1,1,-2) + 3(1,1,2)$$
$$\psi(6,1,\bar{6})_L = 9(1,2,1) + 9(1,2,-1). \tag{6}$$

PROTON STABILITY. The reasons behind the stability of he proton in this model are:

- All the fields in the model have a well defined baryon number B:
  - Gauge fields: The 70 gauge fields in $SU(6)_L \times SU(6)_R$ have B = 0; in $SU(6)_C$ there are 9 leptoquarks with B = 1/3 and 9 with B = -1/3, the other 17 gauge fields have B = 0.
  - Fermions: quarks have B = 1/3, antiquarks B = -1/3 and leptons B = 0.
  - Higgs bosons: Since the fundamental representation of G has well defined B numbers, all irreducible representation of G will also have well defined G.

- B can be written in the fundamental representation of $SU(6)_C$ as

$$B = Dg(1/3, 1/3, 1/3, 0, 0, 0).$$

B is not therefore a generator of G. The model, specified by the corresponding Lagrangian density $\mathcal{L}$, has a $U(1)_\chi$ global symmetry associated with the extension of $SU(6)_C$ to $U(6)_C$. Its generator may be written in the fundamental representation of $SU(6)_C$ as

$$\chi = dg(1,1,1,1,1,1)/\sqrt{12}$$

and as

$$\chi = 1/\sqrt{12} \quad \text{for } \psi(\bar{6},6,1)$$

$$\chi = 0 \quad \text{for } G(1,35,1)$$

and

$$\chi = -2/\sqrt{12} \quad \text{for} \phi(1,\bar{15},15).$$

- On the other hand $SU(6)_C$ has a generator of the form

$$B' = Dg(1,1,1,-1,-1,-1)/\sqrt{12}$$

and therefore

$$B = (\chi + B')/\sqrt{3}.$$

- Now, we can always arrange that the components of Higgs bosons that acquire vevs have B = 0,

$$B < \text{Higgs bosons} >= 0$$

Therefore B is conserved and the proton is stable.

$\chi - B'$ is broken. The corresponding Goldstone boson is however eaten by the gauge field corresponding to B'.

Since the proton is (perturbatively) stable in this model, the mass of leptoquarks needs not to be high and a search for low lying leptoquarks at Hera, Tevatron and LHC makes sense.

VI OUTLOOK

The near future of high energy physics looks very promising, with new accelerator machines within reach and powerful detector techniques under development. The search for the mechanism responsible for the breaking of the electroweak symmetry will have the highest priority in the coming years. In the way we may encounter supersymmetry or maybe technicolor, or even some strange combination of ideas or, which could be the best, something completely unexpected.

VII ACKNOWLEDGEMENTS

This work was supported in part by Conacyt. The authors enjoyed illuminating conversations with G. Altarelli and Monica Pepe. Finally we acknowledge the invitation of Gerardo Herrera to participate in this School.

REFERENCES

1. R. Plano et al., hep-ex/9710016, Talk given at the 5th International Conference on Physics Beyond the Standard Model.
2. H1 Collaboration, S. Aid et al., Physics Letters B.75 (1995) 1006.
3. M. Paulini, Proceedings of the 24th Annual Slac Summer Institute on Particle Physics, 1997.
4. LHC Report 36, K. M. Potter, july 1996, Presented at the ICHEP 1996.
5. F. Abe et al.. (CFD) Nucl. Instr. Methods A271,387 (1988).
6. S. Amato, et al.. (E791) Nucl. Instr. Methods, A 324, 535 (1993).
7. D. M. Gringrich, et. al. (ATLAS) Nucl. Instr. Methods A 364 290, (1995).
8. The D0 upgrade, Fermilab Pub-96/357-E
9. Jay Hauser, Proceedings of Beauty'96 Conference, Rome, 1996 (Submitted to Nucl. Instr. Methods, CMS TN/96-140
10. K. Rybicki, Proceedings of the 28th International Conference on HEP, Warsaw, 1996.
11. F. Piuz, Nucl. Instr. Methods A 371, 96(1995)
12. G. Altarelli, The Status of The Standard Model, hep-ph/9710434;
    J. Timmermans, Proceedings of LP'97, Hamburg, 1997;
    S. Dong, *ibid*
    The LEP Electroweak Working Group, LEPEWWG/97-02 (1997).
13. J.S. Conway, Recent Results of Searches for New Particles in CDF. . FERMILAB-CONF-97-311-E, Aug 1997. Talk given at Workshop on Physics Beyond the Standard Model: Beyond the Desert: Accelerator and Nonaccelerator Approaches, Tegernsee, Germany, 8-14 Jun 1997.
14. Abe F *et al.* (CDF Collaboration), Phys. Rev. Lett. 74,2626 (1995); Abachi S. *et al.* (D0 Collaboration), Phys. Rev. Lett. 74, 2632 (1995).
15. Adolf C. *et al.*
16. C. Dionisi, Proceedings of LP'97, Hamburg, 1997.
17. A. Zepeda, Physics Letters, 132B, 407(1983); 195B, 623E (1987); A. Hernandez-Galeana and A. Zepeda, Zeitschrift für Physik C40, 125 (1988).
18. W. Ponce and A. Zepeda, Phys. Rev. D48, (1993) 240 (1993); R. Gaitan, W. Ponce and A. Zepeda, Phys. Rev. D49, 4954, (1994); W. Ponce, A. Zepeda and J. B. Flórez. Phys. Rev. D49, 4958 (1994); A. Perez-Lorenzana, W. A. Ponce and A. Zepeda, Rev.Mex.Fis. 43, 737 (1997).

Cryogenic Detectors for Dark Matter

Dan McCammon[1]

Physics Department, University of Wisconsin
Madison, WI 53706 USA

Abstract. There is ample observational evidence that the Universe is dominated by dark matter that does not radiate or absorb electromagnetic energy at any observed wavelength. Somewhat more tenuous arguments require that the majority of this dark matter does not consist of baryons. The nature of the "non-baryonic dark matter" is unknown, but one conjecture is that it consists of WIMPs, or Weakly Interacting Massive Particles, with the favored candidate being the lightest supersymmetric particle, the neutralino, with a mass somewhere between 1 – 1000 GeV. Given a mass, the density and velocity are constrained by astrophysical observations, so it becomes reasonable to attempt direct detection of the particles. The primary interaction with matter is expected to be elastic scattering from nuclei, and one detection approach is to measure the recoil energy of the nucleus by the increase in detector temperature that it produces. This is practical in detectors with masses of hundreds of grams operated at millikelvin temperatures, and at least four major efforts along these lines are well under way.

INTRODUCTION

The title of this talk begs a number of questions. What *is* dark matter? Why do we think it exists? Do we know what it should be made of? What is a cryogenic detector, and how does it work? In the following section I will briefly describe the astrophysical "missing matter" problem, the resulting definition of dark matter, and what it *might* be made of. The next section discusses the requirements for detectors for one proposed component of this material, Weakly Interacting Dark Matter (WIMP) particles, why cryogenic detectors are suited to this problem, and an introduction to how they work. The final section shows how two of the current searches are implementing these detectors.

---

[1] mccammon@wisp.physics.wisc.edu

DARK MATTER

Observations of the rotation of spiral galaxies and of the confinement of hot gas in clusters of galaxies require gravitational potentials much larger than can be produced by the matter that is observed in the form of ordinary stars, diffuse gas, and finely-divided dust. Spiral galaxies generally have at least two to three times more gravitational mass than can be accounted for by the stars, gas, and dust they contain, while large clusters of galaxies have 10 – 50 times more than the sum of the mass of the galaxies and hot gas in the cluster.

The total mass in the universe is commonly referred to in terms of Ω, the ratio of the average density to the critical density that would just stop the current expansion after an infinite length of time. The density in stars amounts to $\Omega = .01$, while the inferred gravitational potentials require Ω about 0.3 or more. Inflationary cosmologies, attractive because they provide a natural answer to the questions of why the universe is as flat and isotropic as it is, usually require that $\Omega \equiv 1$.

It is not difficult to conceive of rather ordinary dark matter that we could not see. Putting all of the missing material into uniformly distributed bricks would make it essentially impossible to observe. (We do not know how such bricks could be made, but this could be simply a lack of imagination.) Searches for more conventional structures that could have escaped earlier observation, such as old neutron stars and very faint, low-mass stars, appear to show that they are not present in the required quantities. There is also indirect evidence arguing that no ordinary solution to the "missing matter problem" will be found. This comes from models for the synthesis of light elements in the first few minutes of the big bang. The predicted helium abundance is almost independent of the average density, and is in good agreement with the observed value for all values of Ω from .01 to more than 2 (above which the age of the universe definitely becomes shorter than the ages of the oldest stars). This lends some confidence to the model, in which the deuterium abundance is a strongly decreasing function of the baryon density. The best current observations of deuterium then require that the total baryon density correspond to $\Omega_{Baryon} < 0.1$. This is arguably consistent with $\Omega_{Grav.} > 0.3$, given the large systematic uncertainties in many quantities, but would be very difficult to reconcile with inflation ($\Omega_{Grav.} = 1$).

The solution to this is that much, perhaps most, of the gravitational matter in the Universe must consist of some new kind of particle. Super-symmetrical theories predict a whole family of new particles that are paired with standard-model particles. The lightest of these, the neutralino, is one candidate for astrophysically abundant dark matter.

DETECTION

Much of the information needed to design a direct detection experiment for WIMPs such as neutralinos that would provide the missing mass in a spiral galaxy can be inferred from astrophysical observations. Rotation curve measurements give the mass density distribution. The particles should be gravitationally bound to the galaxy, and, having at most very weak interactions with the normal matter, will be falling through the disk with approximately Keplerian velocities (about 300 km s^{-1} at the position of the Sun). The particle mass is a free parameter, but for any assumed value, the particle number density and flux are then given.

The largest uncertainty is the interaction cross section, but minimal supersymmetric theory predicts that it should be similar to that for neutrinos. The largest cross sections are for coherent scattering from nuclei, and predict rates on the order of one event per day per kilogram of target material for WIMPs in the 2 – 300 GeV mass range. The typical energy transferred in a scattering event goes from a few hundred eV to several keV over the same range of particle masses.

Because of the low rates, it is clear that detectors with active masses of at least a few kilograms and very low backgrounds are required. The low background is particularly important because the signal cannot be turned on and off, so the only modulation available to differentiate extraneous events is the small annual flux variation due to the changing direction of the earth's 30 km s^{-1} orbital velocity relative to the 300 km s^{-1} motion of the sun. The arriving flux is quite asymmetrical due to the large solar velocity, however, and if a *directional* detector could be developed, the diurnal variation of the laboratory orientation of this vector due to the earth's rotation would provide a very strong indication that a real signal had been detected.

Conventional ionization detectors are less than ideal for this application. In the required large sizes, they have high electrical capacities and therefore relatively large readout noise which doesn't allow low enough energy thresholds to be sensitive to the lower end of the interesting range of WIMP masses. This problem is exacerbated by the fact that the W-value, the average deposited energy required to produce an ionization, is about three times larger for low-energy nuclear recoils than it is for electron kinetic energy. The actual threshold is thus a factor of three higher than it would be for normal charged particle or photon events in the same detector.

Cryogenic Detectors

In this section, we will discuss the operation of *phonon-mediated* detectors, where the product of the event that is measured is not the charge nor the light produced, but the phonons, which are lattice vibrations with typical quantum energies of 1 μeV – 20 meV for thermal excitations at temperatures between 10 mK and 300 K. Phonon mediated detectors are usually operated at very low temperatures to minimize the background density of thermal phonons against which the signal must be measured, and this is why they are referred to as "cryogenic detectors". (Normally this term also includes superconducting tunnel junction detectors, which are actually charge detectors that operate very similarly to semiconductor detectors, with the ~1 meV superconducting band gap taking the place of the ~1 eV semiconducting band gap. These are not phonon mediated devices, and we will not consider them further here.)

We further break down phonon mediated detectors into two classes: equilibrium detectors (true calorimeters and bolometers), and athermal, or non-equilibrium

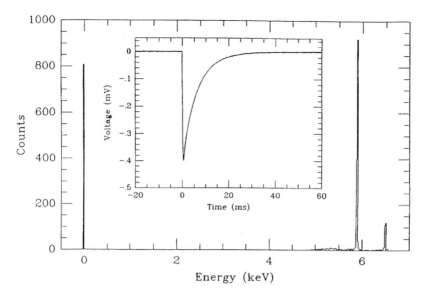

FIGURE 1. Pulse height spectrum of ^{55}Fe obtained with a small calorimeter operated at 50 mK. The inset shows the temperature pulse produced by a single 5.9 keV photon.

devices that look at the super-thermal phonons produced by particle interactions before they are thermalized. Bolometers (where the steady-state temperature rise due to the absorbed power level is measured) of course have a long history as detectors of infrared radiation. In the past ten years or so there has been an increasing interest in the use of physically similar devices in the calorimetric mode to measure the energy of single photon or particle interactions or decays (1–4).

Figure 1 shows the pulse-height spectrum from one of these calorimeters when it is illuminated by soft x-rays from a ^{55}Fe source. The energy from each photon is thermalized in a time short compared to the ~3 ms cooling time constant of this small (~10 μg) detector, and the resulting temperature rise and exponential decay are measured by an attached thermometer. Variations in the pulse heights from monoenergetic x-rays are caused by a combination of the thermodynamic fluctuations in the baseline energy content of the detector at its 50 mK operating temperature and the Johnson noise of the thermometer. The 7.5 eV FWHM resolution is about a factor of twenty better than a state-of-the-art Si(Li) device, and more than an order of magnitude better than the theoretical limit for a Silicon or Germanium detector. There is currently much interest in such detectors for high resolution astronomical and laboratory x-ray spectroscopy. Since the absorber does not need to have good electronic transport properties, a wide range of materials can be used, making thermal detectors particularly useful for neutrinoless double beta decay searches and other investigations where true calorimetric measurements with a variety of materials are required. The calorimeter response and resolution also remain the same for energy deposited in nuclear recoils, while the resolution of a conventional detector is further degraded by the much lower ionization efficiency of these events.

Energy Resolution of Equilibrium Calorimeters

A calorimeter consists of an absorbing element in which the energy is deposited connected to a heat sink by a weak thermal link. A thermometer attached to the absorber measures the temperature rise produced by the deposited energy, while the thermal link conducts the excess energy to the heat sink in a characteristic time $\tau_{th} \equiv C/G$, where C is the heat capacity of the absorber, and G is the thermal conductivity of the link. The deposited energy can be determined from the peak temperature rise: $E = C \Delta T$.

This energy excursion, however, must be measured in the presence of thermodynamic fluctuations in the energy content of the absorber that are produced by the random transport of energy carriers through the thermal link. The magnitude of these fluctuations is given by an elementary calculation in classical statistical mechanics as $\Delta E_{rms} = \sqrt{k_B T^2 C}$ (5). For a dielectric absorber, this can be thought of as the expected size of Poisson fluctuations in the number of phonons contained in the absorber:

$$\overline{N} \equiv \text{average \# phonons} \approx \frac{E_{total}}{\overline{E}_{phonon}} \approx \frac{CT}{k_B T}, \text{ and} \qquad (1)$$

$$\Delta E_{rms} \approx \Delta N_{rms} \overline{E}_{phonon} \approx \sqrt{\overline{N}} \cdot \overline{E}_{phonon} \approx \sqrt{\frac{CT}{k_B T}} \cdot k_B T = \sqrt{k_B T^2 C}. \qquad (2)$$

Note that the magnitude of these fluctuations is independent of the conductivity G of the thermal link but their frequency distribution is not.

The magnitude of these fluctuations does not in itself determine the potential energy resolution. To see why this is the case, we consider their power spectrum, which is flat below a corner frequency ω_0 equal to the reciprocal of the thermal time constant CG^{-1}, and falls as f^{-1} at higher frequencies. The signal pulse from an instantaneously deposited energy will have a sharp rise in temperature followed by an exponential decay with this same time constant. The power spectrum of this pulse has the same shape as that of the thermodynamic fluctuations, as shown in Figure 2. Considering only these quantities, the signal to noise ratio is independent of frequency, and, since any frequency bin gives an independent estimate of the signal, the energy resolution could be made arbitrarily good by using an arbitrarily large bandwidth.

The usable bandwidth is limited by other considerations, the most fundamental of which is noise in the thermometer. We consider an ideal resistive thermometer, with resistance R and logarithmic sensitivity $\alpha \equiv d\log(R)/d\log(T)$. The maximum usable bandwidth is determined by r, the ratio of the fluctuations noise at low frequencies to the Johnson noise of the thermometer. It can be shown that r^2 is proportional to $\alpha^2 b$, where $b \equiv (T - T_0)/T_0$ is the fractional increase of the detector temperature T above the heatsink temperature T_0 produced by the power used to read out the thermometer resistance. The value of α is limited by thermometer

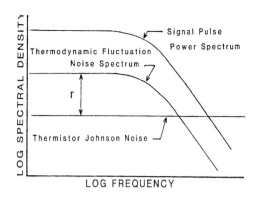

FIGURE 2. Signal pulse and thermodynamic fluctuation noise power spectra.

technology, but b clearly has an optimum value: a very small readout power will give a small signal voltage (for a given temperature and resistance change) relative to the fixed Johnson noise voltage, while a large readout power will raise the temperature significantly and greatly increase the level of the thermodynamic fluctuations. The exact calculation is complicated by the temperature rise, b, (since the fluctuations must be recalculated for the resulting temperature gradient in the thermal link), by the variation with temperature of both the heat capacity of the absorber and the thermal conductivity of the link, and by the electrothermal feedback produced by variations in the bias power during a pulse. This problem has been solved exactly for the ideal case outlined so far (2), with the net detector noise for optimized bias power ($b \approx 0.12$ for most cases) and signal filtering given by:

$$\Delta E^2 = \xi \sqrt{k_B T_0^2 C_0} , \qquad (3)$$

where T_0 and C_0 are the heatsink temperature and the heat capacity of the detector at this temperature, and ξ is a dimensionless factor that depends primarily on the thermometer sensitivity, and weakly on the temperature dependence of the heat capacity and the thermal link conductivity. This result is still independent of the conductivity G, and therefore of the thermal time constant. For values of thermometer sensitivity obtainable with practical semiconductor thermistors ($\alpha \approx 3 - 5$), ξ is about 2, and scales as $\alpha^{-1/2}$ for higher sensitivities.

There are additional limiting factors for semiconductor thermistors. They lose sensitivity to the lattice temperature as the bias power density is increased, and they have an apparently fundamental source of 1/f noise. Both of these effects rapidly become worse as the operating temperature is lowered, and below 0.3 K they provide increasingly severe limits on how small the heat capacity of the thermometer can be made, and on how fast the thermal time constant can be (6,7). The power density restriction puts a lower limit of about 1 ms on the thermal time constant for detectors operated around 100 mK. However, it is still possible to make large detectors with existing thermometer technology that have much better energy thresholds and resolution than conventional solid state detectors of the same size. The relatively slow speeds are not a huge disadvantage for a very low rate experiment, and the phonon-mediated detection provides full response to the energy of low-velocity nuclear recoils (8).

WIMP SEARCHES

There are now at least four major searches for Weakly Interacting Massive Particles getting underway using cryogenic detectors. I will describe the approach of the two largest collaborations here. Both have spent several years on the fundamentals of detector development, and are now deploying detectors of useful size at low-background sites.

The CRESST Experiment

The major collaborators in the Cosmic Rare Event Search with Superconducting Thermometers include the Max Planck Inst. for Physics in Munich, the Technical University of Munich, and Oxford University (9). Their detectors consist of large sapphire cylinders on which are deposited small Tungsten strips. The heat sink is cooled by a dilution refrigerator to a point well below the superconducting transition temperature of the Tungsten (~10 mK). A voltage bias is placed across the Tungsten strip, and the voltage level adjusted so that enough power is dissipated to raise the detector temperature to the transition point when the resistance is some fraction of the normal resistance. The negative feedback inherent to this system makes it relatively easy to keep the strip in a partially superconducting state where the resistance is a very steep function of temperature, as shown in Figure 3a. The current through the strip is monitored with a D.C. SQUID (Superconducting Quantum Interference Device) amplifier, and the reduction in current caused by a temperature increase due to energy deposited in the crystal can be detected. A pulse

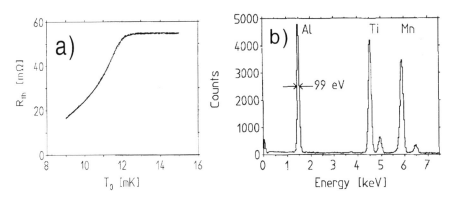

FIGURE 3. a) Superconducting transition of Tungsten strip thermometer. b) Pulse height spectrum of a 32 g sapphire crystal equipped with a similar thermometer and illuminated by several x-ray lines.

height spectrum obtained by illuminating a 32 g sapphire crystal with several x-ray lines is shown in Figure 3b. The current detectors are 262 g, but the resolution is still about 230 eV FWHM. A prototype 1 kg array consisting of four of these crystals is being installed in a heavily shielded coldbox attached to a dilution refrigerator in the Gran Sasso underground laboratory.

The special strength of this experiment is its very low threshold, making it sensitive to very low energy nuclear recoils from WIMPs with masses as low as 2 GeV. The coldbox has room for up to 100 kg of sapphire crystals.

The CDMS Experiment

The Cryogenic Dark Matter Search is run by the NSF Center for Particle Astrophysics. The principal collaborators in this experiment are the University of California, Berkeley, and Stanford University (10). Two types of detectors are being tested in the same coldbox in an underground site on the Stanford campus. The background at this site is low enough to get useful scientific results during the initial phases of the experiment, but once detector testing is complete, a larger array of detectors will be set up in a much deeper site.

One of the detector types is a near-equilibrium calorimeter, with a 165 g high-purity Germanium absorber and NTD (neutron transmutation doped) Ge semiconductor thermometers. The thermal signal is measured with a resolution of about 650 eV FWHM. The Ge crystal also has electrodes on opposite faces with a very small applied field of ~1 V/cm. At low temperatures, this field is sufficient to efficiently collect the charge produced in the event, so the device functions as a conventional solid state detector with a resolution of ~1 keV FWHM at the same time as it measures the thermal signal. The great value of this is shown in Figure 4, where the amplitude of the thermal signal is plotted against the collected charge. The lower trajectory is produced by gamma-rays, which interact with the electrons in the detector. The upper trajectory, where the charge signal is about a factor of three smaller relative to the temperature rise, is due to nuclear recoil events, produced in this case by neutron bombardment. The key point is that in a deep underground site, almost all background events are due to beta and gamma interactions with electrons, while the WIMP signal is from nuclear recoils. The only significant source of nuclear recoil background is neutrons produced by nearby interactions of atmospheric neutrinos, so this efficient separation can reduce the detector background by a factor of more than 100, greatly increasing the sensitivity of the experiment.

The second type of device being tested for CDMS is an athermal or non-equilibrium phonon detector. It uses a Silicon or Germanium disk about 76 mm in diameter and 10 mm thick where one surface is almost completely covered by phonon sensors. These sensors consist of superconducting Aluminum "paddles" connected to Tungsten superconducting transition edge thermometers. Phonons produced in a scattering event in the crystal propagate quasi-ballistically to the surface. If they strike one of the Aluminum paddles, and their energy is larger than the Aluminum superconducting bandgap (about 1 K), they will break up one or

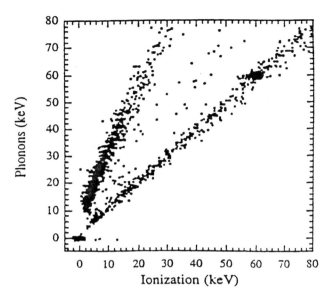

FIGURE 4. Size of phonon signal (temperature rise) relative to the amount of collected charge in a Germanium detector. The lower trajectory is produced by gamma interactions with electrons. The upper one is due to neutron scattering off nuclei.

more of the Cooper pairs in the Aluminum, and the resulting quasiparticles can diffuse to the point where the paddle is attached to the Tungsten strip. The quasiparticles give up their energy to the electron system in the strip, heating it and driving it more normal. The resulting reduction in current through the strip is detected by a SQUID amplifier. These devices also use electrodes on opposite faces of the disk and a small applied field to collect the charge signal.

One advantage of the non-equilibrium phonon detector is that it is much faster than an equilibrium calorimeter of this size. The athermal phonon signal can be collected in about 100 μs, about the same as the ionization signal, as compared to tens of milliseconds for the thermal signal in an equilibrium detector. On the other hand, the athermal phonon sensors must cover a large fraction of the surface, so they detect the phonons on their first hit on the surface, before they are thermalized. The thermometer on an equilibrium calorimeter can be placed anywhere, since it is sampling a phonon density that should the same everywhere. There is also some hope that the propagating phonons initially emitted in the scattering event will show some asymmetry that reflects the direction of the nuclear recoil. If this were the case, then it might be possible to make a directional detector, with the tremendous attendant advantages described in the Detector section above. So far, however, there is no hint that such an asymmetry has been detected.

The CDMS experiment will soon have ~1 kg each of Si and Ge detectors running in the Stanford tunnel, and will eventually have ~10 kg total at a deep underground site. The limits on WIMP fluxes that could be placed by the CDMS experiment are shown in Figure 5, along with the predictions of the minimal supersymmetric models.

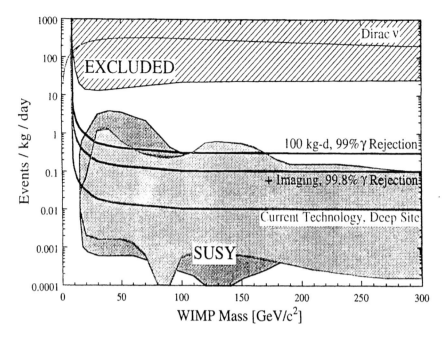

FIGURE 5. Expected limits on WIMP flux as a function of mass from the CDMS experiment, compared with model predictions.

REFERENCES

1. Fiorini, E., and Niinikoski, T.O., *Nucl. Instr. and Methods.*, **224**, 83 (1984).
2. Moseley, S.H., Mather, J.C., and McCammon, D., *J. Appl. Phys.*, **56**, 1257–62, (1984).
3. McCammon, D., Juda, M., Reeder, D.D., Kelley, R.L., Moseley, S.H., and Szymkowiak, A.E., "A New Technique for Neutrino Mass Measurement," in *Neutrino Mass and Low Energy Weak Interactions: Telemark, 1984*, eds. V. Barger and D. Cline, 1985, pp 329–43.
4. Wollman, D.A., Irwin, K.D., Hilton, G.C., Dulcie, L.L., Newbury, D.E., and Martinis, J.M., "High-energy-resolution microcalorimeter spectrometer for x-ray microanalysis," (submitted to *J. Microscopy*, 1997).
5. Reif, F., *Fundamentals of Statistical and Thermal Physics*, New York: McGraw-Hill, 1965, pp 213, 242.
6. McCammon, D., Cui, W., Juda, M., Morgenthaler, J., and Zhang, J., Kelley, R.L., Holt, S.S., Madejski, G.M., Moseley, S.H., and Szymkowiak, A.E., *Nucl. Instr. and Methods*, **A326**, 157–65 (1993).
7. Zhang, J., Cui, W., Juda, M., McCammon, D., Kelley, R.L., Moseley, S.H., Stahle, C.K., and Szymkowiak, A.E., "Non-ohmic Effects in Hopping Conduction in Doped Silicon and Germanium Between 0.05 and 1 Kelvin," *Phys. Rev. B*, (submitted).
8. Alessandrello, A., Camin, D.V., Fiorini, E., and Giuliani, A., *Physics Letters B*, **202**, 611–14 (1988).
9. http://wwwvms.mppmu.mpg.de/cresst/Welcome.html
10. http://physics7.berkeley.edu/home.html

Neutrino Detectors

Kenzo Nakamura

*KEK, High Energy Accelerator Research Organization
Oho, Tsukuba, Ibaraki 305, Japan*

Abstract. Various types of neutrino detectors are reviewed. They are classified as radiochemical detectors, liquid scintillation detectors, water Cherenkov detectors, detectors for neutrino experiments at high-energy accelerators, and other types of detectors. Epoch-making experiments in neutrino physics achieved with each types of the detectors are also mentioned.

INTRODUCTION

Neutrinos interact with matter very weakly. Consequently, neutrino detectors are required to be massive. Because of this, the neutrino detectors are usually simple and economical. However, the detection of low-energy neutrinos requires a technical challenge to overcome rapidly increasing background as the threshold energy of detection lowers. This is particularly true for the detection of solar neutrinos using νe elastic scattering where the only means to suppress the background is to purify the detector materials extraordinarily. If fine tracking is required to identify decay vertex of short-lived particles in the massive detector, this causes another technical challenge.

Various techniques are employed to detect neutrinos. At low energies, typically a few hundred keV to ~10 MeV, radiochemical detectors are used for the detection of solar neutrinos. Also, liquid scintillator is widely used up to several tens of MeV for the detection of cosmic neutrinos as well as those from power reactors. Water Cherenkov detectors have a wide dynamic range, ~5 MeV to a few GeV, and are suitable for the detection of solar neutrinos, supernova neutrinos, and atmospheric neutrinos. At high-energy accelerators, standard neutrino detectors consists of a target calorimeter and a muon spectrometer. Though bubble chambers are not used any more, gigantic chambers were extensively operated for neutrino experiments in the 1970's. For fine tracking in the target region, emulsion technique has now become a standard method.

The upper bounds of the detectable neutrino energy given above are only representative of the currently operating detectors. In fact, they depend on the size of the detector (except radiochemical detectors). Very high-energy neutrinos such as those conjectured to come from active galactic nuclei can be detected by a large underground or underwater/ice detectors using upward through-going muons produced near the detector by charged-current (CC) reactions above an energy threshold determined by the detector configuration.

Flavor identification and energy measurement of the charged leptons produced in CC interctions are important ingredient among others of the neutrino detectors. Calorimeter-type detectors can measure the energy of charged leptons only when these leptons or lepton-initiated showers are fully contained in the fiducial volume. Below the kinematical threshold of muon production, only ν_e can initiate CC interactions. Therefore, the particle identification capability required for low-energy neutrino detectors is only that for electrons.

In the following sections, various types of neutrino detectors are reviewed along with the description of epoch-making experiments in neutrino physics achieved with each types of the detectors.

RADIOCHEMICAL DETECTORS

Radiochemical detectors utilize neutrino capture reactions $A(\nu_e, e^-)B$, where A represents the target nuclide which is transmuted into B as a result of neutrino capture. So far, these detectors have been used for the observations of solar-neutrinos. The famous chlorine detector in the Homestake experiment [1] successfully observed solar neutrinos for the first time around the year of 1970, using the reaction $^{37}Cl\,(\nu_e, e^-)\,^{37}A$ with a threshold of 814 keV.

The number of solar neutrinos N_{obs} detected in unit time is given by

$$N_{\text{obs}} = N_{\text{atom}} \sum_i \int_{E_{\text{th}}} \phi_i(E)\sigma(E)dE \quad (1)$$

where N_{atom} is the number of target atoms contained in the detector, E_{th} the reaction threshold, $\phi_i(E)$ the solar-neutrino flux from a fusion reaction (specified by a suffix i) in the Sun, and $\sigma(E)$ the neutrino-capture cross section. The capture rate is given by $N_{\text{obs}}/N_{\text{atom}}$, and is represented by a conventional unit, SNU (Solar Neutrino Unit, 1 SNU = 10^{-36} capture/atom/s).

Some properties of the chlorine experiment are listed in Table 1 along with those of other ongoing or possible radiochemical experiments. The solar-neutrino fluxes, and, therefore, the capture rates are given by calculations based on a standard solar model (SSM) [2,3]. The Homestake chlorine experiment uses 615 tons of C_2Cl_4 solution, which contains 2.2×10^{30} atoms of ^{37}Cl. The ^{37}Ar production rate expected from the SSM prediction given

TABLE 1. Neutrino capture reactions suitable for radiochemical solar-neutrino detectors.

Reaction	E_{th} (keV)	Half life of product nuclide	Capture rate (SNU)	Abundance of target nuclide (%)	Experiment
$^{37}Cl \to {}^{37}Ar$	814	34.8 d	$9.3^{+1.2}_{-1.4}$ [a]	24.47	Homestake
$^{71}Ga \to {}^{71}Ge$	233	11.43 d	137^{+8}_{-7} [a]	39.89	GALLEX SAGE
$^{127}I \to {}^{127}Xe$	789	36.4 d	36.4 [b]	100.0	Homestake
$^{7}Li \to {}^{7}Be$	862	53.3 d	51.8 ± 5.4 [c]	92.5	
$^{81}Br \to {}^{81}Kr$	471	0.21 My	$27.8^{+5.7}_{-3.7}$ [c]	49.31	

[a] Ref. [2].
[b] Ref. [4].
[c] Ref. [3]. The errors listed here represent 1σ though in the original paper 3σ errors are given.

in Table 1 is only 1.76 atoms/day.[1] This represents the essential point of the radiochemical experiments: how to extract a very small number of produced atoms from overwhelmingly massive detector volume and how to count them.

In the chlorine experiment, the detector is exposed to solar neutrinos for two to three times the half life of radioactive ^{37}Ar. The produced ^{37}Ar atoms are extracted by circulating helium gas through the detector, chemically processed, and introduced into a small low-background proportional counter. ^{37}Ar atoms decay by electron capture with emission of 2.82 keV Auger electrons. The counting lasts for a sufficiently long period to determine the exponentially decaying signal and a constant background.

The principles of gallium experiments (GALLEX [6] and SAGE [7]) and the iodine experiment are similar to the chlorine experiment. In the case of ^{7}Li, it is difficult to count and characterize the electron-capture decay of the produced ^{7}Be atoms because an Auger electron has an energy of only 55 eV. The recoil energy of the nucleus is also ~ 60 eV in 90% of the cases. It has been proposed to detect this total energy release of ~ 110 eV with a cryogenic microcalorimeter [8]. In the case of ^{81}Br, the produced ^{81}Kr atoms have too long a lifetime to count its decay. As a counting method, resonance ionization spectroscopy has been proposed.

The overall calibration of the radiochemical experiment can only be done by using an intense neutrino source with known activity. Both GALLEX [9] and SAGE [10] groups calibrated their gallium detectors with ^{51}Cr source of order ~ 1 MCi. The source was produced by neutron irradiation on ^{50}Cr at a reactor. Electron-capture decay of ^{51}Cr (half life ~ 27.7 d) primarily emits

---

[1] Actually, the observed capture rate was $\frac{1}{3} \sim \frac{1}{4}$. This deficit is the famous solar-neutrino problem. Other solar-neutrino experiments also observe less neutrinos than expected from the SSM calculations. An attractive solution to the solar-neutrino problem is the neutrino oscillation, particularly that in matter [5].

746 keV neutrinos. Both groups demonstrated that the observed ^{71}Ge production rates by the ^{51}Cr source were consistent with the rate expected from the known activity. A possible calibration source of the chlorine experiment is ^{65}Zn (1.35 MeV neutrino is emitted with $\sim$ 50% probability). However, the Homestake experiment has not been calibrated yet.

By nature, radiochemical detectors cannot provide realtime data. They can provide no time information shorter than the exposure period, energy information, nor directional information. They can measure only the number of neutrinos captured in the detector material, which is related to the solar-neutrino flux. Moreover, calibration of the detector is difficult. Though the chlorine and gallium experiments played a historical role in the first-generation solar-neutrino observations, radiochemical detectors are likely not to be used in future.

LIQUID SCINTILLATION DETECTORS

For the detection of low-energy neutrinos, liquid scintillator is used as a total absorption calorimeter. As a low-energy neutrino detector, liquid scintillator has several advantages. It has high sensitivity and, therefore, low threshold; it is easy to purify; it is easy to construct a large detector with an arbitrary shape; and the cost is low.

Conventionally, only $\bar{\nu}_e$'s are detected through the reaction $\bar{\nu}_e + p \rightarrow e^+ + n$ with $E_{e^+} = E_{\bar{\nu}_e} - 1.804$ MeV, and in this case one can use a delayed coincidence technique to reject the background as described below. Recently, however, detection of very low energy (< 1 MeV) solar neutrinos through elastic scattering $\nu_e + e^- \rightarrow \nu_e + e^-$ has been proposed. Only the low-energy recoil electron produces the signal in this case. This requires extremely high level of radiopurity of liquid scintillator as well as other detector materials to suppress the background.

Liquid scintillator is also used for high-energy neutrino detectors as an active element of sampling calorimeters.

Historically, Reines and Cowan discovered neutrino ($\bar{\nu}_e$) at the Savannah River Reactor using 3 modules of 1400-litre liquid scintillator detector [11]. Figure 1 schematically shows the detection principle of this experiment. A $\bar{\nu}_e$ from the reactor interacts with a proton in the water target placed between the liquid scintillator tanks. The produced e^+ annihilates on an electron and two 511-keV γ-rays are emitted back to back, giving a prompt coincidence signal. The neutron is moderated in water and captured by a Cd nucleus dissolved in water in about 10 μs, and the resulting γ's give a delayed signal.

In Table 2 are listed some selected low-energy neutrino experiments using liquid scintillator. Because of the limited space available, not all these detectors but CHOOZ [12], Palo Verde [13], and Borexino [17] detectors are further described below.

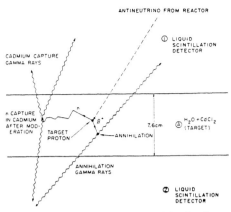

FIGURE 1. Schematic of the detection scheme used in the Savannah River experiment.

The CHOOZ detector [12] is located at a depth of 300 mwe (meter water equivalence). It consists of three concentric regions as shown in Figure 2: a 5-ton central target region of Gd-loaded scintillator contained in a plexiglass vessel, a 17-ton intermediate "containment" region instrumented with photomultiplier tubes (PMTs); and an optically separated 90-ton outer region which vetoes cosmic-ray muons and provides additional shielding against external γ-rays and neutrons. The intermediate region protects the target from PMT radioactivity. Also, the γ-rays from neutron capture in the target are required to be contained in this region. The $\bar{\nu}_e$ signal is a delayed coincidence between the prompt e^+ + two annihilation γ-rays and 8-MeV γ-ray signal

TABLE 2. Some examples of liquid scintillator neutrino detectors.

	Experiment	Segmented?	Scintillator	Fiducial (total) mass (ton)	Ref.
Reactor (ν oscil.)	CHOOZ	No	0.1% Gd loaded	5	[12]
	Palo Verde	Yes	0.1% Gd loaded	12	[13]
Accelerator (ν oscil.)	Kermen	Yes	Gd coated paper between modules	56	[14]
	LSND	No	low scintillator concentration[a]	(167)	[15]
Cosmic Ray (ν burst)	LVD	Yes		(1840)	[16]
Low BG (Solar ν)	Borexino[b]	No	extreme radiopurity	100	[17]
	Kam-LAND[c]	No	extreme radiopurity	1,000	[19]

[a] This allows the detection of both Cherenkov light and scintillation light.
[b] Under construction.
[c] Under construction. This experiment also aims at a study of neutrino oscillations using accelerator neutrinos, detection of neutrino bursts, etc.

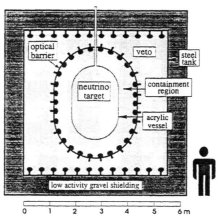

FIGURE 2. Schematic of the CHOOZ detector.

from neutron capture. For the latter signal, the measured energy resolution is 6%.

The Palo Verde detector [13] is located at a 46 mwe depth, much shallower than CHOOZ. This causes larger fluxes of cosmic-ray muons and fast neutrons produced by muon-induced nuclear interactions as well as muon capture by nuclei. To overcome the background, the Palo Verde detector has a segmented structure. A 12-ton fiducial volume consists of 66 acrylic cells, each 9 m × 0.2 m × 0.12 m, viewed by two 5-inch PMTs on each end. The fiducial volume is surrounded by a 1 m water buffer. An active veto layer of liquid scintillator further surround the water buffer. A neutrino event is identified by a prompt triple coincidence produced by e^+ and associated annihilation γ's in neighbouring back-to-back detector elements, followed by a delayed signal from neutron capture.

The Borexino detector [17] is under construction at the Gran Sasso Underground Laboratory. The experiment aims at observing monoenergetic 862-keV ^{7}Be solar neutrinos through νe scattering. The recoil electrons produce a Compton-like spectrum with a kinematical edge at 665 keV. There is no way to discriminate the signal from the background, and, therefore, the background rate must be kept lower than the expected signal rate. The ^{14}C content in the scintillator produces unavoidable background and this sets the threshold of counting at ~ 250 keV. The tolerable contamination level of U and Th is less than $10^{-15\sim16}$ g/g. All other detector materials are also required to have very high radiopurity. The detector design is similar to the CHOOZ detector. The fiducial volume is surrounded by an intermediate buffer region filled with pure pseudocumane, equipped with PMTs. The outer region contains ultra pure water as a passive shield against external radiation. To test the concept of the Borexino experiment, a Counting Test

TABLE 3. Underground water Cherenkov detectors.

Experiment	Location	Depth (mwe)	Total mass (ton)		Threshold (MeV)	Status
IMB	Fairport Harber Ohio, U.S.A.	1,580	H_2O	8,000	~ 30	closed
Kamiokande	Kamioka, Japan	2,400	H_2O	4,500	7	closed
Super-Kam	Kamioka, Japan	2,400	H_2O	50,000	5[a]	operating
SNO	Sudbury, Canada	5,900	D_2O	1,000	5	construction
			H_2O	7,300		

[a] Present threshold is ~ 6 MeV.

Facility (CTF) [18] with 4 tons of liquid scintillator shielded by 1,000 tons of ultra pure water was constructed at Gran Sasso and very promising results have been obtained.

WATER CHERENKOV DETECTORS

The first-generation large water Cherenkov detectors, IMB [20] and Kamiokande [21] were constructed to search for nucleon decay predicted by Grand Unified Theories. Table 3 compares some properties of representative water Cherenkov detectors. Figure 3 schematically shows Super-Kamiokande [21,22]. These water Cherenkov detectors allow tracking of charged particles with moderate precision above Cherenkov threshold, measurement of visible energies, and identification of showering (e and γ) and non-showering (μ and π) particles.

These detectors observe atmospheric neutrinos which at first were considered mere background to the nucleon-decay search. Now the study of atmospheric neutrinos is important in its own right because it provides possible indication of neutrino oscillations [23]. Another important advantage of the water Cherenkov detector is that it can be operated with an energy threshold as low as ~ 5 MeV. Thus, a water Cherenkov detector can be an excellent low-energy neutrino detector. Both Kamiokande and IMB succeeded in detecting a neutrino burst from the supernova SN1987A for the first time [24,25]. The reaction relevant to the detection of supernova neutrinos is $\bar{\nu}_e + p \rightarrow e^+ + n$. This reaction is characterized by a flat e^+ angular distribution with respect to the neutrino source direction.

The fiducial volume of Kamiokande ($\sim 1,000$ tons for the nucleon-decay search and atmospheric-neutrino observation and 680 tons for the solar-neutrino observation) is viewed by $\sim 1,000$ world largest 20-inch PMTs and that of Super-Kamiokande (22,000 tons) is viewed by $\sim 11,200$ PMTs of the same size. The Kamiokande and Super-Kamiokande detectors have sufficiently low thresholds to detect solar neutrinos. Purification of water to remove radioactivity and, in particular, radon is essentially important to

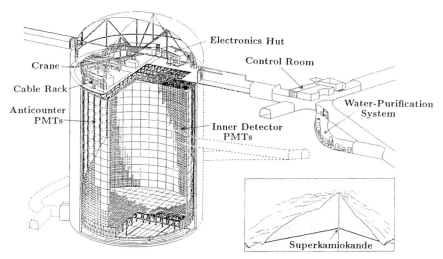

FIGURE 3. Artist's impression of the Super-Kamiokande detector and associated facilities.

lower the threshold. Solar neutrinos are detected using νe elastic scattering. In this reaction, the electron angular distribution shows a sharp forward peak with respect to the neutrino direction. Thus, the solar-neutrino signature is a forward peak on the flat background distribution with respect to the direction from the Sun to the Earth. Because of this directionality, Kamiokande proved that the Sun is really emitting neutrinos. The observed solar-neutrino flux is $\lesssim 1/2$ of the SSM prediction.

Super-Kamiokande [21,22] is a second generation experiment in both nucleon-decay search and solar-neutrino observation. A high-statistics measurement of the recoil-electron energy spectrum of solar-neutrino events is expected to be a key to solve the solar-neutrino problem. The energy calibration of the massive detector required to achieve this goal is $\Delta E/E \sim 1\%$. For this purpose, Super-Kamiokande is equipped with an extremely low-intensity electron linac as an energy calibration source.

Sudbury Neutrino Observatory (SNO) is also a second-generation solar-neutrino experiment [26]. It will use 1,000 tons of heavy water D_2O, shielded by 7,300 tons of light water. The PMTs viewing the central D_2O volume are instrumented in the light-water region to protect the D_2O volume from PMT radioactivity. Solar neutrinos are measured through both CC interctions ($\nu_e + D \to e^- + p + p$ and NC (neutral-current) interactions ($\nu + D \to \nu + p + n$). νe scattering events are also measured. The solar-neutrino observation with NC interactions is particularly important to verify the neutrino oscillation if it is a true solution to the solar-neutrino problem. To detect NC interactions,

MgCl$_2$ will be dissolved in heavy water. Neutrons are captured by ^{35}Cl to produce γ-rays with total energy of 8.6 MeV. Extremely good radiopurity (3.7×10^{-15} g/g of ^{232}Th in heavy water) is required for this scheme to work.

ACCELERATOR NEUTRINO EXPERIMENTS I

In high-energy accelerator neutrino experiments, some common features to be considered in the design of the detector are:
- identification of muons and electrons,
- measurement of muon trajectory and momentum,
- measurement of hadronic energy as well as electromagnetic energy, and
- tracking in the target region.

Depending on the purpose of the experiment, all these features are not necessarily required. The last entry, tracking in the target region, is a rather difficult requirement for the massive neutrino detector. Heavy-liquid bubble chambers were capable of this, but they were not used any more. These old detectors as well as more recent bubble-chamber-quality detectors are discussed in the next section.

The first high-energy accelerator neutrino experiment was performed at the BNL AGS with a primary proton beam energy of 15 GeV by Lederman, Schwarts, Steinberger, and collaborators, and led to the discovery of two neutrinos [27]. The detector used in this experiment was an aluminum-plate optical spark chamber with 10 modules, each consisted of nine 44″ × 44″ × 1″ plates, the total weight being 10 tons. No magnetic field was used. Long-track events were identified as muons and the momentum was determined from the range in aluminum. Electrons were identified by showers; a calibration with electron beams showed that the mean number of sparks was roughly linear with electron energy up to 400 MeV/c. This detector satisfied all the requirements mentioned above.

A popular design of the modern detectors for counter neutrino experiments at high-energy accelerators is a combination of a target calorimeter for the measurement of hadronic energy and a muon spectrometer consisting of magnetized iron plates interleaved with drift chambers. Representative detectors of this category are those for the CDHS and CCFR experiments. If good electron identification capability as well as an electromagnetic energy measurement is required, a target calorimeter must be made of low Z material and have fine granularity. The CHARM-II detector is a representative one.

The CDHS experiment at CERN studied the nucleon structure functions and perturbative QCD, and electroweak physics through both CC and NC inclusive ν_μ ($\bar{\nu}_\mu$) interactions on iron, $\nu_\mu(\bar{\nu}_\mu) + \text{Fe} \rightarrow \mu^\pm + $ anything and $\nu_\mu(\bar{\nu}_\mu) + \text{Fe} \rightarrow \nu_\mu(\bar{\nu}_\mu) + $ anything, in the energy region $30 \lesssim E_\nu \lesssim 300$ GeV. The detector [28] consists of 19 magnetized iron modules, each 3.75

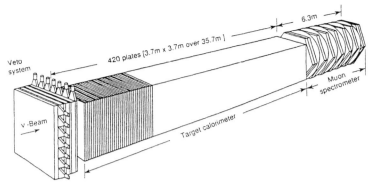

FIGURE 4. Sketch of the CHARM-II detector.

m in diameter and 65 tons in weight. Between the modules are placed drift chamber trackers. The first 7 modules consist of thin iron plates and the last 12 modules consist of thick plates. Plastic scintillator strips are inserted between the plates to sample the hadronic shower. The CDHS detector is a "combined function" detector: the modules function successively as a target, a hadron calorimeter, and a muon spectrometer. The total detector weight is 1250 tons and useful target mass is $\sim$ 750 tons. The detector has good muon identification capability, but it cannot identify electrons. It is also blind to details of the hadron shower.

The concept of the CCFR detector [29] at Fermilab is similar to the CDHS detector, but it is a "separated function" detector consisting of a target calorimeter and a magnetized-iron muon spectrometer. The target calorimeter has a mass of 690 tons, and consists of layers of iron plates, liquid scintillation counters, and drift chambers. The neutrino energy region up to 600 GeV was studied with this detector.

The main purpose of the CHARM-II experiment [30] at CERN was the measurement of $\nu_\mu e$ and $\bar{\nu}_\mu e$ elastic scattering. An average energy of the ν_μ beam was 23.7 GeV and that of $\bar{\nu}_\mu$ beam was 19.1 GeV. A difficult point of this experiment is a 10^4 times larger background rate due to neutrino scattering off nuclei. Hadronic and electromagnetic showers are distinguished by different characteristics of the shower development. Electromagnetic showers are narrow and have characteristic evolution, while hadronic showers are widely spread. The signature of νe scattering is an isolated narrow shower with strongly forward-peaked cone, initiated by a single electron. The calorimeter should have fine granularity to sample the shower energy frequently, and its material should have low Z to reduce multiple scattering for good shower-angle resolution. Thus, the target calorimeter of the CHARM-II detector (see Fig. 4) is constructed out of 420 modules, each consisting of a 48-mm thick glass plate ($\sim 0.5 X_0$) and a plane of plastic streamer tubes with 1-cm wire spacing. The mass of the target calorimeter is 692 tons. the muon spectrometer of this detector is comprised of six magnetized iron modules recycled from the CDHS detector.

ACCELERATOR NEUTRINO EXPERIMENTS II

Large bubble chambers were extensively used in the 1970's for neutrino experiments though they are not used any more. The famous chambers are a heavy-liquid bubble chamber Gargamelle at CERN and two cryogenic bubble chambers, 3.7-m Big European Bubble Chamber (BEBC) at CERN and a 15-ft bubble chamber at Fermilab. Some advantages of the bubble chambers are 4π solid-angle coverage, visibility of all charged tracks, and a fine spatial resolution (50 ~ 150μm). However, the target mass is rather small (~ 1 ton for liquid hydrogen and ~ 12 tons for neon-hydrogen mixture in BEBC, and 18 tons for liquid freon in Gargamelle). For the muon detection and identification, external muon identifiers consisting of iron or zinc absorber and planes of tracking chambers were used for BEBC and the 15-ft bubble chamber.

A historical discovery of neutral currents was achieved by analyses of neutrino events in Gargamelle which was exposed to a neutrino beam produced at the CERN 24-GeV proton synchrotron. A candidate for a NC process $\nu_\mu + e^- \to \nu_\mu + e^-$ was found by searching for a single electron originating in the liquid [31]. Also, neutrino events producing hadrons, but no muon or electron, $\nu_\mu(\bar{\nu}_\mu) + N \to \nu_\mu(\bar{\nu}_\mu) +$ hadrons were observed [32].

A detailed study of charmed particle production in neutrino reactions requires bubble-chamber-quality fine tracking to identify decay vertices of short-lived charmed particles. The experiment also requires a high-energy (several hundred GeV) primary proton beam, but as the proton beam energy increases, the number of background muon tracks increases to an intolerable level for a bubble chamber unless the thickness of a costly muon shield is incresed sufficiently. Here came a Fermilab E531 experiment using an emulsion hybrid spectrometer [33] consisting of an emulsion target, a magnetic spectrometer to measure the charge and momentum of secondary particles, a calorimeter, and a muon counter. The intrinsic spatial resolution of emulsion is ~ 1μm. Thus emulsion can handle accumulation of up to ~ 100 tracks in $100 \times 100\mu m^2$. In the emulsion hybrid experiment, the vertices of neutrino interactions in the emulsion are found by following back the tracks recorded in electronic detectors to the emulsion, and scanning its limited volume predicted by the track reconstruction program. A time-consuming scanning load with a microscope is thus reduced.

Emulsion also provides a means to directly identify τ particles. The CHORUS experiment at CERN searches for $\nu_\mu \to \nu_\tau$ oscillations in the appearance mode using a 800 kg emulsion target in the first run and additional 800 kg

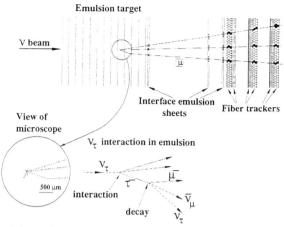

FIGURE 5. Schematic of the front part of the CHORUS detector. A typical signature of a ν_τ CC event is also shown.

in the second run. Figure 5 shows how a ν_τ CC interaction looks in the CHORUS detector [34]. The number of events to be scanned increases with the emulsion mass, but the recent development of fully-automated scanning microscope makes it possible to analyze the data within a reasonable time.

At CERN, another $\nu_\mu \to \nu_\tau$ oscillation experiment NOMAD [35] is operating. The NOMAD is a purely electronic detector and its neutrino target is drift chambers made of low-density and low-Z materials, placed in a uniform magnetic field of 4 kG provided by a recycled UA1 dipole. It may be called an "electronic bubble chamber." The fiducial mass is 3 tons over the volume of $2.6 \times 2.6 \times \sim 4$ m^3. The NOMAD detector is equipped with a transition radiation detector, an electromagnetic calorimeter, a hadron calorimeter, and muon chambers. Thus, it has a good μ/e identification capability. The τ signature is searched for on the statistical basis using various kinematical cuts.

OTHER NEUTRINO DETECTORS

In non-accelerator neutrino physics experiments, a variety of detection technique has been devised. In this section, some other neutrino detectors are discussed.

ICARUS. ICARUS is a liquid-argon time projection chamber [36]. It is also an "electronic bubble chamber." It can track all the charged particles involved with a fine spatial resolution ($\sim 100\mu$m), and measure the ionization loss of these particles all the way along their paths. This feature enables unambiguous particle identification and energy/momentum measurement.

Since the drift distance is very long, very high purity (~ 0.1 ppb electronegative) is required for the liquid argon. A test of a 3-ton prototype was very successful [36]. A 600-ton module is now under construction at Gran Sasso in order to study nucleon decay and atmospheric and solar neutrinos.

HELLAZ. Hellaz [37] is a solar-neutrino detector proposed for the measurement of low-energy pp neutrinos. It is a time projection chamber containing 12 tons of pure helium gas at liquid-azote temperature (77 K) and at 10-bar pressure, well shielded against external radiation. With use of a low-Z material, it enables the complete kinematical reconstruction of νe scattering, by measuring the energy and direction of electrons with an assumption that the Sun is the neutrino source, to obtain the incident solar-neutrino energy. The detection threshold may be as low as 100 keV.

Iron-tracking calorimeters. These are conventional underground detectors for nucleon-decay search and measurement of atmospheric neutrinos. Mass is provided by iron plates interleaved by a fine tracking device. The most recent example is the Soudan 2 detector [38] with a total mass of 963 tons. This type of detector provides tracking, energy measurement and particle identification. However, it is expensive and therefore difficult to scale up.

Deep underwater/ice detectors. For the detection of very high-energy astronomical neutrinos conjectured to come from point sources such as active galactic nuclei, upward-going muons produced by CC interactions near the detector are used. Though underground detectors such as Kamiokande and IMB measured the fluxes of upward-going muons, larger detectors are required to have a sensitivity which is sufficient for the detection of very high-energy astronomical neutrinos. For this purpose, strings of PMTs are deployed deep underwater (for example, Baikal Experiment [39]) or in the Antarctic ice (AMANDA [40]), and penetrating muons are detected utilizing the Cherenkov technique.

CONCLUSION

To conclude this talk, I would like to emphasize that the technical challenge of the neutrino detection is to devise a massive detector with an extremely high radiopurity and uniform response.

REFERENCES

1. Davis, R., Jr., *et al. Phys. Rev. Lett.* **21**, 1205 (1968).
2. Bahcall, J.N., and Ulrich, R.K., *Rev. Mod. Phys.* **60**, 297 (1988).
3. Bahcall, J.N., and Pinsonneault, M.H., *Rev. Mod. Phys.* **67**, 781 (1988).
4. Engel, J., *et al. Phys. Rev.* C **51**, 2837 (1995).
5. Hata, N., and Langacker, P., *Phys. Rev.* D **52**, 420 (1995).

6. Anselmann, P., et al. Phys. Lett. B **285**, 376 (1992).
7. Abdrashitov, J.N., et al. Phys. Lett. B **328**, 234 (1994).
8. Galeazzi, M., et al. Phys. Lett. B **398**, 187 (1997).
9. Anselmann, P., et al. Phys. Lett. B **342**, 440 (1995).
10. Abdrashitov, J.N., et al. Phys. Rev. Lett. **77**, 4708 (1996).
11. Reines, F., et al. Phys. Rev. **117**, 159 (1960).
12. Bemporad, C., *Neutrino '96*, Singapore: World Scientific, 1997, p. 242.
13. Boehm, F., et al. *Neutrino '96*, Singapore: World Scientific, 1997, p. 248.
14. Drexlin, G., et al. Nucl. Instrum. Meth. A **289**, 490 (1990).
15. Athanassopoulos, C., et al. Nucl. Instrum. Meth. A **388**, 149 (1997).
16. Bari, G., et al. Nucl. Instrum. Meth. A **277**, 11 (1989).
17. Alimonti, G., et al. Nucl. Phys. B (Proc. Suppl.) **32**, 149 (1993).
18. Giammarchi, M.G., Nucl. Phys. B (Proc. Suppl.) **35**, 433 (1995).
19. Suekane, F., Preprint TOHOKU-HEP-97-02, 1997.
20. Becker-Szendy, R., et al. Nucl. Instrum. Meth. A **324**, 363 (1993).
21. Nakamura, K., et al. *Physics and Astrophysics of Neutrinos*, Tokyo: Springer-Verlag, 1994, p. 249.
22. Takita, M., *Frontiers of Neutrino Astrophysics*, Tokyo: Universal Academy Press, 1993, p. 135.
23. Fukuda, Y., et al. Phys. Lett. B **335**, 237 (1994).
24. Hirata, K., et al. Phys. Rev. Lett. **58**, 1490 (1987).
25. Bionta, R.M., et al. Phys. Rev. Lett. **58**, 1494 (1987).
26. Ewan, G.T., *Frontiers of Neutrino Astrophysics*, Tokyo: Universal Academy Press, 1993, p. 147.
27. Danby, G., et al. Phys. Rev. Lett. **9**, 36 (1962).
28. Holder, M., et al. Nucl. Instrum. Meth. **148**, 235 (1978).
29. Sakumoto, W.K., et al. Nucl. Instrum. Meth. A **294**, 179 (1990).
30. Geiregat, D., et al. Nucl. Instrum. Meth. A **325**, 92 (1993).
31. Hasert, F.J., et al. Phys. Lett. **46B**, 138 (1993).
32. Hasert, F.J., et al. Phys. Lett. **46B**, 121 (1993).
33. Ushida, N., et al. Nucl. Instrum. Meth. A **224**, 50 (1984).
34. Annis, P., et al. Preprint CERN-PPE/97-100, 1997, submitted to Nucl. Instrum. Meth. A.
35. Rubbia, A., Nucl. Phys. B (Proc. Suppl.) **40**, 93 (1995).
36. Benetti, P., et al. Nucl. Instrum. Meth. A **327**, 173 (1993).
37. Bonvicini, G., Nucl. Phys. B (Proc. Suppl.) **35**, 441 (1994).
38. Allison, W.W.M., et al. Nucl. Instrum. Meth. A **376**, 36 (1996).
39. Belolaptikov, I.A., *Neutrino '96*, Singapore: World Scientific, 1997, p. 524.
40. Lowder, D.M., *Neutrino '96*, Singapore: World Scientific, 1997, p. 518.

LABORATORY SESSIONS

Detection of Cosmic Ray Tracks using Scintillating Fibers and Position Sensitive Multi-Anode Photomultipliers

Muzaffer Atac, Jon Streets and Neal Wilcer

Fermi National Accelerator Laboratory, Batavia, IL 60510 U.S.A.

Abstract. This experiment demonstrates detection of cosmic ray tracks by using scintillating fiber planes and multi-anode photomultipliers (MA-PMTs). In a laboratory like this, cosmic rays provide a natural source of high-energy charged particles which can be detected with high efficiency and with nanosecond time resolution.

INTRODUCTION

The scintillating fibers and photomultipliers (PMTs) were used earlier in a Fermilab experiment, APEX, a search for anti-proton decay. The amplifiers and the scintillating fiber plane couplings were improved for this laboratory experiment.

The majority of the cosmic particles are muons which are mainly from the decay products of charged pions. The mean lifetime of the pions (π^+, π^-) at rest is around 26ns, and they decay into muons and neutrinos 99.987% of the time. Depending on the relativistic γ factor, this lifetime can be much longer. The positively charged muons, μ^+, decay into a positron and a neutrino. The muon lifetime at rest is around 2.2µs. Again, this time can be longer depending on the relativistic γ factor.

We expect the largest cosmic particle flux to reach the earth's surface from the zenith (directly overhead). Tracks from this direction travel through a thinner layer of atmosphere, and so less are absorbed by atmospheric nitrogen and oxygen. When one does this experiment, one will see the display of the charged cosmic ray particle tracks together with their angular distribution after some time (Fig. 1).

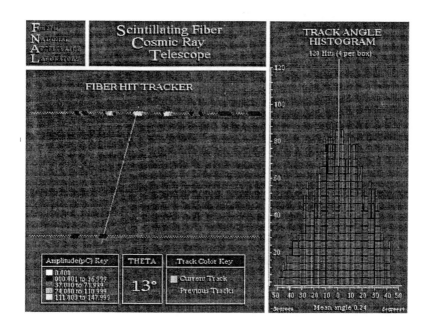

FIGURE 1. Screen display from the cosmic ray telescope.

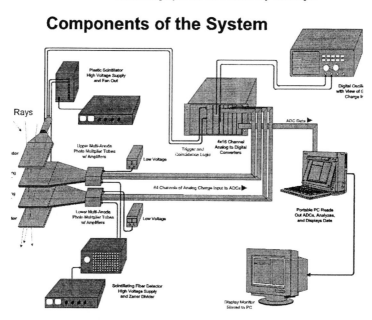

FIGURE 2. Block diagram of the apparatus.

DESCRIPTION

Figure 2 shows the experimental arrangement as a block diagram. The cosmic particles are selected at the given solid angle by a pair of plastic scintillators arranged as a telescope. We use the discriminated signals from these scintillators to provide a trigger to the CAMAC read-out system. Outputs of the discriminators are put through a coincidence module to provide a gate to the LeCroy FERA Analog to Digital Converters (ADCs). The fibers inside each detector are arranged in two layers of staggered rows which are composed of 2mm, singly clad BICRON fibers (Fig. 3). There are 192 fibers in each detector. The fibers are butted onto the face of the MA-PMTs in an 8 by 24 grid. Each fiber is aligned with the matrix of eight strips and 24 wire anodes inside the MA-PMT (Fig. 4). The signals from the 32 anode outputs for each MA-PMT are amplified before being digitized by the ADCs. This reduces the (8×24=) 192 fiber channels to (8+24=) 32 anode signals in each detector. A program running in the PC detects the gate generated by the scintillators, and reads the ADCs. It then decodes the 64 ADC values to reconstruct the hit pattern of the 384 fibers. When studying the display, one should note that often more than one fiber is lit. Sometime this is due to cosmic ray showers where a particle has interacted with the ceiling of the building and produced many secondaries. The most common cause of the multiplier hits in a single fiber plane is due to cross-talk in the MA-PMT. Each wire and strip receives as much as 20% from the charge of its two neighbors. Cross-talk between adjacent wires will appear as hits on neighboring fibers on the event display. Cross-talk between adjacent strips will appear as hits on every eighth fiber.

Scintillating Fiber Paddle Construction

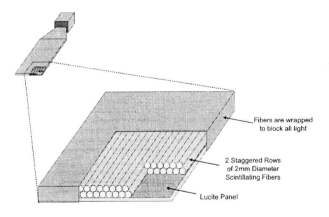

FIGURE 3. Arrangement of fibers in the detector.

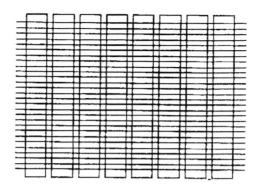

FIGURE 4. Arrangement of 24 wires and 8 strips in MA-PMT.

MATERIALS AND EQUIPMENT

- Butyl PBD and POPOP scintillating fiber planes. The core of the fiber's base material is polystyrene and the clad material is poly-methyl-metacrylate (PMMA), trade name is *Lucite*.
- 64 channels of LeCroy 4300 CAMAC-FERA ADCs.
- Two Hamamatsu R4135 multi-wire anode PMTs.
- Two RCA 8575 PMTs.
- Bias power supplies for the amplifiers on the R4135 signals.
- Portable CAMAC crate with two CAMAC-to-NIM converters for trigger logic.
- LeCroy 4222 gate generator for pedestals.
- LeCroy 622 coincidence unit and LeCroy 621L discriminator unit for trigger logic.
- 100MHz Pentium PC with DSP602 ISA-CAMAC crate interface.

OPERATING INSTRUCTIONS FOR THE SFCRT

Starting Data Taking

1. Turn on the computer.
2. When the computer requests a password, hit CANCEL.
3. Turn on CAMAC crate and power supplies.

4. Attach lemo cable from 4222 OUT1 to the GAI input of the 4301 (for pedestal gate).
5. On the Windows95 screen, there is an icon for the SFCRT. Double-click it. The introductory screen will be displayed while 100 pedestals are taken and the averages downloaded to the 4300s.
6. When the screen changes to the data-taking view (Fig. 1), disconnect the lemo cable from the 4301 GAI input and connect the lemo cable from the 622 OUT to the 4301 GAI input. The system can now trigger on the plastic scintillator coincidence.
7. Hit ENTER to start data-taking.

Getting a Screendump of the Display

Press ALT-PRINT SCREEN. This will save the screen in the computer's paste buffer as a bitmap image. You can then start the PAINT utility that comes with Windows95. Select the PASTE option under the edit window, and the image will appear on the screen. At this point you can save the image to a file or to a printer.

Switch Settings

1. DSP controller: On-line/off-line switch is set to "on-line".
2. Lambda low voltage power supplies for the amplifiers: Both set to "7V".
3. 622 QUAD coincidence module: And/or switch set to "and".

CAUTION: HV supplies must be powered up and down in correct sequence and to correct values. Please do not attempt to change HV setting unless authorized.

4. Fiber detector HV power supply: Set to "NEG 1300V".
5. Fiber detector HV zener divider (COW): Channel 1 peg inserted in the "500 NEG" HV hole. (One detector runs at 1300V; the other at 900V.)
6. Plastic trigger scintillator power supply: Set to "NEG 1700V". Both run at -1700V.
7. Use channels 2 and 3 of the plastic scintillator COW. (Channel 1 is stuck at 100.)

Cabling Hook-up for the SFCRT

Lemo Connections

1. DSP CAMAC controller REQUEST: Set to "GRANT IN".

2. TOP PMT gate to 1st section IN of 621L.
3. BOTTOM PMT gate to 2nd section IN of 621L.
4. Output (lower left) of 1st section of 621L to In of 1st section of 622.
5. Output (lower left) of 2nd section of 621L to the other In of 1st section of 622.
6. Out of 1st section of 622 to IN of 3rd section of 621L.
7. During data taking: OUT of 3rd section of 621L to GAI input of 4301.
8. During pedestal calculation (on start-up): OUT1 of 422 to GAI input of 4301.

Twist-n-flat Cables

1. Cable 1 tube 4 bottom 34-pin connector to 4300 (slot 20) 34-pin connector IN. Pin 1 marking on header up.
2. Cable 2 tube 4 top 34-pin connector to 4300 (slot 21) 34-pin connector IN. Pin 1 marking on header up.
3. Cable 3 tube 4 bottom 34-pin connector to 4300 (slot 22) 34-pin connector IN. Pin 1 marking on header up.
4. Cable 4 tube 4 top 34-pin connector to 4300 (slot 23) 34-pin connector IN. Pin 1 marking on header up.

High Voltage Cables (Red Cables)

CAUTION: HV supplies must be powered up and down in correct sequence and to correct values. Please do not attempt to change HV setting unless authorized.

1. From tube 4 detector to PG1 input of MA-PMT COW (-900V).
2. From tube 9 detector to PG2 input of MA-PMT COW (-1300V).
3. From HV output of model 1570 HV supply to -HV input of COW.
4. From top gate paddle to regulating HV fan-out panel input (3rd from top).
5. From bottom gate paddle to regulating HV fan-out front panel input (2nd from top).
6. From rear connector of 415B to NEGATIVE input of regulating HV fan-out.

Silicon Detectors and Signal Processing

P. Giubellino, M. Idzik

INFN Torino

A. Rudge, P. Weilhammer

CERN, Geneve

In this laboratory the students will familiarize with several aspects of silicon detectors and their use in modern High-Energy Physics experiments. The lab consists of four sections. The first one introduces the general features of the different types of silicon detectors, which are also observed at the microscope, and the standard measurements performed to characterize them, which are executed at the probe station. In the second lab the setup is designed to allow the understanding of the charge collection process and the signal formation, using an infrared diode to generate charge in the detector and very fast electronics to read out the signals. In the third lab a silicon detector coupled with a spectrum analyzer is used to detect particles from radioactive sources, and the problems involved in the detection of X-ray spectra are addressed. The last lab introduces a complex low noise readout system, based on the Viking VLSI chip: in this section the students are confronted with the problems of multichannel readout of a microstrip detector, and can appreciate the excellent spatial resolution achieved and measure the very low noise of the system.

SILICON DIODE AS A PARTICLE DETECTOR, MEASUREMENT OF BASIC PROPERTIES.

During this section we will do what is generally called the characterization of a silicon diode, i.e. we will perform the measurement of its basic properties relevant for its use as a particle detector. Such measurements are the ones routinely performed on any detector which enters a physics lab.

In order to use a silicon diode as a particle detector the following conditions must be fulfilled.
- The diode needs to be reverse biased. In this way an electric field is created in the diode volume, which will move to the opposite electrodes the electrons and holes generated by a ionizing particle. To maximize the collected charge the full thickness of the detector must be depleted, as will be explained later.
- The leakage current of the diode needs to be enough small so that its fluctuations are much smaller than the amplitude of the current pulse generated by a ionizing particle.

In the following sections we give theoretical descriptions for some of the fundamental aspects:
1. Reverse biased *p-n* junction as a detector,
2. Measurement of the depletion voltage by means of C-V curve,
3. Diode's leakage current.
4. Finally we give a brief description of the setup and proposed measurements, which are based on the theory above.

The reverse biased *p-n* junction as a detector

Let us consider a reversed biased *p-n* junction (Fig. 1), with uniform doping concentration N_D in the *n* region and N_A in the *p*, with $N_A \gg N_D$ which is normally the case. We can treat this simple geometry as one dimensional. Without external potential there exists a built-in potential Ψ_0 in the junction, created by the thermal diffusion of electrons and holes:

$$\Psi_0 = V_T \ln\left(\frac{N_A N_D}{n_i^2}\right) \tag{1.1}$$

where:

$$V_T = \frac{kT}{q} \cong 26\,\text{mV} \quad \text{at } T=300\,\text{K} \tag{1.2}$$

and n_i is the concentration of the charged carriers in the silicon $1.5 \times 10^{10} \text{ cm}^{-3}$ (at 300 K).

As an effect, at the boundary between the p and n regions there exists a layer without free charge carriers (depletion layer), but with fixed negative acceptor ions and positive donor ions which create the electric field in it. Any charged carrier generated in this region would be moved by the field in the appropriate direction, opposite for electrons and holes. When an external bias V_R is applied, the total voltage drop in the junction is ($\Psi_0 + V_R$).

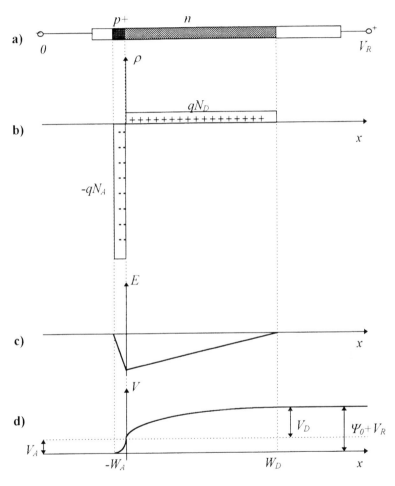

Figure 1. Here are shown 1-dimensional distributions of the following quantities in p-n junction: a) general junction scheme, b) concentration of ionized dopants, c) electric field, d) potential.

If the depletion layer thickness is W_A in the p region and W_D in the n region then one can write the charge balance equation, which expresses the fact that our crystal remains electrically neutral, as:

$$W_A N_A = W_D N_D \qquad (1.3)$$

In the region $-W_A \leq x < 0$ we can write the one dimensional Poisson equation:

$$\frac{d^2 V}{dx^2} = -\frac{\rho}{\varepsilon \varepsilon_0} = \frac{qN_A}{\varepsilon \varepsilon_0} \qquad (1.4)$$

Setting the boundary condition $E(x) = 0$ for $x = -W_A$, from (1.4) one can get (fig. 1c):

$$E = -\frac{dV}{dx} = -\frac{qN_A}{\varepsilon \varepsilon_0}(x + W_A) \qquad (1.5)$$

in this region.
Integrating (1.5) with the boundary condition $V(x) = 0$ for $x = -W_A$, one can get (fig. 1d):

$$V = \frac{qN_A}{\varepsilon \varepsilon_0}\left(\frac{x^2}{2} + W_A x + \frac{W_A^2}{2}\right) \qquad (1.6)$$

for $-W_A \leq x < 0$.
Fixing the potential $V(0) = V_A$, for $x = 0$ we get from (1.6):

$$V_A = \frac{qN_A}{\varepsilon \varepsilon_0} \frac{W_A^2}{2} \qquad (1.7)$$

After similar considerations for the n region one can calculate the voltage drop between the points $x = 0$ and $x = W_D$ as:

$$V_D = \frac{qN_D}{\varepsilon \varepsilon_0} \frac{W_D^2}{2} \qquad (1.8)$$

Summing the expressions (1.7) and (1.8) we get the total voltage drop for the junction:

$$\Psi_0 + V_R = V_A + V_D = \frac{q}{2\varepsilon\varepsilon_0}(N_A W_A^2 + N_D W_D^2) \qquad (1.9)$$

which after replacing (1.3) gives:

$$\Psi_0 + V_R = \frac{qW_A^2 N_A}{2\varepsilon\varepsilon_0}(1 + \frac{N_A}{N_D}) \qquad (1.10)$$

From this expressions we can calculate the thickness of the depletion layer in the p region as a function of the external voltage set and of the doping concentration:

$$W_A = \left[\frac{2\varepsilon\varepsilon_0(\Psi_0 + V_R)}{qN_A\left(1 + \frac{N_A}{N_D}\right)}\right]^{1/2} \qquad (1.11)$$

In a similar way for the n region we get:

$$W_D = \left[\frac{2\varepsilon\varepsilon_0(\Psi_0 + V_R)}{qN_D\left(1 + \frac{N_D}{N_A}\right)}\right]^{1/2} \qquad (1.12)$$

From (1.11) and (1.12) it is seen that the thickness of the depletion layer decreases with growing doping concentration and it is proportional to the square root of the set voltage. In practice abrupt junctions are used, i.e. a very thin layer of $p+$ region is produced (few microns) with very high doping concentration, on a thick n layer (few hundred microns) with low doping concentration (few orders of magnitude less then $p+$). So finally the total depletion thickness equals:

$$W = W_A + W_D \approx W_D \qquad (1.13)$$

In order to get ohmic contact between n region and metal electrode, thin intermediate $n+$ layer is added.

Measurement of the depletion voltage by means of the C-V curve

The depletion voltage is the value of the reverse detector bias voltage for which the whole detector volume is depleted i.e. the *p-n* junction or the electric field fills the whole detector volume. In the following we show that the measurement of the capacitance of the diode versus the reverse applied voltage allows the correct estimation of the depletion voltage.

Since the applied voltage affects the charge distribution in the junction one can define the junction capacitance per unit area as:

$$C_j = \frac{dQ}{dV_R} = \frac{dQ}{dW}\frac{dW}{dV_R} \qquad (2.1)$$

Having in mind the charge balance equation (1.3) one can write:

$$dQ = -qN_A dW_A = qN_D dW_D \qquad (2.2)$$

taking the donor part and then differentiating (1.12) one gets:

$$\frac{dW_D}{dV_R} = \left[\frac{\varepsilon\varepsilon_0}{2qN_D\left(1+\frac{N_D}{N_A}\right)(\Psi_0+V_R)}\right]^{1/2} \qquad (2.3)$$

Substituting equations (2.2) and (2.3) into (2.1) we can finally obtain:

$$C_j = \left[\frac{q\varepsilon\varepsilon_0(N_A N_D)}{2(N_D+N_A)(\Psi_0+V_R)}\right]^{1/2} \qquad (2.4)$$

The same result is obtained using the acceptor part of (2.2) and subsequently (1.11).
For the usual case when $N_A \gg N_D$ one obtains:

$$C_j = \left[\frac{q\varepsilon\varepsilon_0 N_D}{2(\Psi_0+V_R)}\right]^{1/2} \qquad (2.5)$$

or taking into account (1.12) one can express capacitance as a function of the depletion layer thickness as:

$$C_j = \frac{\varepsilon\varepsilon_0}{W_D} \tag{2.6}$$

It is seen from (2.5) and (2.6) that measuring the capacitance versus the applied voltage and making the plot of $1/C_j^2$ vs. V_R one can easily find the depletion voltage of the detector. If the depletion layer is smaller then the thickness of the detector this relation should be linear, according to (2.5), while when it reaches the full detector thickness the capacitance should reach a saturation value (2.6) and stay constant.

Diode leakage current

The leakage current of the reverse biased diode includes quite a few components, among which the most important are: diffusion current from minority carriers diffusing from the neutral region outside the junction, generation current due to generation-recombination processes in the junction and surface current due to the surface charges and defects. In the following we will separately address these components.

Diffusion current.

Under reverse bias conditions there exist limited probability of diffusion of electrons generated in the non depleted *p*-type region and holes generated in the *n*-type region to the junction layer. If it happens then they drift to the appropriate electrodes under the effect of the electric field and so form the diffusion current. The diffusion current is expressed by the well known Shockley formula:

$$J_s = \frac{qD_p p_{no}}{L_p} + \frac{qD_n n_{po}}{L_n} \tag{3.1}$$

where D_x is the diffusion constant, L_x is the diffusion length for electrons and holes respectively and n_{po}, p_{no} are the equilibrium densities of electrons and holes on the p-side and n-side respectively. For the abrupt p^+n junction only the left component (hole current) is significant. Taking also into account that $L_p = (D_p \tau_p)^{0.5}$ and $p_{no} = n_i^2/N_D$, where τ is the lifetime of the minority carriers, one can write:

$$J_s = q\sqrt{\frac{D_p}{\tau_p}} \frac{n_i^2}{N_D} \tag{3.2}$$

The diffusion component is the main one at room temperature for semiconductors with high n_i like germanium ($n_i = 2.4*10^{13}$ cm^{-3}) while for silicon ($n_i = 1.4*10^{10}$ cm^{-3}) it is less important. For a fully depleted detector, and so without the regions with free carriers, the diffusion current is negligible.
The main temperature dependence of the diffusion current is through n_i^2 and so one can approximately write:

$$J_s \approx e^{\frac{-E_g}{kT}} \qquad (3.3)$$

Generation current.
The current generated in the depletion layer may be expressed by the simple formula:

$$J_g = qgW \qquad (3.4)$$

where q is the unit charge, W is the thickness of the depletion layer and g is the generation-recombination rate, which is derived from the Shockley-Read-Hall model. Having in mind that W is proportional to the square root of the applied voltage, one could try to estimate the depletion voltage from the I-V curve (the current versus the applied voltage or rather its square root) for the case where the generation current is the most significant one. To get the temperature dependence of the generation current in a very rough approximation one can assume that the generation-recombination rate is proportional to n_i and so:

$$J_g \approx e^{\frac{-E_g}{2kT}} \qquad (3.5)$$

which gives roughly an increase by a factor of two for a temperature increase of 8 K. Therefore any measurement of the leakage current should be coupled with the precise measurement of the temperature, and in general executed in controlled temperature conditions close to 20°C.

Surface current.
The currents generated in the region of the detector surface are of very complex origin. The main source of these currents are the ionic charges on and outside the semiconductor surface which affect the electric field distribution close to the surface. The absolute value of the surface leakage current is not analytically well determined and strongly depends on the technology and on the geometry of

the detector. To limit the surface current in real detectors several protecting guard rings are usually added.

Description of the setup and proposed measurements

The lab setup includes: probe station, microscope, the instrument for C-V and I-V measurements (which includes a power supply, a capacitance meter and a current meter) and different silicon detectors. An important part of the measurements is to get a minimum of experience using a probe station. For this reason some of the detectors are naked (without bonding wires) and so need to be mounted and measured directly on the probe station making contacts to the instrument with a probe needle. The proposed plan of the measurements is the following:

- Observation of different type of the silicon detectors (diodes, strip detectors, pad detectors, drift detectors) under the microscope. Attention should be given to guard ring structure as a common feature of all kinds of detector design,
- Acquiring of minimum experience in using the probe station as a basic instrument for handling the naked silicon structures during the measurements,
- C-V measurement of simple diodes. According to the previous discussion find the depletion voltage of the detector plotting $1/C^2$ versus the applied voltage. Notice and think out the capacitance behavior during the measurement under daylight and for covered detector,

I-V measurements. Perform the measurements and plot the curves of the current versus voltage (and square root of the voltage). Repeat the same measurement with and without biasing the guard ring structure and measuring separately diode current and guard ring current (mainly surface current). Observe again the current with and without daylight

DIRECT OBSERVATION OF THE SIGNALS GENERATED BY IONIZING RADIATION IN SILICON DETECTORS

In order to directly observe the current pulses from silicon detectors one needs to read out the signal with the amplifiers with rise time much shorter than charge collection time in the detector (current sensitive amplifiers), in order to amplify but not distort the current pulse induced by the motion of electrons and holes in the detector.

Here we give a theoretical description for the following aspects:
1. Theory of signals generated by moving charges in a detector (Ramo theorem),

2. Example of current pulse calculation for a simple diode,
3. Basics about current sensitive amplifiers.

Other theoretical aspects are covered by different labs/lectures and so they are not discussed here. After that we pass to the experimental part of the lab which includes:
4. A description of the lab setup,
5. The proposed measurements/observations to be performed during the lab.

Theory of signals generated by moving charge in the detector

Formulation of Ramo theorem

Let us consider a volume V, inside which M electrodes kept at fixed potentials F_{ak}, where $k=1...M$, surround a region with uniform dielectric constant e (fig. 2).

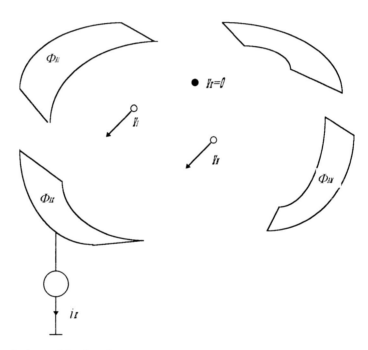

Figure 2. *General scheme of the detection system.*

We define the potentials and electric fields coming only from the supplied voltages at the electrodes as:

$$F_a(r) = \sum_k F_{ak} f_k(r)$$

$$E_a(r) = -\sum_k F_{ak}\, grad f_k(r) \qquad (1.1)$$

where $f_k(r)$ is the potential at point r from the k-electrode, with the assumption that this electrode is kept at fixed potential $F_{ak}=1$, all the other electrodes are grounded, and there is no charge in the detector volume. It is clear that $f_k(r)$ depends only on the geometry of the detection system.

We define the charge density distribution in the detector volume as:

$$r(r,t) = \sum_{i=1}^{N} q_i\, d(r-r_i) \qquad (1.2)$$

Then the total potentials and fields in the volume may be written as a superposition of the ones from the potentials on the electrodes (F_a) and from the charge within the detector volume (F_r).

$$F(r) = F_a(r) + F_r(r)$$

$$E(r) = E_a(r) + E_r(r) \qquad (1.3)$$

Knowing the charge motion (i.e. q_i, $r_i(t)$, $v_i(t)$, where $i=1...N$) one can write the Ramo theorem.

The current induced on the electrode k by the known motion of the charge is expressed as:

$$\boxed{i_k(t) = -\sum_{i=1}^{N} q_i\, grad f_k(r_i) \bullet v_i(t)} \qquad (1.4)$$

Here we would like to stress again that the field vector $f_k(r)$ is of pure geometrical nature while the velocity $v_i(t)$ depends on the electrode's potential as well as on the presence of other charges inside the detector volume. In the Ramo theorem it is assumed to be known.

Proof of Ramo theorem

The Gauss law:

$$div[ee_0 E] = r \qquad (1.5)$$

for the potentials F_r, F_a, may be written as:

$$-\text{div}[ee_0 \, \mathbf{grad}F_r(\mathbf{r},t)] = r(\mathbf{r},t) \quad (1.6)$$

with boundary conditions: $F_{rk}=0$, $k=1...M$ \hfill (1.7)
and for the potential F_a:

$$-\text{div}[ee_0 \, \mathbf{grad}F_a(\mathbf{r},t)] = 0 \quad (1.8)$$

with fixed potentials F_{ak} on the electrodes.
Using the Green theorem for the potentials F_r, F_a one can write:

$$\int_V \{F_a(\mathbf{r},t)\text{div}[\mathbf{grad}F_r(\mathbf{r},t)] - F_r(\mathbf{r},t)\text{div}[\mathbf{grad}F_a(\mathbf{r},t)]\} dV =$$
$$= \int_S [F_a(\mathbf{r}_s) \, \mathbf{grad}F_r(\mathbf{r}_s,t) - F_r(\mathbf{r}_s) \, \mathbf{grad}F_a(\mathbf{r}_s,t)] \bullet d\mathbf{S} \quad (1.9)$$

where V is the volume of the region considered, without the M electrodes, S is the area of the electrodes, and $\mathbf{r}_s$ is the vector on the electrodes surface.
Substituting equations (1.6), (1.7) and (1.8) into (1.9) one can get:

$$-\int_V F_a(\mathbf{r},t) \, r(\mathbf{r},t) \, dV = \int_S F_a(\mathbf{r}_s)ee_0 \, \mathbf{grad}F_r(\mathbf{r}_s,t) \bullet d\mathbf{S} \quad (1.10)$$

Replacing $r(\mathbf{r},t)$ by its distribution (1.3) we may write:

$$-\sum_{i=1}^{N} q_i F_a(\mathbf{r}_i,t) = \sum_{k=1}^{M} F_{ak}(t) Q_{rk}(t) \quad (1.11)$$

where $Q_{rk}(t)$ is the charge induced in the electrode k by the distribution of all N charges, and may be written as:

$$Q_{rk}(t) = \int_{Sk} e(\mathbf{r}_s)e_0 \, \mathbf{grad}F_r(\mathbf{r}_s,t) \bullet d\mathbf{S} \quad (1.12)$$

Now introducing the potentials from (1.1) to the equation (1.11) one can get:

$$\sum_{k=1}^{M} F_{ak}(t) [Q_{rk}(t) + \sum_{i=1}^{N} q_i f_k(\mathbf{r}_i)] = 0 \quad (1.13)$$

To fulfill this expression, the part in the parenthesis must always vanish and so:

$$Q_{rk}(t) = -\sum_{i=1}^{N} q_i f_k(r_i) \qquad (1.14)$$

Differentiating this equation with respect to time ($i = \dfrac{dQ}{dt}$), one can finally get the current (1.4):

$$i_k(t) = -\sum_{i=1}^{N} q_i \, \mathbf{grad}\, f_k(r_i) \bullet v_i(t)$$

Calculation of the current pulse for a diode

Let us consider a reverse polarized (by voltage V) silicon diode of thickness D, and assume that there are N electric charges q_i moving with velocities v_i to the appropriate electrodes. According to the Ramo theorem the instant current induced on the k-th anode is:

$$i_k(t) = -\sum_{i=1}^{N} q_i \, \mathbf{grad}\, f_k(r_i) \bullet v_i(t) \qquad (2.1)$$

Assuming that the size of the diode is much larger then its thickness we can use a one dimensional approximation and replace the product in the Ramo theorem by a simple multiplication. The unit potential $f_k(r)$ in this case may be easily obtained (from Gauss law) as:

$$f(x) = 1 - \frac{x}{D} \qquad (2.2)$$

where x is the distance from the anode. Then its gradient:

$$\mathbf{grad}\, f_k(r) = \frac{df(x)}{dx} = -\frac{1}{D} \qquad (2.3)$$

and finally we can write the expression for the current as:

$$i(t) = \sum_{i=1}^{N} q_i \frac{1}{D} v_i(t) \qquad (2.4)$$

At this point one could think of what will happen if we calculate the current on the cathode, instead of on the anode. Looking at (2.1) we see that the only factor that may change in this case is $f_k(r)$, and in fact one can easily get that:

$$f_{cathode}(x) = \frac{x}{D} - 1 \qquad (2.5)$$

and so:

$$i_{cathode}(t) = -\sum_{i=1}^{N} q_i \frac{1}{D} v_i(t) \qquad (2.6)$$

We see that for a simple diode the current signal on the *n-side* has exactly the same shape as on the *p-side*, but opposite sign. This is not the case for strip or pixel detectors, where there is not such a simple symmetry between the two sides of the detector.

In order to know the full time dependence of the current, one needs to calculate the velocities for all charges. To do it, let us make the additional assumption that the detector under exam is the standard p^+-n diode, and that the charge generation is point like, giving a total charge $-Q$ for the electrons and Q for the holes.

The velocity of the electrons and of the holes may be written as:

$$\frac{dx}{dt} = v(t) = \pm m_{e,h} E(t) \qquad (2.7)$$

where m is the mobility and $E(t)$ is the electric field.
Solving the Poisson equation for this simple case, one gets:

$$E(x) = \frac{2V}{D^2} (x - D) \qquad (2.8)$$

From equations (2.7) and (2.8) we can calculate the velocity:

$$v(t) = \frac{1}{\tau} (D - x_o) \exp[\pm \frac{1}{\tau} t] \qquad (2.9)$$

where $t = \dfrac{D^2}{2\mu V}$

Replacing this equation in the Ramo formula we finally get for electrons and holes:

$$i_e(t) = \frac{2QV}{D^3} m_e (D - x_o) \exp[- \frac{2V}{D^2} m_e t] \qquad (2.10)$$

$$i_d(t) = \frac{2QV}{D^3} m_d (D - x_o) \exp[+ \frac{2V}{D^2} m_d t] \qquad (2.11)$$

Basics about current amplifiers

As readout electronics for semiconductor detectors in high energy physics three configurations of amplifiers a), b) and c), are generally used:
a) charge sensitive configuration; characterized by very good S/N performance, output signal proportional to the charge deposited in the detector, limited bandwidth,

b) current sensitive configuration; characterized by an output signal proportional to the current from the detector (so the shape of the detector signal is conserved), high bandwidth, noise performance worse than for charge sensitive preamplifier,

c) voltage sensitive configuration; which is used for charge measurement, can be also very fast, but has a great disadvantage in the requirement of a stable input capacitance (detector capacitance), which is usually not the case for silicon detectors (and so this configuration is rarely used)

All these configurations are usually used with negative feedback loops at input stages. The only exception is the current sensitive configuration, which is also used with a grounded base connection. In this configuration the input transistor's base is kept at fixed potential, while the input signal is connected to the emitter. As mentioned above the purpose of the current sensitive configuration is to preserve the shape of the detector signal. It is seen that for this configuration the amplifier input impedance is of resistive nature. Using the criterion of maximum power transmitted to the amplifier (with input resistance R), which might be written as:

$$\frac{dP(R)}{dR} = 0,$$

one can get (for exponentially decaying detector pulses) an approximate condition for R:

$$RC \approx \tau,$$

where C is the total detector capacitance and τ is the characteristic time of the detector pulse. Having in mind standard values of detector capacitance $C \sim$ (1-100) pF and typical collection times in silicon detectors $\tau \sim$ (1-20) ns, one would usually need an amplifier with input resistance $R \sim$ (10-1000)Ohm. Such low values of input resistance are accessible almost only with a grounded base connection. This is one of the reasons why the grounded base configuration is often used. In addition, this configuration has very good high frequency properties, due to substantial reduction of Miller capacitance.

Description of the lab setup

The lab setup consists basically of the following components:
- silicon diodes, with different doping concentration. They allow shining with infrared light on both sides (is it obvious ?). Some of the diodes have been irradiated with hadrons,
- two channels of fast (T_{rise}=600ps) current sensitive preamplifiers (biased at +12V, -6V, 0V), based on a grounded base configuration, followed by operational amplifiers (biased at +12V, 0V),
- red LED diode with trigger input and output with optical fiber,
- pulser,
- oscilloscope.

In order to understand better the relation between signals on p-side and n-side of detector under different irradiation conditions, the detector is read out from both sides. In Fig. 3 the scheme of the detector-readout electronics connections is shown. Since both sides are read out at the same time, separate

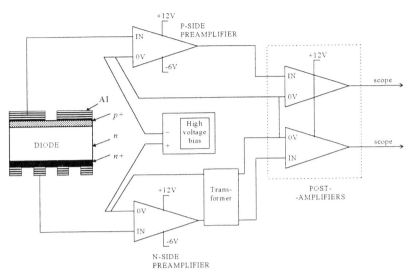

Figure 3. *Scheme of the double sided readout*

and floating bias lines for each amplifier are used. As it is seen from the scheme, to match DC levels between n-side readout and postamplifier, a transformer was used. It is also well seen that the detector is biased through the preamplifiers.

WARNING!!! Don't switch the bias ON without confirmation from the lab attendants.

Proposed measurements/observations

During the lab the following observations and measurements should be performed:
- Understanding different hardware issues: the way the signal from the detector is read out, one side readout, double side readout, what is the reason for having floating power supply, difference in readout scheme for DC-coupled and AC-coupled detectors.

- Observation of signals from p-side and n-side of the detector: basic properties of the signal like shape and amplitude, how they are connected to the physical processes inside the detector, relation between signals on p-side and n-side of the detector, what is the role of electrons and holes in signal formation, what would be the difference in pointed above topics for strip or pixel detectors, ...
- Observation of signals shining the infrared light on the p-side and n-side of the detector: are the signals different for different lighting conditions, if so then why, how they change for different bias voltage of the detector, ...
- Estimation of depletion voltage for different diodes: how to do it, what is the physical meaning of depletion voltage, ...
- Comparison between signals from different diodes under different radiation conditions (lighting on p-side or n-side) and different detector bias voltage, what is the difference between diodes, how can one check if the diode was highly irradiated (passed inversion), what are the effects of radiation on silicon detectors, how they can be measured.

ENERGY SPECTRA

Among the main reasons for the success of silicon detectors are the fact that they produce a signal that is proportional to the energy deposited in the detector volume, which is generally referred to as "being linear", and the fact that the energy needed to produce a charge pair is very small (3.6 eV in silicon), which translates in a very good ratio signal/energy deposit. These characteristics have very important implications on possible applications and on the performance of the detectors. One should remember, though, that the signal released by a traversing particle in a thin silicon detector is generally very small, and even smaller for X-rays, so the possibility to profit of these features will depend strongly on the noise of the readout electronics connected to it.

ENERGY DEPOSITION and RESOLUTION

The energy resolution σ^2 which can be achieved is given by the quadratic sum of two factors: the noise of the readout chain[1] and the statistical fluctuations in the generation of charge carriers. Two main cases should be considered: low-energy regime, in which essentially all of the particle energy is deposited in the

---

[1] It is important to remember that for small signals the main "noise" source could be downstream of the frontend electronics, i.e. in the digitisation phase. The width of the lowest significant bit in the Analog-to-digital conversion introduces an error (quantization error) which is often comparable to the noise in the analog chain before it.

detector volume, and high-energy regime, in which the particle traverses the detector losing a small fraction of its energy.

Low Energy

σ is in this case simply the square root of the number of generated charges, i.e. E/δ, where E is the energy of the detected particle and δ is the average energy needed for the creation of an electron-hole pair. In fact, this naive expression must be multiplied by a further factor F, called Fano factor, which depends on the specific material and takes into account the fact that the charges are not generated in a fully uncorrelated way. The value of the Fano factor is between 0 and 1.

For charged particles most of the energy loss is due to ionization. For low-energy photons (X-rays) most of the energy loss is due to the photoelectric effect, so if a photon interacts it loses its full energy, giving a gaussian-shaped signal. For higher energies (for silicon starting in the tens of KeV) the Compton effect becomes important, and gives rise to a shoulder of signals lower than the peak, for which the secondary photon has escaped detection (this is very likely, since the detector volume is small, and the photon interaction probability is consequently also small).

The spectrum observed looks therefore like in the following Figure 4 :

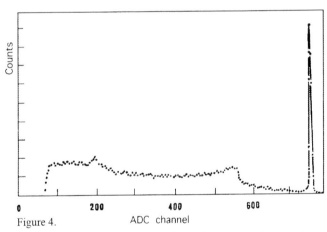

Figure 4.   ADC channel

At even higher energies, when the threshold for electron-positron pair creation is reached (about 1 MeV), this becomes the dominant process. In this lab, we observe X-rays of energies from the 13.4 KeV of the K-line of Rb to the 60

KeV of the Am X-rays. Only for this last one we observe the Compton shoulder appear.

High Energy

The average energy loss of relativistic particles other than electrons (high-energy electrons lose most of their energy through bremsstrahlung) is described by the Bethe-Bloch formula:

$$-\frac{dE}{dx} = Kz^2 \frac{Z}{A} \frac{1}{\beta^2} \left[\frac{1}{2} \ln \frac{2m_e c^2 \beta\gamma^2 T_{max}}{I^2} - \beta^2 - \frac{\delta}{2} \right]$$

Where
z is the charge of the incident particle,
Z is the atomic number of the medium,
A is the atomic mass of the medium,
β and γ are the kinematic variables of the incident particle,
$m_e c^2$ is the electron mass,
I is the mean excitation energy,
T_{max} is the max. kinetic energy which can be transferred to a free electron in one collision,
δ is a density correction factor,

The resulting curve is shown below (dashed curve). Its characteristic features are readily visible: for low energies (β less than 1) the energy loss decreases as β^2, and after a minimum it grows slowly. Particles with energy above the so-called $1/\beta^2$ region are called minimum ionizing particles (MIPs). A significant fraction of the energy loss goes to knock-on electrons of relatively high energy, which escape from a thin detector. Therefore the energy deposited is in general lower than the energy lost. The energy deposited is expressed by the restricted energy loss, for which the largest energy transfers are excluded. In the plot one can compare the energy loss in silicon (dashed curve) and the restricted one (full line): it is clear that the rise at high energies is fully due to the high energy transfers, and is not visible with a thin silicon detector.

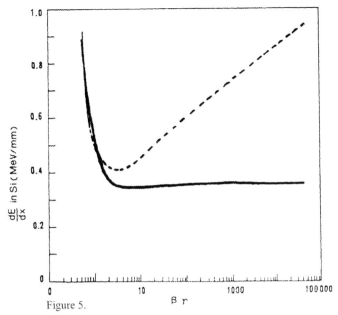

Figure 5.

For a MIP, the statistical fluctuations in the energy deposited in a thin silicon detector follow the distribution shown below.

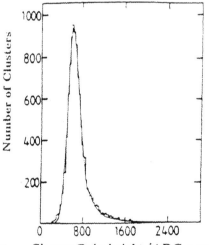

Figure 6.   Cluster Pulseheight [ADC counts]

This distribution is usually referred to as a Landau curve, but it is a bit wider, and physically much more complex, than the one calculated from the Landau formula. In our lab, you can observe the energy deposition of β particles from a

^{107}Ru source (3.54 MeV), which is very close to that of a MIP, and exhibits the typical Landau-type behavior.

APPLICATIONS

TRACKING

One of the most important applications of the linearity of silicon detectors lies in the possibility of improving the spatial resolution of detectors which are segmented to a pitch comparable to the spatial spread of the collected charge. In this way, the charge is shared among several electrodes, and impact point is reconstructed as the centroid of the signals. This method can be very powerful when the noise is low: experimentally resolutions down to few microns have been achieved even in large experimental setups.

In addition, when double-sided microstrip detectors are used, the measurement of the deposited charge allows one to correctly match the signals on the two sides when more than one particle hit the detector at the same time, thus removing ambiguities.

PARTICLE IDENTIFICATION

From the energy deposition plots below, it is clear that once the momentum of a particle is known, and it is below the minimum of ionization, the measurement of the dE/dx allows a measurement of the particle velocity. Therefore, known both momentum and velocity, one can derive the particle mass, and thus identify it. In other words, a system of silicon detectors in a magnetic field can provide a full stand-alone spectrometer, measuring the trajectory of the particle, the momentum and the mass. The following figure shows the momentum-dE/dx correlation for particles with momenta in the few-hundred MeV region.

It is clear, though, that an individual energy deposition follows the Landau curve shown before, so an individual measurement will fall anywhere on the curve, giving a very rough measurement of the *average* energy loss of a particle of that momentum. What is normally done is to use a few measurements (which comes naturally if the silicon detectors form a tracking system, in which each particle crosses several detector layers) and perform a "truncated mean" of them, i.e. take the average of the lowest ones. This method improves the result very rapidly, and already with just three or four measurements one gets good resolution, as shown in the following plot.

From the Bethe-Bloch formula we see that the energy deposition is proportional to the square of the charge of the crossing particle, so it is straightforward to identify particles with charge more than one (ions). This method is used, for example, for experimental setups which fly on satellites or will fly on the space shuttle. They use silicon detectors to study the composition of primary cosmic rays, using for example (NINA experiment) a simple telescope of silicon microstrip detectors to get the charge, the range (and thus the energy), and the angle of incidence of nuclei in the cosmic rays. The following figure shows an example of the response of a silicon detector to ions of different charge (picture courtesy of G. Raciti).

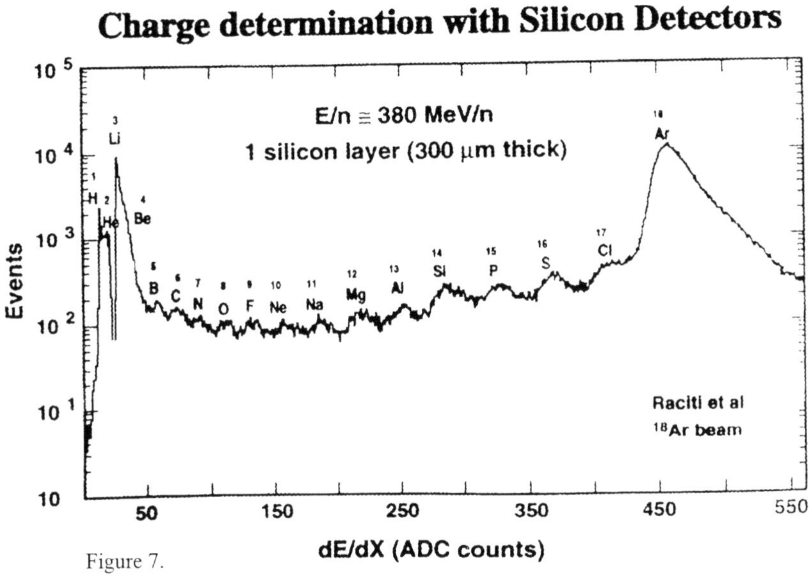

Charge determination with Silicon Detectors

Figure 7.

SPECTROSCOPY

When one has to measure an x-ray line, the quality factors are *resolution* and *statistics*, both essential to the determination of the peak.
The resolution depends on the factors observed before: the noise of the readout system
and the statistics of charge carriers. The second factor gives a small advantage to Germanium relative to silicon (2.96 eV/pair against 3.62), but the excellent low-noise electronics which has been developed for silicon detectors can offset this (and the electronics *you* are using *does*).

The reason for which Ge, although rather cumbersome to use, since it needs to be cooled to
77 K, is still the material normally used comes from statistics: the cross section for the photoelectric effect is proportional to the charge of the medium nuclei to the power 4 to 5, so Ge is 40 times more efficient than silicon in detecting low-energy photons. In addition, it is easier to fabricate Ge detectors of large volume, again important for efficiency.
One could think of offsetting these problems by simply increasing the time of the measurement. This is not only annoying, but often impossible, since the measurement has to be performed in a time short compared with mean life of the source, which for cases of practical interest can be very short.

In this lab you will directly measure the spectrum of X-rays from ^{241}Am, which have an energy of 60 keV. You will also observe, from the memorized spectra, other measurements which were done at CERN with the identical setup. These come from a special source which is often used to calibrate a system, since it provides several lines of known energy in the range from 13 to 60 keV. The source we used is rotary one, which allows to irradiate, with the α from a ^{241}Am source, several target materials. The alpha particles collide with nuclei of the material and excite them to unstable states from which they decay via the emission of X-rays. Through the measurement of the position of these peaks, we can derive a response function of our system, so that its energy response is known accurately to later measure unknown sources.
The width of the peaks is not their natural width, but reflects the noise of our readout+detector system: you can easily appreciate the excellent performance of it! The memorized spectra are reproduced in the following Figure 8.

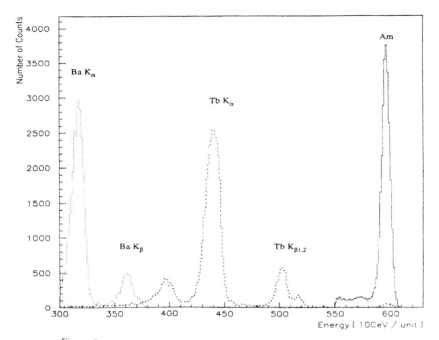

Figure 8.

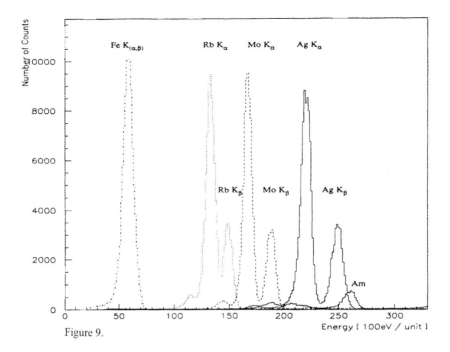

Figure 9.

A last comment on the setup you are using: to measure the X-rays from the Americium source, we stop in a thin copper foil the α particles emitted by the source, which otherwise would saturate our readout system (remember that the system has been designed for MIPs, which on average deposit in 300 μm of silicon about 87 keV, so the signal from the 5-MeV α is 50 times our typical signal).

If you remove the copper foil and observe the signals on the scope, you will notice that there is NO difference switching ON or OFF the bias of the detector. The reason is that the α penetrates very little into the surface of the detector, and the built-in potential of the p-n junction is enough to collect most of the deposited charge!

Amplifiers and Electronics

G. Hall and R. G. Payne

Blackett Laboratory, Imperial College, London SW7 2AZ

CONTENTS

1. Introduction

2. Background material
2.1 Amplifier characteristics
2.2 The MSD2
2.3 Bipolar transistor circuits
2.4 Operational amplifiers
2.5 Analysis of the MSD2 circuit

3. Experimental programme
3.1 Measuring the amplifier characteristics
3.2 Construction of an amplifier

References

Appendices
I. Calibration using test pulses
II. Bipolar transistor operation
III. A fast current sensitive preamplifier ...
IV. Transistor data sheets
V. Technical details

1. INTRODUCTION

This experiment is intended to provide an introduction to the electronics used by particle physicists for amplifying signals from detectors. It is deliberately simple, recognising that this subject will be new to many. It will concentrate on the electronics, rather than the detectors, but will emphasise a non-mathematical approach to the circuits and concentrate on experiments, hopefully guided by some intuition. The object is to try to gain enough confidence to apply this knowledge in practical situations and perhaps to stimulate further interest in going beyond this simple start.

Not everyone who uses a particle detector wishes, or needs, to become an electronic expert. However to observe particle signals it is usually necessary to use an amplifying circuit. Therefore it is essential to understand the influence of the amplifier on the observations. (This also holds for other parts of the measuring system, such as the oscilloscope, which are employed.) Many physicists perhaps do not feel as confident in electronics as they do in other areas, such as computing. Therefore, here we intend to use an amplifier and observe its behaviour, then to try to interpret it and understand how best to use it. Finally, we shall build some amplifying elements to try to enhance the performance and build on what we have learnt.

We are trying to cater to widely varying levels of background knowledge, so those with prior experience may skip those parts they already know. If you know everything, maybe you don't need to do the experiment and should try one of the others! For those who want to go further from what they learn here, the classic text on electronics is the book by Horowitz and Hill [1]. This is highly recommended - in fact, so much that if you only study one book on electronics in your entire life this should be the one. Even a few chapters will pay dividends!

2. BACKGROUND MATERIAL

2.1 Amplifier characteristics

From the point of view of the potential user, there are a few characteristics that need to be known when starting work with a new amplifier. These mainly concern:

> the amplifier type - current, charge or voltage
> input and output impedance
> gain and linearity
> frequency response
> noise level or signal to noise ratio.

The noise level is a very important parameter but is deliberately placed last on the list above. Before estimating the noise in the system it will be important to understand exactly what is being measured. The other characteristics listed can influence the result.

Amplifier type

Before going too far, it is wise to have figured out what quantity is being amplified. The output signal may be measured in volts but it may well depend on input current or charge and not an applied voltage.

Input and output impedance

The impedance of the amplifier as seen by a signal is vitally important - this is the input impedance. In many situations amplifiers are designed so that they look like very low impedances to ground to measure currents, which is generally achieved using the feedback techniques explained in section 2.4. The precise value could be very important in determining how signal current pulses are shared between different paths. Not all signal paths might lead into an amplifier. Conversely, a voltage amplifier is usually designed with a high input impedance to reduce the influence of any resistance associated with the signal source.

The output impedance is important when observing the results of the amplification, either directly on an oscilloscope or when coupling the amplifier into subsequent electronics. Many amplifiers are not designed to drive the 50Ω loads typical of standard cables used for signal transmission. In contrast an observation terminated in 50Ω can sometimes look very different from one terminated in 1MΩ because of different RC time constant circuits created by the termination which integrate the output signal differently.

Gain and linearity

For our purposes, this is a measure of the ratio of output and input signals. It may not always simply be a non-dimensional number, for example if the amplifier converts currents to voltages. It will usually be necessary to know what the gain of the system is so that the magnitude of the observed pulse produced by a given input signal can be used to calibrate the system.

The a.c and d.c gains may be different. That is, a circuit may amplify fast pulses differently than it does slowly varying or constant current or voltage levels. This has some practical importance in case there are non-zero levels associated with your signal.

The amplifier may also have some limit to the range of signals which it can handle. Most of the amplifiers in common use are linear, which means that there is a linear relation between the magnitudes of input and output signals. Signals outside the linear range may be distorted or truncated. It is obviously important to be prepared for this.

Frequency response

Although it may not be immediately apparent, the bandwidth of the amplifier being evaluated has important consequences. If it is required to measure pulses with risetimes in the few nanosecond range it is clearly necessary to have an amplifier with good performance at frequencies well into the MHz range. This is a challenge and most amplifiers display reduced gain at high frequencies, usually in a predictable fashion. However, to study signal pulses which have durations of a few tens of nanoseconds it may not always be necessary to use the entire frequency range of the signal and, in fact, bandwidth limitation plays an important role in optimising signal to noise performance.

Despite its importance, this aspect of the amplifier performance is not being given much quantitative emphasis in this experiment. We are relying on the amplifier designers to have achieved this for us (see Appendix III). However it will be worth observing the time development of the signals under different experimental conditions. It will be discussed in the lectures on signal processing as well as in another experiment.

Noise level

Ultimately, this is a very important quantity since it determines the smallest observable signal. It often needs care to measure accurately since many of the amplifiers now in use are intended to measure small signals, eg of a few femtocoulombs.

2.2 The MSD2

The experiment is based around a widely used amplifier, the MSD2, designed some years ago [2] for the readout of silicon microstrip detectors at CERN but which has been employed on a large scale also for read out of gaseous and other types of detector. Although it does not use the most up to date technology, it is still a very useful amplifier for many present applications and it is an excellent device to illustrate some of the principles of amplifier use.

The MSD2 design is based on bipolar transistors and so an introduction is given to some aspects of their behaviour. Not all circuits use bipolar transistors, but it is very easy to extend an understanding of bipolar circuits to other types of transistor, and also to integrated amplifiers or op-amps.

2.3 Bipolar transistor circuits

The bipolar transistor is a three terminal device of two possible types - npn or pnp, referring to the doping of the three regions of the device. There is a very good description of the useful operating principles in Horowitz and Hill [1]. To a first approximation, it is not necessary to worry about how the transistor physically works, just to know some simple rules. Of course the rules have some justification [see Appendix II].

To proceed with the first part of the experiment, it is not essential to have much familiarity with the internal workings of the MSD2 so, if you wish, you can defer a detailed reading of the following section until you need to understand a bit more. By the end of the experiment it will be necessary to have read this, because some of the interpretation of the results will only be possible with this information.

Fig 2.1. Circuit schematics of npn and pnp transistors

The circuit diagram of an npn and pnp transistor is shown in Fig. 2.1. For pnp transistors the bias polarities are inverted. The arrow in each case indicates the direction of conventional current flow in the device. The terminals are labelled emitter, collector and base. The rules for npn transistors are:

1. Collector voltage is more positive than emitter
2. the base emitter junction behaves like a conducting diode, while the base collector diode is reverse biased
3. Base current is proportional to collector current:
   $I_C = \beta I_B$ where β (or h_{fe}) is known as the current gain.

In typical transistors $\beta \sim 100$, although it depends on operating conditions and can be larger or smaller by orders of magnitude for different types of transistor. It is therefore not a good parameter to base a design around. However, since β is generally large and, by current conservation,

$I_B + I_C = I_E$

then since $I_C \gg I_B$, $I_C \approx I_E$

In normal operation, when the transistor is conducting, rule 2 implies that

$$V_B = V_E + V_{BE} \approx V_E + 0.7V$$

This is the way in which the transistor is usually operating in an amplifier circuit. Alternatively, the transistor can be operated as a switch. For example, if $V_B \approx V_E$ then current will not be conducted and the transistor is off.

There are two very common circuit configurations which employ the bipolar transistor: the emitter follower and the common emitter amplifier.

The emitter follower

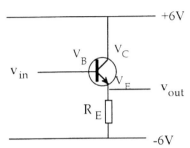

Fig.2.2. The emitter follower circuit configuration

The input to the circuit is at the base of the transistor while the output is taken from the emitter.

First, examine the quiescent or d.c. conditions, assuming the voltages in the illustration

$$V_E \approx V_B - 0.7V$$
$$I_E = (V_E + 6V)/R_E \approx (V_B + 5.3V)/R_E$$

So the current in the transistor is controlled by the base voltage, which will be chosen for that purpose.

Now let's look at the response to a small signal,

$v_{in} = v_b = \Delta V_B$ (the notation $v = \Delta V$ is a common convention)

$$= \beta(V_E + 0.7V) = \beta V_E$$
$$= v_e = v_{out}$$

This is an amplifier with a gain of 1, since $v_{in} = v_{out}$, which may not seem very useful. However, if we estimate the input impedance of the circuit

$$R_{in} = v_{in}/i_{in}$$
$$i_{in} = \beta I_B = \beta I_C/\beta = i_c/\beta$$

Since $i_c = v_e/R_E$
$$i_{in} = v_e/\beta R_E = v_{out}/\beta R_E = v_{in}/\beta R_E$$
and $R_{in} = \beta R_E$

so, even allowing for a range of β values, this is a circuit with a high input impedance.

In some circuits the emitter resistance R_E is omitted which would appear to cause problems. In fact there is an intrinsic dynamic resistance associated with the base emitter junction which can be estimated to be (see Appendix II)

$$r_e = kT/qI_C \approx 25/I_C \; (½)$$

where I_C is expressed in units of mA. Thus in all the expressions above the resistor R_E is replaced by $R_E + r_e$ and the input impedance is more accurately expressed as

$$R_{in} = \beta (R_E + r_e)$$

The output impedance can also be estimated. The base is connected to the signal source v_s through an associated resistance R_S.

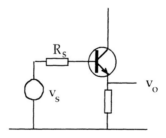

Fig. 2.3. Estimation of emitter follower output impedance

Then, when no signal is applied and a test voltage v_o is applied at the emitter terminal, we find

$$v_b = i_b R_s + i_e r_e$$
$$i_c = \beta i_b$$
$$v_o = v_b$$

so $R_{out} = v_o/i_e \approx R_s/\beta + r_e$

In the normal case, the output impedance is quite small.

The function of the emitter follower now is more clear. It has unity gain, high input and low output impedance which makes it an ideal buffer circuit for many applications.

Common emitter amplifier

A typical circuit is shown in fig. 2.4. The input to the circuit is at the base of the transistor while the output is taken from the collector.

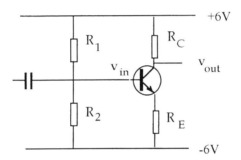

Fig. 2.4. A common emitter amplifier

First, examine the quiescent or d.c. conditions, assuming the voltages in the illustration

$$V_C \approx 6V - I_C R_C$$
$$V_E \approx -6V + I_E R_E \approx -6V + I_C R_E$$

The current in the transistor is again controlled by the base voltage selected.

The small signal response is,
$$v_{out} = v_c = -i_c R_C$$
$$v_{in} = v_b = v_e = i_c R_E$$
so
$$v_{out}/v_{in} = -R_C/R_E$$

This is an amplifier with voltage gain. The input impedance is easily calculated by referring to the emitter follower. An a.c. signal sees the bias resistor network in parallel with the input transistor, thus
$$R_{in} = R_1 \parallel R_2 \parallel R_E/\beta$$

while at the output we see a reverse biased diode (the base-collector junction) in parallel with the resistor R_C. Thus typically
$$R_{out} \approx R_C$$

and the output impedance may be high and therefore not suitable for driving all loads.

Amplifiers with feedback

The bipolar circuits discussed above are voltage amplifiers. However, using feedback, current or charge amplifiers can be constructed. To see this it is convenient to introduce the concept of the operational amplifier which is a building block used in many circuits whose properties can be exploited without detailed knowledge of its internal structure. Many circuits can be approximated as op-amps to interpret their behaviour.

2.4 Operational amplifiers

The ideal op. amp is a very high gain voltage amplifier with a single output but dual, differential inputs. The bare amplifier gain, A, is referred to as the open loop gain. The voltage at the output is the product of the gain and the voltage difference between the inputs.

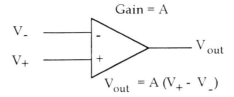

Fig. 2.5. An operational amplifier schematic

In reality the op-amp is never used in this simple way but almost always in a feedback mode where a fraction of the output voltage is fed back to the inverting (v_-) input and subtracted from the input signal. This is called negative feedback and explains why the inputs are labelled as + and -: the output goes positive when the non-inverting input (v_+) goes more positive than the inverting input(v_-), and vice-versa. The gain of the amplifier with the feedback network connected is referred to as the closed loop gain.

The open loop gain of op-amps is made so high and they are designed to draw such small input currents that circuits with them can be analysed with two simple rules:

i) the output attempts to set itself so that the voltage difference between the inputs is zero
ii) the inputs draw no current

Applying these rules to the circuit below:

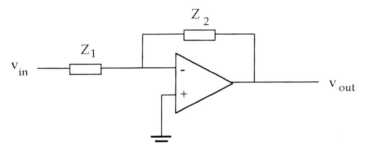

Fig. 2.6. An inverting amplifier

$$v_- = v_+ = 0$$
$$(v_{in} - v_-) / (v_- - v_{out}) = Z_1/Z_2 \quad \text{-->} \quad v_{out}/v_{in} = -Z_2/Z_1$$

which could be an alternative to the common emitter amplifier in some cases. The impedances in the networks above could be resistors, capacitors or combinations of both which can produce interesting and useful results.

In this case, as in many others, the non-inverting terminal is held at ground potential forcing the inverting terminal to ground potential also. This is then referred to as a *virtual ground*. It should be clear that this presents an ideal arrangement for a current amplifier since the input impedance is practically zero and this is often exploited, as in the following example.

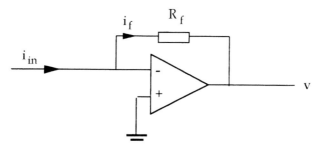

Fig.2.7. A current sensitive amplifier

By omitting the input resistance of the inverting voltage amplifier and making Z_2 a resistor R_f the gain of the circuit is most usefully written

$$v_{out}/i_{in} = -R_f$$

which provides current to voltage conversion. You may also notice that the current in the feedback loop is identical to the input current (notice that there is no current flowing into the amplifier) so

$$i_f = i_{in}$$

which provides another useful way of analysing many feedback circuits.

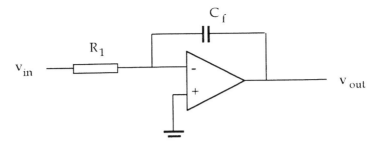

Fig. 2.8. An integrator

If the feedback element is a capacitor C_f and the Z_1 is a resistor (fig. 2.8) the circuit becomes an integrator with $i_{in} = v_{in}/R_1$ and

$$v_{out} = -(1/R_1 C_f) \int v_{in} \cdot dt$$

The integrator is widely used without R_1 (which would be a noise source) as a charge sensitive preamplifier. Working in the frequency domain we therefore write

$$v_{out}(\omega) = -i_{in}/j\omega C_f \quad \text{since} \quad Z_2 = 1/j\omega C_f$$

which may be recognisable as the Fourier transform of a step function in time, or, if not,

$$v_{out}(t) = -Q/C_f \quad \text{in the time domain, for a } \delta \text{ function input.}$$

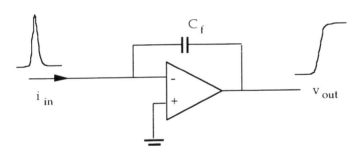

Fig. 2.9 A charge sensitive amplifier

The operational amplifier concept is useful also in the analysis of circuits which are not strictly op-amps and it is often helpful to analyse complex circuits by breaking them down into building blocks based around op-amp configurations. The MSD2 is analysed in this way in the next section. The same ideas are also exploited in modern VLSI circuit design where a concept known as the Operational Transconductance Amplifier (OTA) is often used. This is essentially an op-amp with a high output impedance in which the output signal is a current. It can be turned into the voltage amplifier used in the examples above by adding a buffer with a low output impedance.

Other consequences of feedback

In the examples above, the operational amplifiers are assumed to have ideal characteristics in the derivation of the transfer function: the ratio of output to input signal. It is a useful exercise to compute the impedances of the amplifier configurations given, assuming this ideal behaviour. Of course, real amplifiers may approach but not reach the ideal case.

In a real operational amplifier the (open loop) gain is finite, perhaps 10^5 or more at a few Hz. It falls with frequency, with important consequences for the risetime of output signals, but over a wide frequency range the gain can be regarded as constant. The input impedance of the amplifier with feedback is

$$Z_{in} \approx Z_f/A \quad \text{for } A \gg 1$$

which is usually a significant reduction, and thus advantageous for many situations. A drawback in some circumstances is that the feedback network, if resistive, acts as a noise source. In any case there is a compromise to be made between sensitivity and input impedance, which both increase with Z_f.

Practical limitations

In reality the ideal op-amp does not exist and details of the performance of these circuits are affected by the deviations from ideal behaviour. In particular the speed of an op-amp is not usually as high as simpler, purpose designed, circuits so they are not normally used in circumstances where very fast response is required.

In nuclear pulse amplifiers, the op-amp approximation is worse than integrated circuits designed as operational amplifiers. The d.c open loop gain will typically be much less than a real operational amplifier (perhaps 10^3-10^4) but the emphasis will be on high gain at high frequencies (say > MHz) to ensure that the amplifiers can follow fast input signals. The details of how this is achieved are beyond the scope of these notes.

All active components, transistors and op-amps, have operating ranges of voltage and current which are not intended to be exceeded. In the experiment everything should have been arranged here so that operation is well within the limits.

2.5 Analysis of the MSD2 circuit

A circuit diagram of a single channel is given in fig.2.10

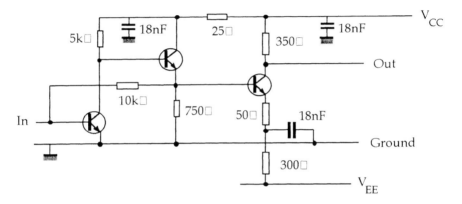

Fig.2.10. One channel of the MSD2

It should be possible to identify the basic elements of the circuit (fig.2.11). There are three: a voltage amplifier, a unit gain buffer with feedback to the input and a further stage of gain. The last stage has some subtle features which may become evident in the experiment. Appendix III, by the designers, explains the circuit in detail and why the components and operating conditions were chosen.

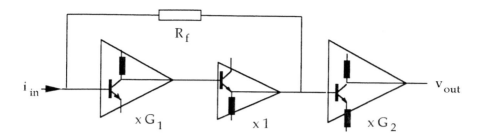

Fig. 2.11. Block diagram of the MSD2

3. EXPERIMENTAL PROGRAMME

Setting up

The equipment for the experiment is already mounted on a board for easy use. There is a diagram (fig.3.1) showing an outline of the layout of the board. *You will probably find it helpful to refer to this during the experiment, especially when you have to change connections on the board.* The MSD2 amplifier is in the form of a thick film hybrid, which is the green square object.

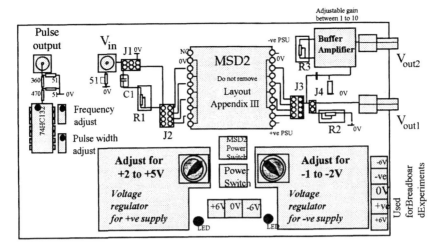

Fig. 3.1. The layout of the experiment circuit board

Before starting the measurements it will be worth spending a few minutes examining the board. The on-off switches are in the middle, one for power and a second for cutting the power to the MSD2. The basic power supply to the board is not critical and intended to be in the range (±) 6-9V. The voltage levels used by the MSD2 are adjustable on the board. The detailed results of the experiment will depend on the settings chosen. Suggested values are ~ +3V, -1.5V. They may already have been set but it will be useful to check and record them.

The MSD2 is a four channel hybrid, of which we will employ one channel only in the experiment. The channels are all practically identical but if need arises it is possible to switch to another channel by moving the jumpers J2 and J3.

The jumpers J1 and J4 will be switched between different positions during the experiment. J4 should be removed at the start of the experiment. Use a pair of fine nosed pliers to remove and re-insert the connecting wire. It's probably a good idea to switch off the power on the board while making these changes. The potentiometers R1 and R2 are adjustable and will be altered during the experiment by screwdriver. R3 controls the gain of the op-amp buffer and can also be adjusted.

There is one input connection (V_{in1}) which is the lemo socket on the left of the board. On the right hand side are two lemo output connectors. The lower one

(V_{out1}) allows observation of the output from the MSD2 directly. The upper one (V_{out2}) provides a buffered output from the MSD2 via a fast op-amp, with a gain adjustable by means of the potentiometer R3. There is a circuit diagram of the arrangement in fig. 3.2; many of the components are hidden on the underside the board in surface mount form but the potentiometer is on the top of the pcb.

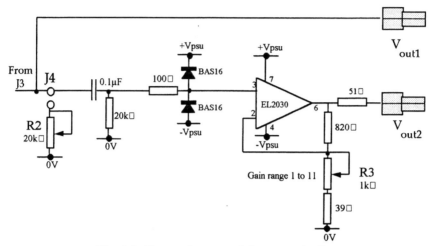

Fig. 3.2. Circuit diagram of the output buffer

The input signal to the experiment is provided in the form of a test pulse with a square waveform. The pulse generator is on the board, in the form of a digital chip mounted on the left of the board. The frequency and pulse width are adjustable in the ranges ~1kHz - 1MHz and ~100nsec - 20μsec but the amplitude is fixed. The rise and fall times are ~5nsec. The circuit has been designed so that the pulser output goes directly to the V_{in} input through the lemo connectors. You can choose the repetition rate and width as you wish. Some suggested values are given in the course of the experiment. It will be necessary to measure the amplitude of the input signal from time to time.

Bear in mind that none of the observations should be considered *right* or *wrong*, so feel free to experiment - eg with different terminations of cables and cable lengths. All of them should be useful in understanding the amplifier and, within reason(!), it should be safe against damage. Finally, try to keep a good record of your observations, including accurate sketches of waveforms. This will be of great help when reviewing your results.

3.1 Measuring the amplifier characteristics

3.1.1 *Pulse response*

Set the pulser to produce a square waveform with a frequency around 100kHz; the amplitude should be about 20mV. Insert a capacitor (eg 10-20pF) in the **central** jumper position of J1. Connect the scope so that on one channel you can see the waveform produced by the pulser. It will probably be best to set the scope to trigger on the pulser waveform.

This is the method for injecting pulses simulating detector signals into the amplifier which is described in Appendix I (It will be worth taking a few minutes at this point to read this explanation). For 18pF and 20mV the signal injected will be equivalent to a charge $\sim 2 \times 10^6$ electrons.

Connect the other channel to V_{out1} and examine the waveform. You should be looking at the response of the amplifier to detector-like signals.

Look on different timebase settings at the output from the amplifier, with different cable lengths and terminations (ie 50½/1M½).

Compare outputs from leading and trailing edges of the test pulse train.

Try to find explanations for your observations.

You can also choose to use the output from V_{out2} if you wish. Make sure you adjust R3 for a suitable gain of the buffer circuit (fig. 3.2). V_{out2} has a slightly different behaviour with respect to variations in cable length and termination. Why?

Once you feel you understand your observations, vary the amplitude of the input signal and measure the amplitude of the output pulses. You can alter the input signal magnitude using different test capacitances to cover the range you wish.

Make a plot of output signal vs input signal magnitude. Over what range of signals does the amplifier behave in a linear fashion?

You might already be interested in making a rough estimate of the noise in the system.

After this point you will find that you will gradually require to understand a little more of the internal workings of the MSD2 to answer all the questions which are raised. It will be worth consulting the circuit diagram and block diagram, and referring back to the section on the bipolar transistor above. You can also carefully probe the MSD2 using the digital voltmeter if you wish. This may be useful to verify operating voltages throughout the circuit and component values.

3.1.2 Input impedance

To estimate the input impedance of the amplifier remove the test capacitor from J1. Insert a wire jumper **across the left hand sockets** so that the input is connected in series to the capacitor C1. Observe the response to a square wave at the input.

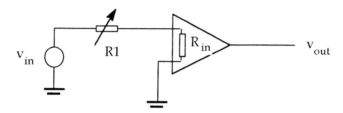

Fig.3.3 *Schematic of measurement of input impedance.*

The input waveform is not connected to the MSD2 directly but passes through a 1μF capacitor (C1) and a variable resistor (R1). Initially set the resistor to its minimum value. Turn off the power at the switch while you measure the resistance using a voltmeter. There are a pair of sockets at the side of the potentiometer to make this easier.

In this mode the amplifier can be viewed as a voltage amplifier with resistive feedback. Try and sketch the schematic circuit diagram.

What is the gain you expect to measure?

What is the purpose of the 1μF capacitor and does it have any influence on the measurements? You can check this experimentally if you wish.

Observe the output waveform. (If you have set the pulse rate high it may be useful to reduce it to observe the waveform clearly.) It is not simply an image of the input waveform scaled up. Can you suggest why not?

You should now measure the gain of the system by making observations for different values of R1.

Plot the results on log-log paper and compare with your expectation. If you are not sure of the value of the expected gain, try and deduce the feedback resistor value from your data. It can be directly measured on the MSD2 to check your result.

You now should have enough data to estimate the input impedance of the circuit. What is it?

The initial part of the output waveform has an exponential decay time constant associated with it. Try and measure the time constant by plotting the amplitude of the output pulse versus time on log-linear paper. If you can, try and explain the shape of the pulse and the value of the time constant. This is not simple if you are analysing the circuit for the first time so don't worry if you can't yet do this. However, try and return to this question later if you don't have the answer.

3.1.3 Output impedance

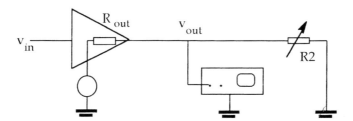

Fig. 3.4. Schematic of measurement of output impedance.

Return to equipment set up used for the pulse measurements of section 1. Insert the jumper J4.

Connect the output V_{out2} to the scope and select an amplitude in the linear range of the MSD2 which is not too small.

Measure the amplitude of the output pulse as a function of the resistance R2. (Measure the resistance using the voltmeter between the jumper J4 and a suitable ground, eg the lemo socket. Switch the power off while you do so.)

Plot your results and deduce the output resistance of the MSD2.

You should now be able to explain all of the observations made above and understand the function, and perhaps design, of the buffer amplifier. An alternative to the op-amp buffer is to build a simple transistor circuit to perform the buffer and gain function.

3.2 Construction of an amplifier

On the right of the board on which the MSD2 is mounted is a blue "breadboard". This can be used to prototype circuits constructed from discrete components and op-amps prior to building up a circuit with soldered components. Components can be mounted by inserting them into the holes in the board which are connected together in groups. If it's not clear, probe with a voltmeter to see how the holes are connected together.

The goal of the final part of the experiment is to use the knowledge you've now gained to build a buffer amplifier suitable of driving a 50½ load. It is proposed that this should be composed of two stages:

  (i) a common emitter amplifier stage with a gain of ~5-10 which can be driven from the MSD2 output directly (V_{out1}), followed by

  (ii) an emitter follower to drive the load.

In case you have limited time, omit the gain stage and just build the emitter follower.

The design parameters are that it should run from the board power rails, ie ~±6-9V, and that it should use npn transistors, with any of the resistors and capacitors available. Just take a look what is provided, which should cover a wide range. Start by making a circuit design on paper before you begin inserting components.

Acknowledgements

We thank all those who have read and made comments on the experiment, especially Pierre Jarron.

REFERENCES

1. Horowitz, P. and Hill, W. *The Art of Electronics*, Cambridge: Cambridge University Press, 1980.

2. Jarron, P. and Goyot, M., *Nucl. Instr. Meths.* **226**, 156 (1984).

3. Sze, S. *Physics of Semiconductor Devices*.New York: John Wiley and Sons, 1981.

APPENDIX I: CALIBRATION USING TEST PULSES

There are a number of ways to calibrate the response of a detector-amplifier system. One of the most useful is to apply a voltage step of a known amplitude to a capacitor of defined value at the input of the amplifier. This is usually done by means of an arrangement like that sketched below:

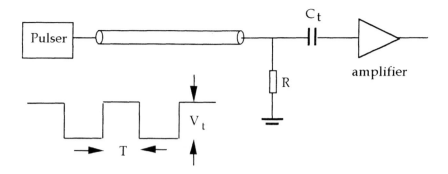

Fig. 1.1 Schematic of test pulse calibration with a square wave

The input signal, usually a train of equal positive and negative going steps, is sent from a pulser via a cable terminated in its characteristic impedance, R. Even if the cable is too short to be considered as a transmission line it can be detached and terminated in the same resistance at a scope (i.e. 50½ here) for proper observation of the signal.

On arrival of a positive voltage step at the cable end the test capacitor is charged by a very rapid flow of current which simulates the signal response of a detector. The amount of charge placed on the capacitor is simply

$$Q_t = C_t.V_t = N.e$$

The system response to this series of positive and negative current flows is therefore a series of positive and negative going train of pulses, each one of which represents the impulse response of the amplifier system.

Since the charge injected into the system is well defined, then the response of the system can be calibrated by measuring the amplitude, at its peak, of the output pulse v_{out}. The value of the ratio v_{out}/Q is an *arbitrary* quantity (ie its absolute value has no significance) which depends on the actual gain of the system. However it can now be used to estimate the sizes of signals from interactions in the detector in useful units of numbers of electrons.

APPENDIX II: BIPOLAR TRANSISTOR OPERATION

It is often helpful to have a slightly more detailed description of bipolar transistor. For circuit design models of device behaviour are important, especially at the integrated circuit level. For many applications much simpler models can suffice.

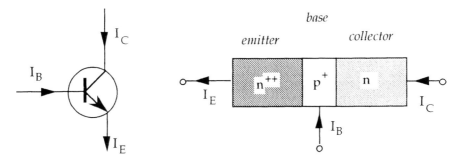

Fig. II.1 *The circuit diagram and schematic diagram of a bipolar transistor*

Physically the bipolar transistor is comprised of three regions:
- *emitter* - heavily doped,
- *base* - moderately doped (with opposite type),
- *collector* - lightly doped with the same type as emitter

so that the final device is n^{++}-p^+-n or p^{++}-n^+-p, which are usually designated as npn or pnp.

The Ebers Moll model

A basic model of the transistor which is useful for many simple circuit analyses is known as the Ebers-Moll model. It is based on the idea of the transistor as two diodes back to back. The currents in the diodes are assumed to have ideal characteristics [3].

Since the emitter base junction is a forward biased pn diode then
$$I_E = I_{E0}(e^{qV_{BE}/kT} - 1) \sim I_{E0}\, e^{qV_{BE}/kT}$$

where I_{E0} is the saturation current of the diode. The base collector junction is reverse biased, typically by several volts and the current at the collector is composed of two components, the base collector leakage current, I_{BC}, and a component due to diffusion of the emitter current across the base region, $I_{diffuse}$. Since

$$I_{BC} = I_{CO}(e^{qV_{BC}/kT} - 1) \sim I_{CO}$$

which is normally quite small, the diffusion component dominates.

The diffusion current is a consequence of majority carriers from emitter diffusing across base. However in the, lightly doped, base region a small fraction of them combine with majority carriers there. If a fraction α of the carriers recombine in the base, then

$$I_C = \alpha I_E$$

and $I_B = (1-\alpha)I_E$ by current conservation.

Thus $I_C = \alpha/(1-\alpha) I_B = \beta I_B$ with $\beta = \alpha/(1-\alpha)$

For example, if $\alpha = 0.99$ $\beta \sim 100$, which gives rise to the amplifying action of the transistor, whereby a small change in the base current gives rise to a large change in collector current.

It is useful to verify that there is a finite dynamic resistance associated with this current. If V_{BE} is allowed to vary, then

$$i_c/v_{be} = \partial I_c/\partial V_{BE} = I_0(q/kT) e^{qV_{BE}/kT} = I_C(q/kT)$$

or $\quad i_c = v_{be}/r_e$ with $r_e = (kT/q)/I_C$ or $r_e(½) Å 25(mV)/I_C(mA)$

The resistance often appears in equations in the inverse form of a conductance, called the transconductance of the transistor, g_m, thus

$$g_m = 1/r_e$$

Noise characteristics

Since amplifier noise is an important consideration when using detectors it is useful to identify the basic sources of noise in bipolar circuits. There are three principal origins:

shot noise in collector current $\quad i_c^2 = 2eI_C Æf$

shot noise in base current $\quad i_b^2 = 2eI_B Æf$

thermal noise in base $\quad e_b^2 = 4kT \cdot r_{bb'} \cdot Æf$

where $r_{bb'}$ represents the usually small but finite resistance of the base and contacts.

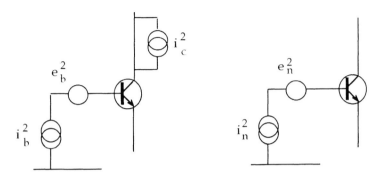

Fig. II.2. *Noise sources in a bipolar transistor*

In noise analyses, it is conventional to lump all the noise sources at the input of the circuit. This can be achieved by recognising that the relation between changes in base voltage and collector current can be written

$$i_c = g_m \cdot v_{be} = I_C \cdot (q/kT) \cdot v_{be} \qquad r_e = 1/g_m$$

so, the collector current appears as a *series* noise source (voltage) at the input

$$e_n^2 = (4kTr_{bb'} + 2qI_C r_e^2)Æf = (4kTr_{bb'} + 2(kT)^2/qI_C)Æf$$

while the base current contributes *parallel* noise

$$i_n^2 = 2eI_B Æf$$

Thus, because of the correlation between base and emitter currents, there is a trade off between series and parallel noise which sets a limit to bipolar circuit performance. For example, e_n^2 can be reduced by increasing I_C but i_n^2 is increased. In practice when numerical values are inserted it is found that bipolar transistors give best performance for high speed applications but do not match

Field Effect Transistors, particularly Junction FETs, for low noise if speed is not of prime importance.

APPENDIX III:
A FAST CURRENT SENSITIVE PREAMPLIFIER

See P. Jarron & M. Goyot., Nucl. Instr. Meth. **226,** 156 (1984).

APPENDIX IV: TRANSISTOR DATA SHEETS

2N3904   **NPN**
2N3906   **PNP**

These are general purpose transistors. See the data handbooks from many manufacturers.

APPENDIX V: TECHNICAL INFORMATION

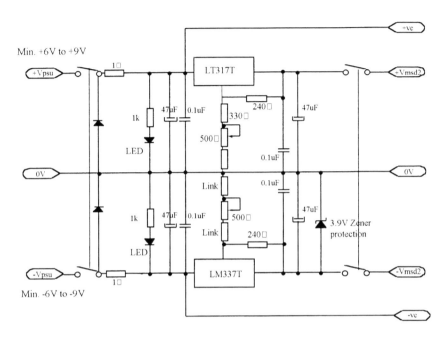

Fig V.1 The power supply regulation

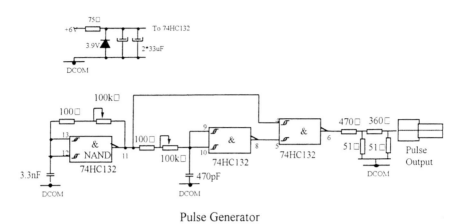

Pulse Generator

Fig V.2. The pulse generator circuit

Use of De-Randomizing Buffers in a Data Acquisition System [1]

Marvin Johnson

Fermilab, P.O. Box 500, Batavia, IL, 60510, USA

Marleigh Sheaff

Departamento de Fisica, CINVESTAV-IPN, Apdo. Postal 14-740, 07000, Mexico, D. F., MEXICO

Abstract. This is the course manual for a laboratory course that was designed to demonstrate a basic technique employed in data acquisition systems for the purpose of reducing deadtime and thereby increasing throughput. A custom CAMAC module was designed and built at Fermilab for use in these measurements. The use of queueing theory to predict the performance of the system is discussed, including some equations which can be used for that purpose.

INTRODUCTION

Modern high energy physics experiments typically require sophisticated, complex hardware triggers at first level to select events for subsequent processing, since the events of interest represent only a very small fraction of the total cross section. Although the rejection factors provided by these triggers can be large, $\sim 10^4$ or more in some cases, the data throughput to higher level triggers still must have large bandwidth to keep the deadtime at an acceptable level. Since this has to be achieved within rather stringent financial constraints, the design of the data acquisition path plays an important role when planning a new (or upgraded) experiment. The purpose of this laboratory experiment is to measure the significant reduction in deadtime (or, equivalently, increase in bandwidth) that can be achieved by adding de-randomizing buffers to each channel of front end electronics. A custom CAMAC board has been designed and built at Fermilab for these measurements.

---

[1] Laboratory Course presented at the *1997 ICFA School on Instrumentation in Elementary Particle Physics*, Leon, Guanajuato, Mexico, 7-19 July, 1997.

An example of the use of this technique is the SVXIII chip chosen for the CDF Upgrade, which contains four buffers, i.e., four sample-and-hold devices, per channel. The D0 experiment will use an earlier model, the SVXII, which contains only a single sample-and-hold stage. The difference in cost is approximately $1M. The use of 4 buffers instead of just one is expected to increase the data throughput by a factor of approximately 5, i.e., 50,000 events per second out of level 1 for CDF versus 10,000 out of level 1 for D0 at the expected luminosity of 2×10^{32}. Since the physics goals of D0 are somewhat different from those of CDF, the use of the more expensive chip was felt to be a less effective use of the funds available for the D0 upgrade than other detector upgrades more specific to those goals.

In a high energy physics experiment, both trigger rate and event length (and thus, output rate) are almost always randomly distributed. It is possible to predict the behavior of a data acquisition system for such an experiment using queueing theory. The results of the measurements made using the CAMAC board designed for this course can be compared to the predictions made for the system implemented on the board, which is a single queue containing a variable number of buffers. The match between the throughput predicted and that actually achieved may not be exact because of other factors, for example noise pickup in the system, that alter the input and/or output rate distributions so that they are no longer Poisson-like.

BACKGROUND

A queueing process is one in which events accumulate at some input stream, are serviced by a facility or process, and then exit in an output stream [1]. A simple example is a line at the grocery store. The customers in line form the queue, and the check-out cashier acts as the service facility. As customers pay for their goods, they successfully exit the queue and the line advances. Measures of effectiveness are important even to the customers in the grocery store. How long must they wait before receiving service? As one can imagine, the problem becomes more complicated when several cashiers are involved or when a person with an extremely large number of groceries is in line. The customer has to decide which line appears to provide the fastest service, and the store manager has to decide how many check-out registers need to be open to satisfy the customers and yet keep costs down.

Even the simplest data acquisition system represents a queueing process. Information is gathered at various detectors and transmitted to a central processing point, where it is assembled and then recorded onto a storage medium. An understanding of the queues formed at each stage of a data acquisition system is critical for design optimization. Rarely will the interarrival time of data be intrinsically well-matched to the computer readout time. Since physics processes tend to occur randomly, the exact arrival time of each event can not be known in advance. The inter-arrival times will follow a probability distribution instead. Event length differs from event to event, and thus processing (read out) time is also randomly distributed. A queueing model provides a means of studying a proposed data acquistion system. The goal is to minimize the idle time of the processors for a given system configuration and thus to maximize the throughput it can achieve taking into account the costs involved. Often general trends and steady-state behaviors can be predicted, and, in some instances, even transient solutions can be obtained explicitly.

Six characteristics are required to adequately describe a queue. Both the distribution of arrival times of events into the queue and the distribution of processing times for events must be specified. The number of parallel service facilities affects queue dynamics as well. (A store with three open check-out lines can service more customers per unit time than a store with only one.) In practical queues, there is usually a restriction on the system capacity. After a certain number of events have gathered in the queue, all subsequent events are ignored or rejected until the next event exits the queue. In some cases, the queue discipline can be important. Typically, the queue is first in, first out (FIFO) but there are many situations where certain events take priority upon arrival and some where service takes place on a last in, first out (LIFO) basis (i.e. a stack).

Using the standard shorthand notation, the data acquisition system implemented on the custom CAMAC board is an example of an M/M/1/(1-255)/FIFO queue. The first M refers to the exponential distribution of the interarrival times of the cosmic ray events (equivalently, a Poisson distribution in the arrival rate [2]). The second M denotes the approximately exponential distribution of the read out times, 1 specifies the number of parallel servers in the queue, and (1-255) is the user-specified maximum number of spaces (buffers) in the queue. When the number of buffers is set to 3, for example, no more data can be collected as soon as there are three events in the queue. All incoming events are rejected until the event currently being processed has been read out and the queue advances. The queue discipline is traditional first in, first out (FIFO). This queue is a very common model, which has well-known analytic solutions.

Let the event arrival rate and readout rate be Poisson distributed with mean rates λ and μ, respectively. Then, the inter-arrival times and processing times follow exponential distributions with means $1/\lambda$ and $1/\mu$. The ratio $\rho = \lambda/\mu$ is known as the utilization factor. Let K be the maximum number of events the queue can contain before further events are rejected, i.e., the total number of buffers. Using this notation, we can determine p_n, the probability of n events being in the queue at time t. (For details of the mathematics, see [3].) We do this by first writing down the set of difference equations in both n and t for the steady-state system. Then, taking the limit of small intervals in time, we solve equations differential in t to produce a set of relations of the form:

$$p_n = (1-\rho)\rho^n/(1-\rho^{K+1}) \qquad (1)$$

for $\rho \neq 1$ and $(n \leq K)$. Note that

$$\sum_{n=0}^{K} p_n = 1 \qquad (2)$$

and that the fractional deadtime is just the probability that the system is in the state where all K buffers are full, i.e., the probability that there are K events in the queue. Substituting K for n in the above equation, we see that this is:

$$p_K = (1-\rho)\rho^K/(1-\rho^{K+1}) \qquad (3)$$

EXPERIMENTAL SETUP

The Equipment

The overall experimental setup is shown schematically in Figure 1. A list of the equipment used for the experiment follows.

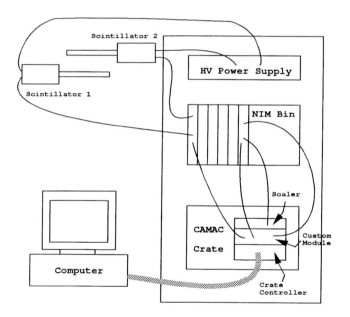

FIGURE 1. Experimental Setup

Scintillation Counters:

- Plastic scintillator
- Magnetic shields
- High voltage zener divider
- Photomultiplier tubes
- High voltage power supply

The scintillation counters contain organic plastic compounds doped with materials that scintillate. The passage of a charged particle excites the dopant molecules, which then de-excite by fluorescence. The photomultiplier registers the flourescence photons on a vacuum sealed photocathode optically bonded (with optical epoxy or optical grease) to the plastic scintillator and amplifies the resultant photoelectrons through a series of high voltage dynodes.

NIM Electronics:

- NIM bin with power supply
- NIM cooling fan
- Linear fan in/fan out
- Discriminator
- Majority logic unit
- NIM/TTL level translator
- Pulser
- Delay and interconnect cables

NIM is an international system of logic levels and module standards. Each of the following components can be plugged into a standard NIM bin. A linear fan in/fan out accepts an input signal and reproduces it at multiple output ports. A discriminator sends out a NIM true pulse when the input signal is above a certain voltage threshold (set by using a small screwdriver to adjust a potentiometer through a port on the face of the module). The majority logic unit produces a NIM pulse when a user-selected number of the four possible inputs are NIM true. The user sets the requirement by inserting a pin into one of four small ports marked 1-4 on the front face of the module. TTL is another system of logic levels. The level translator converts between NIM and TTL, both ways. The pulser unit sends out pulses of user defined frequency, amplitude, and width.

CAMAC System:

- CAMAC crate with power supply
- DSP crate controller
- Custom CAMAC module
- CAMAC 12-channel 24-bit blind scaler module

CAMAC is an acronym. It represents an agreed-upon standard for computer-hardware interfaces. The crate is controlled by a crate controller which issues commands on the backplane (the DATAWAY) of the CAMAC crate. Each module plugs into an 86-pin card-edge connector in the crate's backplane (called a station), and communicates with the controller. CAMAC is primarily a command protocol package. It provides a system of commands that can be issued, an addressing scheme, and timing diagrams that must be adhered to, but there is no restriction on what an addressed module can do in response to a command. The specific action performed when a command is received and decoded by a custom module is totally at the discretion of the designer of the module. The custom CAMAC module listed above was designed at Fermilab for this laboratory course. It contains an integration circuit (which produces a voltage proportional to the integrated charge of the input signal), an analog-to-digital converter which converts that voltage into a number (integer), a clock and countdown counter, and scaler displays. The unit is controlled by a Field Programmable Gate Array (FPGA). The FPGA is a very large array of small transistor gates that can be programmed externally to perform a defined set of logical state machine operations. The FPGA on the custom CAMAC board has been programmed to provide the functionality of the M/M/1/(1-255)/FIFO queue described above. Each channel of the blind scaler contains a counter which increments by one every time its input transitions from NIM false to NIM true. The unit can be gated off by an external veto.

Computer:

- Compaq Portable III 386 computer
- DSP Technologies backplane card
- 40-pin interface cable

Details of Experimental Arrangement

Figure 2 shows details of the NIM electronics used to form the trigger and the data for the experiment. The delay cable shown is approximately 300 ns. Other cables are of appropriate lengths to correctly "time in" the trigger.

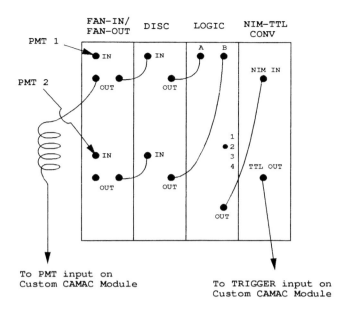

FIGURE 2. Detail of NIM electronics and wiring between modules

The events collected in this experiment are cosmic rays. The event trigger is made by forming the coincidence of the signals from two scintillation counters mounted one above the other. A piece of steel of the same size as the counters inserted between the counters ensures that the particles that trigger will be of high enough energy to be minimum ionizing in the upper counter. Anode signals from the two photomultiplier tubes (PMT's) attached to the counters are sent to the NIM fan in/fan out module. Each pulse is then sent through a discriminator. When the pulses are above the set threshold, the discriminators send a NIM true pulse to the coincidence (majority logic) unit. The coincidence unit (with the pin placed at 2 to indicate that both input signals are required to be true in order for the output to be true) performs a logical AND of the two inputs. The output is a NIM true and it is used to form the gate (i.e., event trigger) for the CAMAC board. The integrator on the board is of the switched capacitor design. Since it uses TTL switches, the level translator is needed to switch the trigger from NIM to TTL logic. The TTL signal should be connected to the input port labeled "Trigger" on the custom CAMAC board.

The anode pulse from the PMT of the upper counter is routed through the linear fan in/fan out module to the input labeled "PMT" on the custom CAMAC board. This signal must be delayed by approximately 300 ns. to ensure that it arrives in time with the gate for the integrator. (The switches described above are very slow.) The delay is imposed by adding a long cable between the fan in/fan out and the PMT input on the custom board. The PMT pulse is charge-integrated by the op amp circuit within the CAMAC module. The integrated pulse size (charge = pulse area = height x width) is then digitized by an 12-bit ADC. This 12-bit integer is the "data" entered into the queue (providing there is an available buffer) for each event. The queue itself is actually created by the FPGA in its own memory space. The number of buffers in the queue is set by the user using the upper of two dip switches on the face of the custom CAMAC module.

Since each event contains exactly one word of data, the processing time would not be random if the data word were then simply read out. Instead, the random event processing time expected in a real experiment is modeled using the pulses themselves. The charge collected due to minimum ionization loss is actually Landau rather than Poisson distributed, but that is close enough to give an approximately exponential distribution of read out times. The 12-bit ADC value is loaded into a countdown counter. The on-board clock is used for the count down. As the counter is clocked, the train of pulses that results is output through the output port labeled "AMP" (amplitude) on the front face of the custom CAMAC module. This should be input to channel 0 of the CAMAC blind scaler. Since the train of pulses output by the custom CAMAC module is TTL, use the TTL/NIM converter to convert to NIM before inputting to the blind scaler. When the counter reaches zero, the module issues a CAMAC look-at-me to notify the computer that the data in the scaler is ready to be read out. After the computer reads out the scaler and resets it to 0, it writes a data word to the custom module. The module responds by advancing the FIFO by one and loading the next data word (if at least one is waiting for processing) into the countdown counter. The countdown times can be made longer by using the second (lower) dip switch on the face of the custom CAMAC module. The switch setting effectively multiplies the intrinsic countdown time by a factor equal to the value set.

If, during the processing time for one event, another cosmic ray passes through the scintillation counters, it will generate a trigger and the pulse from the upper counter will be digitized. Then, if there is at least one buffer available in the defined queue, the data for the new event will be placed in an available buffer. If all of the buffers in the queue are full, however, the new event must be ignored, because there is no place to store it. There will only be space for a new event in the queue after the computer has read out the leading event in the FIFO and the queue advances. When the queue advances, the next event in the FIFO is loaded into the countdown timer, and the process repeats.

Figure 3 is a block diagram which shows the main features of the custom CAMAC module. The module is connected to the Dataway in the CAMAC crate through an 86-pin card edge connector through which it receives power and receives and sends CAMAC commands. There are several scaler displays on the front panel of the CAMAC module. They are "Accepted" and "Rejected" event counters, a display of the current event's amplitude and the current number of full buffers in the queue. There is also a "Reset" button and a "Run/Stop" switch, which will enable you to operate the board manually.

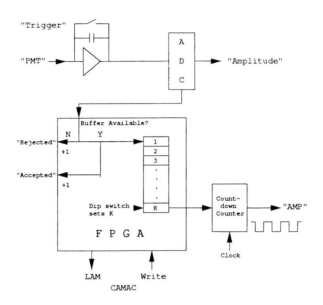

FIGURE 3. Block Diagram of Custom CAMAC board

PROCEDURE

Before beginning to experiment with this setup, be sure to read through this manual, and to familiarize yourself with the hardware. Check to make sure that the NIM electronics is correctly set up to form the trigger. Also, be sure the PMT signal you are digitizing comes from the upper counter so it will be certain to come from a minimum ionizing signal. The relative positions of the two scintillation counters should be checked to make sure you are maximizing the coincidences.

You may have to boot the computer before beginning work; the switch is located on the rear left corner of the machine. The central data acquisition program is located in the c:icfa97 directory. To run the program, simply type "readdaq". Enter a suitable output file name when prompted. At the end of a data taking run, this file will contain a listing of amplitudes for the events that were accepted and written to disk and will also contain some summary information at the end.

The main goal of this experiment is to measure the decrease in deadtime that can be achieved by increasing the number of buffers in which data can be stored while it is waiting to be processed. You can select the depth of the queue to use (in the range 1-255, although you will find that there is not a lot of change once you get past 10 or so). At the end of each experimental "run", defined as the time between a "Reset/Start" and "Stop", record the number of accepted and rejected events. The experimental deadtime can then be calulated as:

$$(Rejected)/(Rejected + Accepted) \qquad (4)$$

Select the number of buffers for each run by using the upper set of dip switches on the face of the module. It is best to start with one buffer and to set the lower dip switches, if necessary, to get a deadtime between 10 and 20% for the one buffer case. Then, perform a series of runs increasing the number of buffers by one each time. Since the number of rejected events decreases rapidly as the number of buffers is increased, be sure to take a long enough run each time to get the statistics you need to keep the errors sufficiently small for a meaningful measurement of each data point. A plot of the deadtime versus the number of buffers should indicate a trend that can be explained in terms of the queue model described above.

A plotting program, called "plot" has been provided that will enable you to plot a histogram of the pulse sizes. These are expected to follow a Landau distribution. You can also use the external NIM pulser unit set to a frequency much greater than the event rate to measure the event processing time distribution. Just put the output of the pulser into the input of the blind scaler instead of the "AMP" output of the custom module. The processing times will be seen to be approximately exponentially distributed. Set the number of buffers to a high enough value that the processor deadtime is negligible for this measurement.

ADDITIONAL OPTIONS

If there is time remaining, there are some additional aspects of the setup that can be varied. The NIM pulser unit can be used to form the trigger and data for the events instead of the cosmic ray telescope. The pulser represents a deterministic arrival time distribution, since the pulse frequency is constant in time. By varying the size of the pulses (amplitude times width, both also constant) issued by the generator, you can control the digitized pulse size (total charge) which is loaded into the countdown timer. In effect then, the pulse size will control the processing time, which will also be the same for each event. The larger the pulse size, the longer the processing time. Using the pulser in this manner can be a good demonstration of what can happen with a poorly matched set of arrival and service time distributions.

Another option is to lengthen the countdown time and thus the average readout time for events by using the lower set of dip switches. The pile-up of events in the queue will grow larger and the loss of data will become more dramatic as the run progresses. This is an example of a transient effect. Once the queue fills up, data will be lost at a constant rate. Try this with a relatively large number of buffers for several different choices of this number. At the other extreme, if the service time is short and the interarrival time between events is much longer than this, events will never accumulate in the queue. These limiting cases should be observable experimentally.

For given arrival and service time distributions, i.e., a fixed value of ρ, increasing the number of available buffers decreases the number of events rejected by reducing the idle time of the processors. This is the meaning of the term term "de- randomizing" buffers. With an insufficient number of buffers, the arrival time of events at the processing facility is random, the processors remain idle after reading out one event while they wait for the next, and the system is inefficient. Adding buffers smooths the random arrival times, with the result that the processors are kept busy nearly full-time, as if they were presented with a stream of events matched to the processing rate.

ACKNOWLEDGEMENTS

This work was made possible by support from the Consejo Nacional de Ciencia y Tecnologia, Mexico, and the U. S. Department of Energy and National Science Foundation. The authors also wish to thank Rafael Gomez and Timothy Meyer for their contributions to hardware and software for this project as well as to parts of this manual.

REFERENCES

1. D. Gross and C. M. Harris, *Fundamentals of Queueing Theory*, John Wiley and Sons, New York, 1979, Chapters 1-2.
2. Ibid, pp. 23-29, discusses the equivalence between a Poisson distribution of arrival rate and an exponential distribution of interarrival times.
3. Ibid, pp. 64-71, discusses the steady state solution for a M/M/1/K/FIFO queue.

Parallel Plate Avalanche Counters

A. Martínez-Dávalos and R. Alfaro-Molina

Instituto de Física, UNAM, A.P.20-364, 01000 Mexico D.F.

Abstract. The aim of this laboratory session has been to introduce some of the main characteristics and basic principles of operation of a family of gaseous detectors known as Parallel Plate Avalanche Counters (PPACs). The construction of a simple position-sensitive PPAC is described, and a set of measurements to test its performance is suggested. Some results of the measurements done by the students during the school are presented.

INTRODUCTION

As its name indicates, PPAC detectors are based in the process of avalanche multiplication in gases. In its simplest form a PPAC is comprised of two parallel electrodes made of thin metallized plates or fine mesh grids separated by a gas-filled gap. The application of a high voltage between the electrodes generates an intense and uniform electric field inside the detector sensitive volume. By using small gaps and/or low pressures the resulting reduced electric field (E/p, in $V/m \cdot atm$) is such that avalanche multiplication can be started immediately by primary ionisation electrons [1,2]. For this reason PPACs normally have very good time resolution and count rate capability [3-6].

The main application of these counters has been in the field of heavy ion physics, particularly for fast timing studies and as ΔE (transmission) detectors. They are normally required to be very thin (i.e. low pressure and thin electrodes,) to operate at high gains with good counting rate capability, and to have good radiation hardness. Besides, some experiments require to cover large areas, and to have some form of position sensitivity.

For these reasons a family of hybrid detectors was developed using the design characteristics of Multiwire Proportional Chambers and of PPACs [4-7]. The multiwire structure not only provides the required position information, but also introduces a second step of gas multiplication due to the high intensity of the electric field in the vicinity of the wires. This double mechanism of avalanche formation (first in the constant field region and then near the wires) is very useful and allows to obtain large gains (10^4-10^5) with fast charge collection times. Time resolutions down to ~100 ps at counting rates of the order of 10^4 s^{-1} mm^{-2} have been reported [5]. The main drawback of these chambers is that energy resolution tends to be poor (~20%) due to the different contributions to the avalanche process from the heavily ionising particles, which produce a mixture of amplification factors in the output pulse.

Some recent uses of PPACs reported in the literature include a detector proposed for the determination of energy and direction of jets in the very forward regions of the Large Hadron Collider [8]. They have also been used in experiments based on kinematic coincidence methods for heavy ion reaction studies [9] and as part of position sensitive particle-identification telescopes for forward angle heavy ion elastic scattering measurements [10].

EXPERIMENTAL PROCEDURE

Description of the detector

Since one of our aims was that the students could build a simple PPAC during the lab session, we chose a design that, although easy to assemble, could also demonstrate position sensing capabilities. A set of 5 detectors was prepared for this school. One was a ready-build chamber and 4 others were provided as kits for the students to assemble and test them on-site.

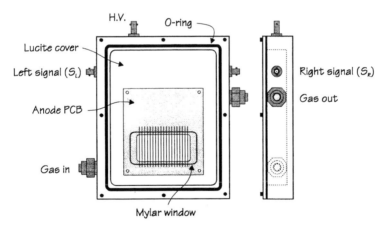

FIGURE 1. Schematic diagram of the PPAC detector.

The chambers consisted of a central anode wire plane placed between two cathodes. The anode was formed by 15 Cu-Be wires (50 µm diameter) with a 2.5 mm pitch, soldered to individual conductive strips of a circuit board used for electrical connections and mechanical support. The front cathode was made of a thin (200 µg/cm^2) aluminised Mylar foil strongly stretched and glued to an epoxy frame. The back cathode was a single 4×2 cm^2 Cu strip laid down on a printed circuit board (PCB), except for the ready-build chamber, where a second Mylar foil was used to demonstrate the PPAC in transmission mode.

The detector was enclosed in a 10×12×2.5 cm^3 gas-tight, high-density polyurethane container (see fig. 1). The box had a transparent Lucite cover to show the inner parts of the chamber once it was assembled, and a 5×3 cm^2 entrance window made of 1.5 µm thick Mylar. The position along the anode was determined using the charge division readout method [11], due to its low cost and ease of implementation. The methods consists in building a resistive chain along one end of a given electrode (in this case the anode wire plane); the position is determined by the ratio of the charge collected at one end of the chain (terminated into a low impedance) and the total charge.

Assembling the chambers

Each kit consisted of the box for housing the detector, anode and cathode PCBs, 50 µm Cu-Be wires, resistors, capacitors and soldering equipment. The detector box was already fitted with appropriate feed-throughs for the high voltage (HV) input, signal outputs and gas connectors (fig. 1). In order to simplify the construction of the chamber the students were provided with ready-tensed wires mounted on a PCB frame. In this way the anode plane could be quickly build by placing the frame directly on top of the anode PCB and soldering the wires using the anode tracks as guidelines (fig. 2.a).

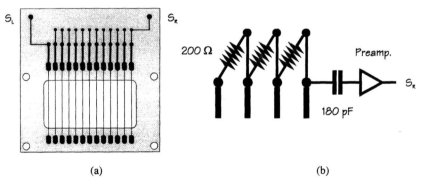

(a) (b)
FIGURE 2. (a) Anode plane formed by 50 µm diameter Cu-Be wires. (b) Section of the resistive chain for readout by charge division.

Once the wire plane was completed, the chain for the charge division readout was made by soldering a series of 200 ohm resistors on the back of the anode

PCB (fig. 2.b). The next step was to assemble the anode and cathode planes using plastic screws and spacers. At this point some variations could be done by, for example, using different separations between anode and cathode (typically 3 to 5 mm); also, the chamber could be operated with or without the front cathode.

The chamber was mounted on the plastic box and each end of the readout chain was connected through a 180 pF capacitor to their respective signal outputs (S_L and S_R on fig. 1); one end of the anode plane was soldered to the HV connector through a high value resistor (1-10 Mohm) in order to limit current in case of breakdown, and the cathodes were connected to a common ground. Final testing of electrical connections was made before placing the Lucite cover of the detector box with the corresponding O-ring. After this procedure was completed the detector was ready for connection to the gas handling system.

Initial Testing

As with all gaseous detectors, the choice of gas depends on specific experimental requirements. When working at low pressures (~10 Torr) isobutane (iC_4H_{10}) is normally used [3-10]. However this requires to enclose the chamber in a vacuum vessel and to operate it with a differential gas pumping system.

For simplicity, during the lab sessions all the tests were carried out with an Argon-CO_2 (80-20%) mixture flowing steadily at atmospheric pressure, using a ^{241}Am source (5.48 MeV α-particles). The source was mounted on a holder with two selectable collimators (0.5 and 2 mm diameter) and placed directly on top of the entrance window of the chamber. The collimators were used either to improve position determination or for higher counting rates, respectively.

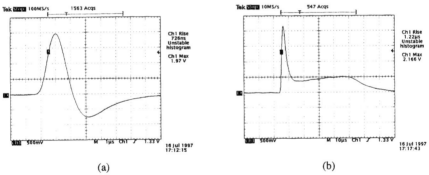

FIGURE 3. Signal output (128 sample mean) as seen on a digital oscilloscope. (a) Fast bipolar and (b) Unipolar output.

Each detector output was connected to an Ortec 142C charge-sensitive preamplifier and to one input of an Ortec 855 dual shaping amplifier, operated with a fixed time constant of 0.5 µs. After flushing the detector for several minutes with the gas, the HV power supply was switched on and the bias slowly increased, making sure to start from 0 V and not to exceed 3 kV.

When the bias was of the order of 1.5 kV the signals were clearly visible in the 1 µs/div, 100 mV/div scale of the oscilloscope. Figure 3 shows typical signals obtained using either the bipolar or the unipolar output of the amplifier. A first activity was to analyse the effect of the readout electronics and to give a physical interpretation of these signals. Measurements of rise time and the effect of voltage on the slow component of the signals (see fig. 3.b) were also made.

SAMPLE MEASUREMENTS

With a digital oscilloscope

Measurements of the pulse height *vs.* voltage characteristics of the chamber were made using the same setup as described above. In this case, with the ^{241}Am source positioned near the centre of the chamber and for a given voltage, one had first to make sure to trigger on the bigger signals in order to register only those events of maximum energy deposition. For this purpose it was useful to operate the oscilloscope in averaging mode (64 sample mean or higher). Figure 4 shows a log-linear plot of the results obtained with one of the chambers; notice the exponential growth of gain in this regime, as indicated by the fit (solid line).

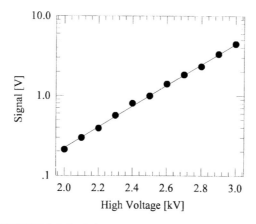

FIGURE 4. Pulse-height *vs.* Voltage measured on the anode plane.

The use of the digital oscilloscope also helped to illustrate, at least qualitatively, the position sensing capabilities of the detector. In order to verify that any difference between S_L and S_R was due to a change in position of the source, each readout line was first calibrated with an Ortec 419 pulse generator, in such a way that they produced the same output for a single given input signal. Then, each output of the detector was connected to the oscilloscope using the amplifier in fast bipolar mode. Each signal was averaged at least 128 times in order to reduce noise, and the resulting waveforms were subtracted from each other using the Math channel of the oscilloscope.

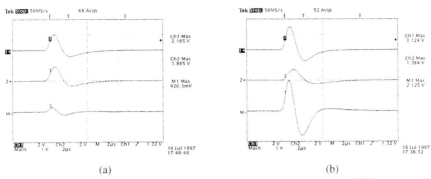

(a) (b)

FIGURE 5. Determination of position using the digital oscilloscope. ^{241}Am source placed: (a) near centre, (b) towards right.

Figure 5 shows two screenshots of the oscilloscope when the ^{241}Am source was placed in two different positions. In both cases the waveforms correspond to S_R, S_L and $(S_R - S_L)$ from top to bottom, respectively. In the first case the source was placed near the centre; since the magnitudes of both signals were approximately the same, the difference tends to cancel out. In the second case the source was moved towards the right end of the chamber and, as it would be expected, the magnitude of S_R was higher than that of S_L. This difference is obviously reflected in the Math waveform. Similar measurements were made with the source in different positions and the waveforms were stored on floppy disks for later analysis. Some additional activities were suggested to the students, such as to obtain a rough calibration of position, to determine the position resolution, and to observe the effect of the applied voltage on these parameters.

Data acquisition with a CAMAC system

A more precise determination of position required the use of a CAMAC system. The setup for CAMAC acquisition also demonstrated the use of NIM electronics for signal conditioning and trigger generation (gate generators, linear gate stretchers, delay amplifiers,) and permitted to obtain data in event mode to be stored as computer files for off-line analysis.

The system was based on a LeCroy CAMAC-GPIB 8901A interface and an Ortec AD811 analog-to-digital converter. Several programs were developed under National Instrument's LabView to perform on-line analysis, to plot 1D and 2D spectra of the digitised signals and to store event data on disk. A simple sorting program written in Visual Basic was also used to illustrate off-line analysis.

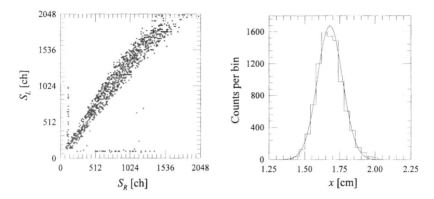

FIGURE 6. (a) 2D plot of the PPAC output, S_L vs. S_R. (b) Histogram of the position determination from the event mode acquisition data.

A run of up to 10^5 counts could be obtained in a few minutes under typical operating conditions. The signals (S_R and S_L) were digitised with 11 bit precision and stored as 2-byte integers (4-byte events). Since they represent the charge collected at the ends of the readout chain, the total energy for a single event is proportional to $(S_R + S_L)$. The position of the traversing particle, as measured from the left end of the chamber, would be given by

$$x = l \cdot \frac{S_R}{(S_R + S_L)}$$

where l is the total length of the chamber along x, approximately 4 cm in our case. Figure 6 shows a 2D plot (S_L vs. S_R) for a 10^4 event run, and the corresponding histogram for position determination when x was evaluated event by event using a lower threshold of 80 AD units in order to reduce the contribution from electronic noise. The position of the source, $\bar{x} = 1.7$ cm, was obtained from a Gaussian fit to the data (solid line) with a FWHM resolution of 2.2 mm, which is roughly equivalent to the wire pitch. These data were taken with the source as close as possible to the chamber and using the 0.5 mm collimator, therefore they represent the best resolution we could get during the lab session, with the chamber working at atmospheric pressure. Additional activities were to analyse data acquired with a bigger collimator and/or with the source further away from the chamber.

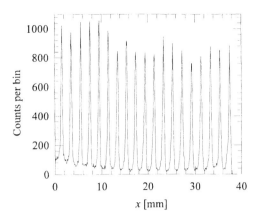

FIGURE 7. Position spectrum for a 20 wire, 2 mm pitch chamber with isobutane at 13 Torr.

In order to illustrate how the PPAC performed when working at low pressure, event data acquired with a 20 wire, 2 mm pitch chamber filled with isobutane at 13 Torr, were also available for sorting. In this case an uncollimated ^{241}Am source located at 94 cm. from the chamber was used. Figure 7 shows the position spectrum for these data; it can be observed that the individual wires are clearly resolved. The sorted data could be used to determine the average FWHM resolution, uniformity of peak position, etc.

ACKNOWLEDGEMENTS

Part of the preliminary work for the school was carried out with support from DGAPA-UNAM, grant 100196. We gratefully acknowledge the help of Mr. Mario Rangel Guzmán in the construction of the kits, and of Mr. Arcadio Huerta Hernández with the LabView programming.

REFERENCES

1. Knoll G.F., *Radiation Detection and Measurement*, New York: John Wiley and Sons, 1989, ch. 6, pp. 160-198.
2. Sauli F., Principles of operation of multiwire proportional and drift chambers, CERN Internal Report 77-09 (1979).
3. Breskin A. and Zwang N., Timing properties of parallel plate avalanche counters with light particles, *Nucl. Instr. and Meth.* **144**, 609-611 (1977).
4. Eyal Y. and Stelzer H., Two dimensional position sensitive transmission parallel plate avalanche counter, *Nucl. Instr. and Meth.* **155**, 157-164 (1978).

5. Breskin A., *et al.*, Properties of very low pressure multiwire proportional chambers, *IEEE Trans. Nucl. Sci.* **27-1**, 133-138 (1980).
6. Mazur C. and Ribrag M., Position sensitive multiwire proportional parallel plate detector for heavy ions, *Nucl. Instr. and Meth.* **212**, 203-208 (1983).
7. Steltzer H., Multiwire chambers with a two-stage gas amplification, *Nucl. Instr. and Meth.* **A 310**, 103-106 (1991).
8. Arafiev A. *et al.*, Parallel plate chambers: a fast detector for ionizing particles, *Nucl. Inst. and Meth.* **A 348**, 318-323 (1994).
9. Beghini S., *et al.*, A compact parallel plate detector for heavy ion reaction studies, *Nucl. Instr. and Meth.* **A 362**, 526-531 (1995).
10. Alfaro-Molina R. *et al.*, An inexpensive PSD for heavy ion elastic scattering measurements, *Nucl. Instr. and Meth.* **A**, in press (1997).
11. Alberi J.L. and Radeka V., Position sensing by charge division, *IEEE Trans. Nucl. Sci.* **23-1**, 251-258 (1976).

Muon Lifetime Measurement

Luis Villaseñor

Instituto de Física y Matemáticas, Universidad Michoacana, Edificio C3 Ciudad Universitaria, Morelia, Mich., 58060, México

Abstract. A simple experimental setup to measure the muon lifetime is presented. The muon detector consists of a sealed container with liquid scintillator coupled to a 2.5" photomultiplier (PMT). A home-made electronics module controlled by the parallel port of a personal computer (PC) digitizes the time interval between two consecutive PMT pulses in a time window of 25.6 μs. The muon lifetime is obtained by analysing thousands of double-pulse events in which the first pulse corresponds to a cosmic ray muon that stops inside the detector and the second to the decay electron coming from the weak decay of the muon. The background noise comes from random coincidences of pulses due to muons crossing the detector within the same time window. The PC is used as the data adquisition (DAQ) and data analysis computer. In addition to the muon lifetime, the charge ratio of cosmic ray muons and the capture rate of negative muons by carbon nuclei can be measured if the number of events is sufficiently high.

INTRODUCTION

One of the simplest experiments that can be used to illustrate the basic elements of a modern high energy physics experiment is the measurement of the muon lifetime. This experiment demonstrates the use of the following techniques: scintillation detectors; NIM electronics; logic pulse coincidence and timing electronics; on-line data adquisition with a PC and off-line analysis of the data [1]. The muons used for this experiment are cosmic ray muons which come from the decay in flight of secondary pions and kaons, the latter result from collisions of primary cosmic rays with nuclei high in the earth atmostphere. At sea level the muon flux is about 180 $m^{-2}s^{-1}$ and it corresponds to approximately 75% of the flux of charged cosmic rays. The muon flux has a mean energy of 2 GeV and a differential spectrum falling as E^{-2} up to a few TeV [2]. In this energy range the angular distribution is approximately $cos^2\theta$, where θ is the zenith angle. The plus to minos ratio is 1.25 - 1.30, where the positive muons exceed the negative ones because the particles in the primary collisions are mostly positive.

The muon lifetime at rest has been measured to be 2.19 μs, and the fact that muons produced at heights of tens of kilometers reach the ground before decaying is a consequence of the relativistic time dilation. The experimental setup along with the home-made timing and trigger electronics are described in detail. It is interesting to note that the charge ratio of cosmic ray muons, and the capture probability per unit time of negative muons by carbon nuclei, can be measured with the same setup provided the number of events collected is sufficiently high.

DESCRIPTION OF THE EXPERIMENTAL SETUP

Figure 1 shows the experimental setup used to measure the muon lifetime. The detector consists of a container filled with liquid scintillator (mineral oil made up of carbon and hydrogen, for example Bicron BC-517H). A PMT is coupled to the bottle exercising care to avoid any light leakage to the interior of the detector. The PMT base has an SHV connector for the high voltage and one BNC connector for the signal. The operating point of the PMT is around -1500 V. A 50 Ω resistor must terminate the PMT output to match the impedance of the RG58 coaxial cable used. Standard NIM modules are used to

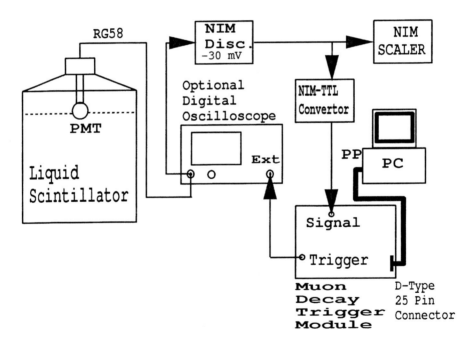

FIGURE 1. Experimental setup for the muon lifetime measurement.

discriminate and to convert the logic pulse from NIM to TTL. The TTL pulse is connected to the adquisition electronics which performs the trigger logic and the measurement of the time interval between two consecutive pulses between the time window of 25.6 μs.

The adquisition electronics is referred to as the muon decay trigger module; it interfaces with the PC through the parallel port of the latter. The DAQ program that runs on the PC can be written in Basic, C or any other computer language (including graphics programming packages such as LabView). For simplicity we give a sample program written in Basic. The use of the digital oscilloscope is optional to record the signal of the first and second PMT pulses and therefore study quantitatively the scintillation light signals of crossing and stopping muons as well of the decay electrons.

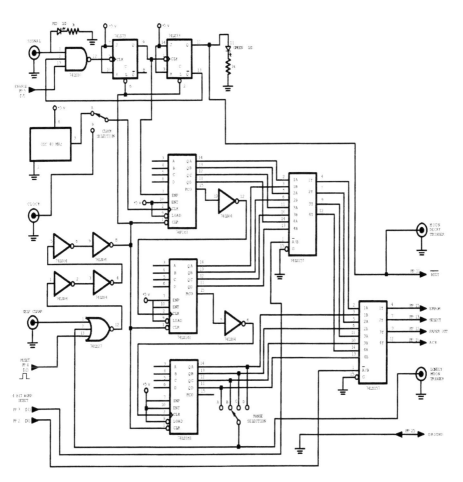

FIGURE 2. Schematic diagram of the Muon Decay Trigger Module.

DESCRIPTION OF THE ADQUISITION ELECTRONICS

Figure 2 shows a complete schematic diagram of the muon decay trigger module. It consists of two J-K type flip-flops in toggle mode, 74LS73, which allow the clock signal from the 40 (or 20) MHz oscillator, or from the external clock, to activate the three 4-bit counters, 74LS161 cascaded to form a 12-bit counter [3], only during the time window formed by two consecutive TTL pulses coming from the SIGNAL input BNC connector. The time window

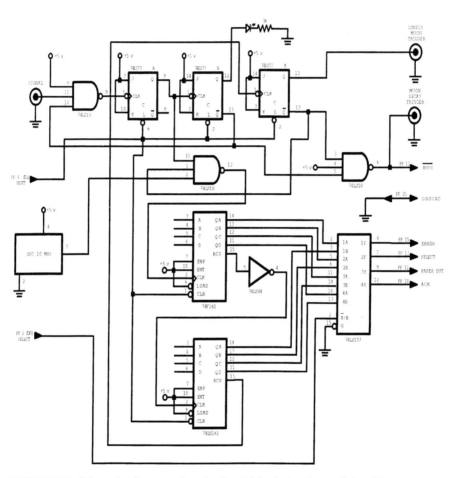

FIGURE 3. Schematic diagram of a simpler DAQ electronics module. The computer polls the BUSY line until a transition is detected; the DAQ program decides whether the transition is caused by two pulses within the time window, or by a single pulse and the end-of-window pulse.

is pre-selected by means of the RANGE SELECTION switch located to the right of the third counter; for a 40 MHz clock, i.e., the CLOCK SELECTION switch in position A, the positions marked A, B, C and D correspond to maximum time ranges of 51.2 μs, 25.6 μs, 12.8 μs and 6.4 μs, respectively. These numbers are the products of the clock period, normally 25 ns, times the maximum count range for 11, 10, 9 and 8 bits, respectively. In case there is no second pulse in the preselected time window after the first pulse, the carry-like pulse from the third counter clears and resets all the counters and flip-flops.

The parallel port is commonly found on the back of the PC as a D-type 25 pin female connector (the D-type 25 pin male connector corresponds to the serial RS-232 port). The parallel port is widely used for interfacing home projects to the PC; it allows the input of 9 or more bits and the output of 12 bits. For wider compatibility we use the so called uni-directional mode of the parallel port, in opposition to the bi-directional mode; one of the problems is the lack of standards for operation of the parallel port before IEEE-1284-1994 [4]. Table 1 shows the pin assignments of the parallel port connector and table 2 shows two of the three software registers; the third is the Control Register which we do not use.

Every time two consecutive TTL pulses separated by a time interval lower than the maximum range occur, the MUON DECAY TRIGGER output transits from low to high, and the green LED lights and stays lit until the module is reset by the PC. All signals are either TTL or TTL compatible. Under normal oparation mode, the SELF CLEAR input must be connected to ground and the module must be enabled by sending D3=1 through the data port which is located at the base address of the parallel port. The address locations for LPT1 and LPT2 are typically 378h and 278h respectively (the address 3BCh is not typically used today, but it was once used when the parallel port was contained on the video card). When the PC is first turned on, BIOS (Basic Input/Output System) determines the number of ports and assigns logical addresses to them. This information is briefly displayed on the screen of the PC.

The DAQ program that runs on the PC constantly polls the BUSY signal of the parallel port (pin 11) until it detects a value of 0 corresponding to the occurrance of an event; the latter is defined as two consecutive pulses separated by less than the maximum-range time window. A more elegant way of doing this is by using the parallel port's interrupt request, but this complicates the DAQ program. After an event has been detected, the PC reads out the 12-bit counter, 4 bits at a time, by means of the two multiplexers 74LS157. On the two 4-BIT WORD SELECT lines the PC sends D1=0, D0=0 to select the lower 4 bits, then it sends D1=0, D0=1 to select the middle 4 bits and finally, it sends D1=1, D0=0 to select the upper 4 bits. The readout is done through bits 3 to 6 of the Status Port located at base address + 1. After the readout is complete, the PC resets the module by sending the sequence D2=1, D2=0.

TABLE 1. Pin assignment of the D-type 25 pin parallel port connector.

Pin	Signal	In/Out	Register	Inverted
1	Strobe	In/Out	Control	Yes
2	Data 0	Out	Data	
3	Data 1	Out	Data	
4	Data 2	Out	Data	
5	Data 3	Out	Data	
6	Data 4	Out	Data	
7	Data 5	Out	Data	
8	Data 6	Out	Data	
9	Data 7	Out	Data	
10	Ack	In	Status	
11	Busy	In	Status	Yes
12	Paper Out/End	In	Status	
13	Select	In	Status	
14	Auto Linefeed	In/Out	Control	Yes
15	Error / Fault	In	Status	
16	Initialize	In/Out	Control	
17	Select Printer	In/Out	Control	Yes
18	Ground	Gnd		
19	Ground	Gnd		
20	Ground	Gnd		
21	Ground	Gnd		
22	Ground	Gnd		
23	Ground	Gnd		
24	Ground	Gnd		
25	Ground	Gnd		

A sample DAQ program written in Basic language is shown below; it assumes that the base address of the parallel port is 378h.

```
         OUT &H378, 8                                'sends enable
   10    OUT &H378, 12                               'resets module
         OUT &H378, 8                                'sends enable
   100   IF (INP(&H379) AND 128) = 128 GOTO 100      'waits for event
         'READOUT SEQUENCE ads
         A1 = INP(&H379) AND 120                     'lower 4 bits
         OUT &H378, 9                                'word select
         A2 = INP(&H379) AND 120                     'middle nibble
         OUT &H378, 10                               'word select
         A3 = INP(&H379) AND 120                     'upper 4 bits
         DTIME =0.025*(A1 + A2*16 + A3*256)          'time interval
         GOSUB WRITEOUT                              'saves data to HD
```

```
        GOTO 10                                    'next event
```

The low to high transition of the trigger outputs can be used to trigger other devices such as ADC's, digital oscilloscopes, etc., when the DAQ program is running. Alternatively, the module can be configured in a self clear mode by connecting the MUON DECAY TRIGGER output to the SELF CLEAR input. In this mode the module resets itself after each event and an external scaler can be used to measure the trigger rate of double pulses. Another operation mode of the module can be configured by connecting the signal to the CLOCK input and by placing the CLOCK SELECTION switch in A. In this way the module operates as a scaler that counts the number of TTL pulses between the PC commands of enabling (D3 = 1) and disabling (D3 = 0) the module. A similar module but this time with a CAMAC (Computer Automated Measurement and Control) interface is also available so that the students optionally use a CAMAC based DAQ system.

A simpler version of the adquisition electronics is shown in figure 3; unlike the electronics of figure 2 where the trigger is done in hardware, the trigger decision for the diagram of figure 3 is done in sotware. The price paid for the extra simplicity is the increase of the dead time. A simple DAQ program to use the adquisition electronics of figure 3 is the following:

```
        OUT &H378, 4                               'resets module
    10  OUT &H378, 0                               'resets module
        OUT &H378, 4                               'resets module
   100  IF (INP(&H379) AND 128) = 128 GOTO 100    'waits for event
        'READOUT SEQUENCE ads
        A1 = INP(&H379) AND 120                    'lower 4 bits
        OUT &H378, 5                               'word select
        A2 = INP(&H379) AND 120                    'upper nibble

        DTIME =A1 + A2*16                          '8-bit count
        GOSUB WRITEOUT                             'saves data to HD
        GOTO 10                                    'next event
```

EXPERIMENTAL PROCEDURE

Step 1. First of all you need to determine the 'voltage plateau'. This is the operating point for the high voltage of the PMT. The use of a volatage lower than optimal causes a decrease in the efficiency for the detection of the decay

TABLE 2. Software registers of the PC's parallel port; the Control Port register is not shown.

Offset	Name	Read/Write	Bit No.	Properties
Base + 0	Data Port	Write (Read/Write if port is bi-directional)	Bit 7	Data 7
			Bit 6	Data 6
			Bit 5	Data 5
			Bit 4	Data 4
			Bit 3	Data 3
			Bit 2	Data 2
			Bit 1	Data 1
			Bit 0	Data 0
Base + 1	Status Port	Read Only	Bit 7	Busy
			Bit 6	Ack
			Bit 5	Paper Out
			Bit 4	Select In
			Bit 3	Error
			Bit 2	IRQ (Not)
			Bit 1	Reserved
			Bit 0	Reserved

electron; on the other hand, if the voltage is too high, there will be an excess of background noise due to electronic noise in the PMT. It is important that small changes in the high voltage do not lead to great changes in the pulse rate; for this reason it is convenient to select a stable operating point which lies on the voltage plateau. In order to determine the operating point select a fixed threshold with the NIM discriminator of - 30 mV and take measurements of the pulse rate as a function of the NEGATIVE high voltage of the PMT. Do not exceed -1800 V as the PMT may be damaged. You may use steps of 20 V and use the NIM scaler to count the single pulse rate using appropriate time intervals leading to statistical errors in the pulse rate lower than 5% [5]. Once the operating point is set, make a rough estimate of the muon flux: divide the pulse rate by the effective area of the detector. You should get of the order of 400 $s^{-1}m^{-2}$.

Step 2. You can optionally check the electronics. For this purpose, use two channels of a NIM delay/gate generator module. Connect the inverted output of the first channel to the input of the second, and the inverted output of the second to the input of the first. You can control the oscillation frequency and duty cycle with the knobs on the module. Use the oscilloscope to look at the pulses. Now you can run the sample DAQ program given and check that the time interval between double pulses, as measured by the DAQ system, coincides with the value given by the oscilloscope.

Step 3. If everything works normally up to this point, you are ready to take data. Modify the sample DAQ so that the data are written to a file on the hard disk of the PC. Optionally you can add your on-line display subroutine to the DAQ program so that the time interval between consequtive pulses is histogramed in real time during the data taking process.

Step 4. If you are reasonably expert with a computer language, you can attempt to use the digital oscilloscope to digitize the waveform of two pulses. Connect the MUON DECAY TRIGGER output of the electronics module to the EXTERNAL TRIGGER input of the oscilloscope. Write off-line (or

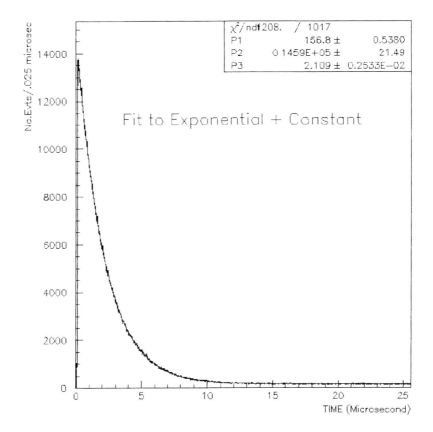

FIGURE 4. Raw data with a fit superimposed. The solid curve is a fit to the data using the function $P_1 + P_2 \exp(-t/P_3)$. From which a measurement of the background noise giving 156.8 events/0.025 μs is obtained. The 1 340 845 events shown correspond to an integrated data taking time of 2 572 h with a 35 l detector.

even on-line) analysis subroutines to integrate the area under each of the two pulses to obtain the PMT charge produced by stopping muons and by the decay electron. You can use the peak positions in these charge distributions to calibrate the detector and to monitor the PMT gain [6,7].

Step 5. After you collect a reasonably large number of events (at least one full day of data taking), you can start with the data analysis [8]. Write a subroutine to fit an exponential plus a constant to the raw data. The decay constant of the exponential gives you directly the muon lifetime; the constant

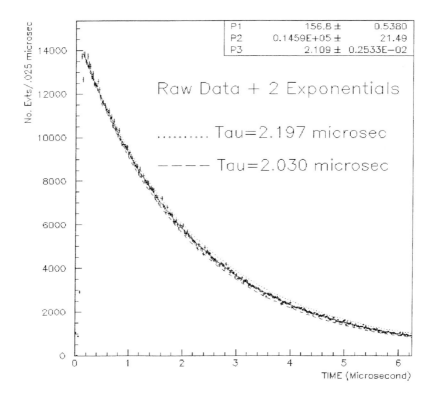

FIGURE 5. Comparison of the experimental data with two exponentials of different decay time. Both exponentials have the same amplitude and the same asymtotic value equal to the background noise of 156.8 events/0.025 μs. The 1 340 845 events shown correspond to an integrated data taking time of 2 572 h with a 35 l detector.

gives you the background noise due to randon coincidences of PMT pulses within the time window of 25.6 μs. Alternatively, the noise rate (number of noise events per bin) can be calculated by the formula $f^2 \Delta T$, where f is the single pulse rate and ΔT is the integrated time of data collection. Explain this formula. The PAW (Physics Analysis Workstation) [9] macro used to obtain Figure 4 [10] is the following:

```
macro 3
set xmgl 3.0
set xlab 2.0
```

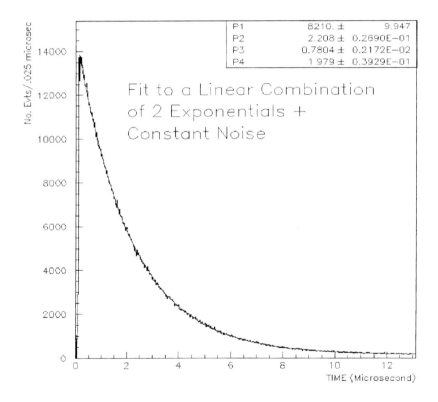

FIGURE 6. Raw data with a fit superimposed. The solid curve is a fit to the data using the function $P_1[(\exp(-t/P_2) + P_3 \exp(-t/P_4)] + 156.8$. The 1 340 845 events shown correspond to a data taking time period of 2 572 h with a 35 l detector.

```
opt fit
set fit 011
/ntuple/create 100 muon 1 ' ' 10000 't'
nt/read 100 data_file.dat
cre/1dhisto 10 'Time Interval between Pulses' 1024 0. 25.6
nt/project 10 100.t
ve/cr par(3) r 156. 145000.   2.2
hi/fit 10(6:1024) fun3.for ! 3 par
atitle 'TIME (Microsecond)'   'No.Evts/.025 microsec'
return
```

Where the file fun3.for contains the following:

```
FUNCTION FUN3(X)
COMMON/PAWPAR/PAR(3)
FUN3=par(2)*exp(-x/par(3))+par(1)
END
```

The muon lifetime that you obtain is smaller than 2.19 μs because the negative muons that stop inside the liquid scintillator fall into K orbitals around the carbon and hydrogen nuclei and can be absorbed through inverse beta decay, namely, a proton and a negative muon interact through the weak force to produce a neutron and a neutrino. Positive muons, on the other hand, are not absorbed as they never cross the nuclei due to their electrostatic repulsive force. As a first correction, you can assume that the ratio of negative to positive muons reaching the detector is one and use the known lifetime of the muons in vacuum (2.19 μs) to estimate the rate of capture of negative muons by carbon nuclei in the detector. The size of this effect is depicted in Figure 5, where two exponentials are compared to the experimental data, one corresponding to negative muons (the one with the smaller decay time) and the other to positive ones. Note that the statistical error bars on the data points are smaller than the gap between the two curves; this is a necessary condition to obtain a direct measurement of the negative-muon capture effect.

Step 6. After several days of data taking you can put together all the data sets and attempt to measure directly the charge ratio of cosmic ray muons and the capture probability per unit time of negative muons by carbon nuclei. A simple PAW macro used to obtain the results shown in Figure 6 is the following:

```
macro 5
opt fit
```

```
set fit 011
/ntuple/create 100 muon 1 ' ' 10000 't'
nt/read 100 data_file.dat
ve/cr par(4)  r 7000. 2.2 .500 2.09
ve/cr step(4) r 1000 .1 .1 .1
cre/1dhist 20 'Time Interval between Pulses' 1024 0. 25.6
nt/project 20 100.t
hi/fit 20(6:500) fun4p.for ! 4 par step
atitle 'TIME (Microsecond) ' 'No. Evts/.025 microsec'
text 2 12000. 'Fit to a Linear Combination ' .55
text 2 11000. '   of 2 Exponentials + ' .55
text 2 10000. '     Constant Noise ' .55
return
```

Where the file fun4p.for incorporates the value measured for the backgroung noise in Figure 4; it contains the following:

```
FUNCTION FUN4p(X)
COMMON/PAWPAR/PAR(4)
FUN4p=156.8+PAR(1)*(exp(-x/par(2))+par(3)*exp(-x/par(4)))
END
```

Step 7. Write and hand in a detailed report of all the work you did related to this laboratory course.

ACKNOWLEDGEMENTS

I would like to thank the students Francisco Alcaraz, Cuauhtemoc Pacheco and Jose Acevedo from the University of Michoacan and also to Arturo Gonzalez from the University of Guanajuato for helping with the construction of the electronics and detectors for this laboratory course. I also wish to acknowledge the support from the University of Michoacan during the realization of this work.

REFERENCES

1. W.R. Leo, *Techniques for Nuclear and Particle Physics*, Berlin, Heidelberg: Springer Verlag, 1987.
2. M. Aguilar-Benitez et al, *Review of Particle Properties*, Phys. Rev. **D50** (1994).
3. P. Horowitz and W. Hill, *The Art of Electronics*, New York: Cambridge University Press, 1989.

4. C. Peacock, *Interfacing the PC: Parallel Port, Version 3.0*, WWW document at http://www.senet.com.au/ cpeacock/parallel.htm (1997).
5. G. Moreno and L. Villaseñor, *Topics in Experimental High Energy Physics*, Rev. Mex. Fis., **36 S1** (1990).
6. L. Villaseñor et al, *Use of Decay Electrons from Stopping Muons as a Tool for Calibration of Čerenkov Tanks of the Pierre Auger Project*, HE 6.1.1 to appear in the Proceedings of the XXV International Cosmic Ray Conference, Durban South Africa (1997).
7. L. Villaseñor et al, *Calibration and Monitoring of Water Čerenkov Detectors with Stopping and Crossing Muons*, Auger Project Technical Note, GAP-97-033 (1997).
8. J. Orear, *Notes on Statistics for Physicists*, CLNS 82/511 (1982).
9. Brun R. et al.,*PAW (Physics Analysis Workstation)*, Cern Computer Center Program Librery, Q121 (1989).
10. J. Estevez, L. Villaseñor, A. Gonzalez and G. Moreno, *Measurement of the Charge Ratio of Secondary Cosmic Ray Muons*, Rev. Mex. Fis. **42**, No. 4, 649-662, (1996).

X-RAY IMAGING CHAMBER

Yu. Zanevsky, S.Chernenko, L.Smykov,
G. Cheremukhina

141 980 Dubna Moscow Region Russia

1. INTRODUCTION

Imaging radiation detectors are widely used for many investigations in biology, medicine, astronomy and so on. The application of position-sensitive detectors based on a gaseous multiwire proportional chamber (**MWPC**)[1]. leads to decreasing the radiation dose and to accelerating research. One and two-dimensional **MWPC**-based devices have successfully been applied to:

- **X**-ray crystallography of protein single crystals;
- **X**-ray structural analysis of amorphous, microcrystalline or powder samples;
- small-angle neutron scattering and thermal neutron structural analysis of powder or biological protein samples;
- two-dimensional autoradiography
- **X**-ray radiology.

The purposes of this laboratory work are the following:
1) experimental work with an imaging detector based on a **MWPC**;
2) study basic detector parameters.

The experimental work is divided into three parts. In the first one, **MWPC** properties will be studied with an oscilloscope. Further on, an imaging system with a **PC** allows us to make and to process some "radiographs". In the third part, the students should look through the files with typical defects of the detector and an example of using a two-dimensional detector for **X**-ray structural analysis.

All measurements will be performed with an X-ray Fe^{55} radioactive source (5.9 keV photons). A standard **P-20** (or **P-10**) gas mixture (argon- methane in the ratio 80 : 20/80:10) will be used.

PLAN OF THE WORK

1. Familiarization with a description of the laboratory work.
2. Switch-on of the detector and investigation of its parameters with the help of a scope.
3. Control of signal commutation according to a block-diagram of the registration system and the timing diagram of signals with a scope.
4. Study of the **SOFTWARE SKETCH** using the program **D2.EXE**.
5. Experimental work with MWPC on-line with a PC/AT using some phantoms as samples. Mastering of the investigation methods of main parameters of the detector (on the basis of our operating files or files from the directory :\SCHOOL\D).
6. Familiarization with typical defects of MWPC and the read-out system using respective files from the directory :\ **SCHOOL** \D.
7. Looking through the files from the directory :\. **SCHOOL** \D. These results have been obtained with the two-dimensional detector used for investigations in the field of protein crystallography (designed by **JINR** for the Shubnikov Institute of Crystallography, Moscow).

One or another point of the laboratory work is done depending on the students interests.

2. IMAGING DETECTORS

The common feature of the above applications is the requirement to display the distribution of particle fluxes entering the sensitive area of the detector. A similar problem is solved in radiology, nuclear medicine, industrial radiography and so on. There are two different approaches to the imaging of ionizing radiation particles: "integrating"(1) and "pulse counting"(2) systems.

(1) an `integral` response proportional to the total particle flux is measured in each image pixel,

(2) the coordinates of each particle interacting in the sensitive volume of the detector are determined separately, and the image is accumulated in the digital form with an electronic imaging system.

As usually, the former approach allows the used equipment to be simpler and cheaper. Moreover, in fact this method has no limitation on a maximum count rate.

In principle, the latter approach is more preferable for imaging of ionizing radiation particles because it allows a supplementary separation of particles according to such parameters as pulse height distribution, time correlation of events, etc. Such separation of particles enhances the image contrast due to better background rejection. Moreover, this approach decreases radiation doses needed for the irradiation of the investigated object by one or two orders of magnitude [2,3]. The applications are limited by the rate capability of the whole imaging system. The available limit is $10^6 - 10^7$ events/sec [4,5].

Considering a statistical character of particle detection, the root-mean-square σ of the average number **N** of particles passing through an image pixel is $\sqrt{N}$. Let our image contain 256x256 pixels and σ be smaller than 1%. Thus, more than 10 minutes are required to get reasonable statistics on the image at a count rate of 10 events/sec. From this it follows that "pulse counting" detectors are more convenient for such applications as diffraction experiments, autoradiography and so on. On the contrary, at present "integrating" detectors are more powerful for radiology or industrial radiography.

2.1 Imaging detectors based on a MWPC

An imaging detector is a complex device involving all processes from detection to displaying. An imaging "pulse counting" detector based on a **MWPC** will be used in our laboratory work [6,7]. Its schematic drawing is shown in fig.1. Particles are registered in the detector. Further on, the coordinates of each event are determined by means of **CAMAC** electronics, while an image is accumulated in a histogramming memory. The obtained images are processed and displayed using a **PC**. The particle interaction takes place in the gas filling the detector. The charge produced by this interaction is amplified in a high electric field around the anode wires. Due to the motion of positive ions from the anode to the cathodes, positive electric signals are induced on the cathode wires close to the interaction point.

There are a few coordinate readout methods in **MWPCs**:
- delay line readout;
- method of charge division on a resistive electrode;

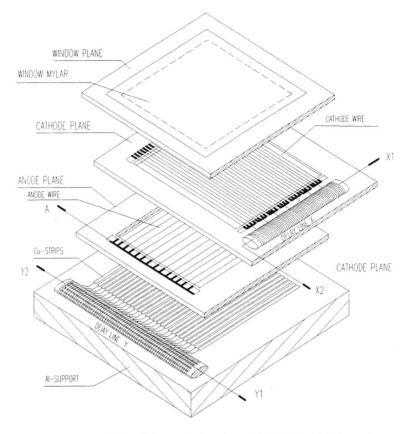

Fig1. Schematic drawing of MWPC with DL readout.

- method of wire cluster center finding;
- method of charge centroid finding;
- direct counting from each wire.

The choice of method is usually a compromise between all requirements for the imaging system (i.e. spatial resolution, rate capability, differential and integral non-linearities, long-term stability and so on.). In the case of delay line readout, separated cathode wires or groups of cathode wires are connected to a delay line in the range from 1 ns to 5 ns per mm. Induced signals are spread from an interaction point to both ends of the delay line, and the difference between the time of signal arrivals at these ends determines the coordinate of each interaction.

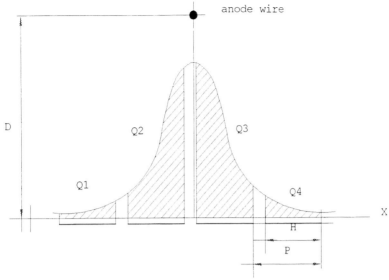

Fig 2. The shape of induced charge on cathode strips.

H-strip's width,   P-strip's pitch

Thus, only five readout channels are necessary to determine two coordinates while the fifth anode channel is used for amplitude event selection. A spatial resolution of 0.1 % of the total coordinate length and a count rate of up to 10 events/s are a reached by delay line readout. The delay line readout method allows us to get a very uniform response of the imaging detector. Anode and cathode signals from the low-noise preamplifiers and amplifiers are fed to a constant fraction discriminator. A veto logic built in the time-to-digital converters (**TDC**) checks proper time relations between all five signals and rejectssimultaneous events. The used TDC's digitizes time intervals with a step of up to 1 ns/ch.

The coordinates of each event (2 x 8 bits) are fed into a histogramming memory of 64K. An image is transmitted to the **PC** from the histogramming memory via the **CAMAC** dataway, the crate controller and the interface.

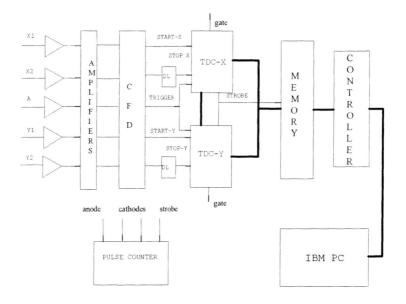

Fig. 3 Schematic diagram of the electronics.

2.2 Mechanical construction of the MWPC

The parameters of the detector depend on mechanical properties and tolerances as well as on electronics. The chamber desctibed below has a typical construction of a small size **MWPC** for X-ray imaging (fig.1). The detector consists of several insulating fiberglass frames stressed between an aluminum support plate and an aluminum frame. The fiber glass frames are machined, and their tightness is guaranteed by compressing a rubber placed in the grooves. The sensitive volume is placed just behind the mylar input window. The cathode made as a single-side printed circuit board is glued on the aluminum support plate. The second cathode and the anode frame are wound with thin wires by a weaving machine. The preamplifiers are placed on the rear side of the support plate. Gas is supplied via plugs and holes drilled in the support plate having the gas inlet and outlet in opposite corners of the sensitive area of the chamber. The main parameters are the following:

-anode-cathode gap	3.15 mm
-cathode-mylar window gap	3.15 mm
-anode wire spacing	2.0 mm
-anode wire diameter	20 mkm
-cathode wire diameter	50 mkm
-sensitive area of the detector	130 x 130 mm

3. EXPERIMENTS

The experiments consist of two stands:
- at the first one, laboratory work is done with an oscilloscope (PART 1);
- at the second one laboratory work is done with the IMAGING system, including read-out electronics, PC and respective software (PART 2 and PART 3).

WARNING:

1.1) Make sure that the gas mixture is flowing through the **MWPC**. The value of the flow must be 30-50 cm per minute.

2) The anode high voltage does not exceed +2.3 kV, and it is recommended not to operate above 2.2 kV for a long period of time.

PART 1

3.1. Measurements with an oscilloscope.

3.1.1 Examination of proper functioning the **MWPC**.

Place the collimator with the Fe^{55} source in the center of the detector. Verify a proper connection of the amplifier output with the oscilloscope (the cable should be terminated to 50 Ohm). Increase slowly the anode potential. You should observe
negative signals (about 200 mV). Notice that the pulse height distribution is relatively narrow; this corresponds to the full absorption of 5.9 keV photons. Only a small part of pulses (about 15%) belongs to the escape peak in argon, i.e. 5.9 - 3.2 = 2.7 keV.

3.1.2 Examination of gas gain coefficient homogeneity.

The **MWPC** (in particular, anode wires) is a delicate instrument and its proper functioning is dependent on mechanical accuracy. Illuminating the overall detector area homogeneously, a uniform response is expected. It is important for imaging applications to ensure a constant pulse height across the overall area of the detector. To verify this, install the collimated **Fe** source in the center of the **MWPC** and adjust the amplitude of signals to be 5 vertical divisions on the scope (between 0%-100% markers). Further, observe the pulse height, moving slowly with the collimator in the x-direction and subsequently in the y-direction. The changes of the pulse height should not exceed ± 15 %. A standard deviation of the average pulse height can be determined measuring the pulse amplitude at many points of the whole area of the detector. For a more precise determination of the standard deviation, a measurement with a multichannel analyzer and a calibrating pulse generator is needed.

3.1.3 Measurement shape of the anode and cathode pulses.

Placing the collimator in the center of the **MWPC**, we can determine the rise time of anode or cathode signals. A rise time is usually defined as the time required for a pulse to go from 10 % to 90 % of the full amplitude. The constants for the anode signal rise time and fall time are 70 ns and 100 ns, respectively, and for the cathode signal-50 ns and 50 ns, respectively. To observe cathode signals, it seems more practical to synchronize the scope using anode signals and to employ a 2 mm collimator. Placing the collimator on the respective opposite side of the chamber, verify the absence of reflections from the ends of the delay lines. Reflections should not exceed 10% of the pulse height while reflections of less than 5% are achievable.

3.1.4 Determination of the delay line attenuation.

Scanning the detector by means of the collimated source, you can observe changes of the signal rise time and signal amplitude. When you observe the signal from one end of the delay line (**X**), move the source to the opposite side. The rise time of
pulses increases and, on the contrary, their amplitudes decreases. Determine the attenuation of the delay line as $\frac{A_0 - A}{A} * 100$ [%], where A_0 is the amplitude when the source is placed on the side of the observed channel and A on the opposite side.

3.1.5 Determination of a specific delay of the delay line

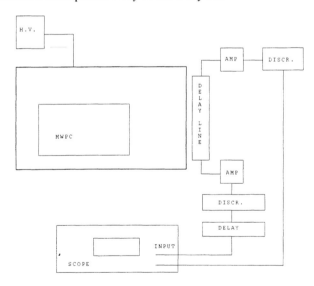

Fig.4. A block diagram of the measurement of specific delay.

Connect cathode signals via the constant fraction discriminators and the delay line to the oscilloscope (see fig.4). Select a 100 ns/div sensitivity on the scope and place the collimator (2 mm in diameter) on the edge of the **MWPC** input window. Fix the position of the pulse on the scope screen and move the collimator to the opposite side. Measure this time interval and determine the specific delay of the delay line (ns/mm). Notice that the coordinate is "measured" on both ends of the delay line, and thus the measured time interval must be divided by 2 to get the real specific delay.

3.1.6 Measurement of the gain curve vs. high voltage.

Connect the detector anode output of the amplifier and measure the full gain curve (pulse height as a function of high voltage). Plot the results in a semilogarithmic scale. According to an exponential growth of the avalanche charge, you get a linear fit with a deviation for the highest amplitudes. Fig.5 shows the measured amplitude vs.anode voltage.

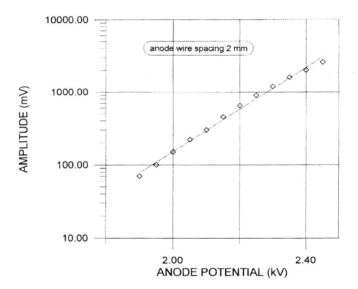

Fig. 5   Pulse height vs.anode voltage

PART 2

3.2 Measurements Using the Imaging System with PC

A block-scheme of the imaging system is shown in fig.3
Make sure that the gas mixture is flowing through the **MWPC**. Study the used program **D2** using its Software Sketch. Three phantoms are available for radiography measurements: a) arrays of holes (3.5 mm in diameter)with a pitch of 10 mm; b) a wedge of mylar-five steps ranging from 0.11 mm to 0.55 mm; c) a "black box". Place the collimated Fe^{55} source in the center of the **MWPC** input window. Set the anode potential to be 2.1 kV and, using the scope, check if negative signals appear at all five outputs of the amplifier. Further, adjust the anode voltage to obtain a 600 mV amplitude of the cathode signals.

3.2.1 Measurement of the efficiency curve of the MWPC.

Place the collimated Fe^{55} source in the centre of the **MWPC** input window. Make connections according to fig.6. Measure the count rate of each signal channel versus anode potential. Then, remove the source and measure the noise count rate. Use the commands EXP and START of the program D2 for setting a time interval and start of data acquisition. Plot the results as shown in fig.7. Pay your attention to the parameter "signal/noise ratio" (signal-average amplitude of signals and noise-full amplitude(6 sigma) of noise signals (at the output of the amplifier)). The space resolution of the detector and stability of its parameters depend on the value of this ratio. For our type of electronics this value should be 20 or more. This corresponds to the end plateau of the efficiency curve. Evaluate the signal/noise ratio using the scope and plot the results.

3.2.2 Homogeneity of the detector efficiency.

Differential nonlinearity σ is a relative root-mean-square deviation of counts in the channels devided by the average value:

$$\sigma = \frac{\sqrt{N_i^2 - N_{iv}^2}}{n \cdot N_{iv}}$$

and it is used to evaluate the homogeneity detector efficiency.

Set a signal amplitude of 600 mv. Install the uncollimated source on a holder at a distance of 200-300 mm from the chamber. After 2-3 minute time exposure, make data acquisition and process them. Determine differential nonlinearity on both **X** and **Y** coordinates using the **X** and **Y** histograms (to increase statistics, sum up the channels in the image between the corresponding markers).

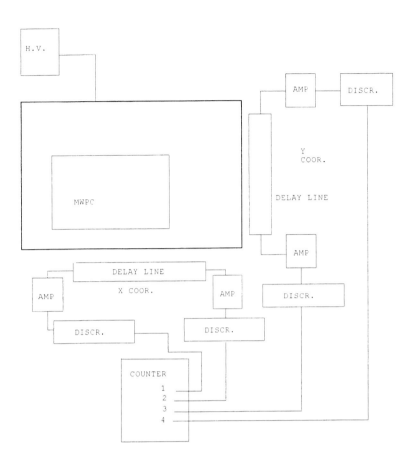

Fig.6 A block-diagram of the measurement of the counting rate plateau.

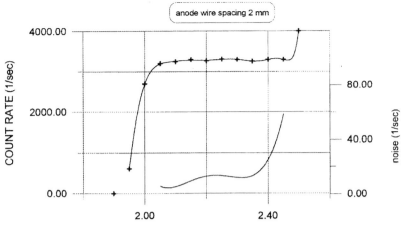

Fig.7 Counting rate plateau and noise rate as a
function of the anode potential

3.2.4 Measurement of the spatial resolution of the detector.

Place the slit collimators (0.16mm in width) on the input window at a distance of 50-100 mm. The collimators should be perpendicular to the X-coordinate. Locate the Fe^{55} source on the collimator and make data acquisition. Then, move the source to the second collimator and do the same. Process the data. Adjust each slit relative to the channels (the distribution must be symmetrical and as narrow as possible). Determine the size of the channel and the value of FWHM in mm (in the histogram is given in the channels)

3.2.5 Measurement of integral nonlinearity of the detector.

Integral nonlinearity is a maximum deviation of the measured coordinate from its true position. Install the phantom containing arrays of holes (3.5mm in diameter, pitch-10mm) on the input window and the uncollimated Fe source on the holder at a 200-300 mm distance from the chamber. Make data acquisition for 5-10 minutes and process them. Use the value of CGP (Center of Gravity of Peak). It is given in channels. Plot the results as shown in fig.8.

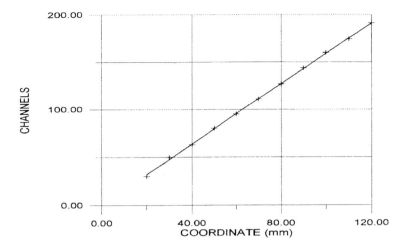

Fig. 8   Position linearity of the imaging detector with MWPC

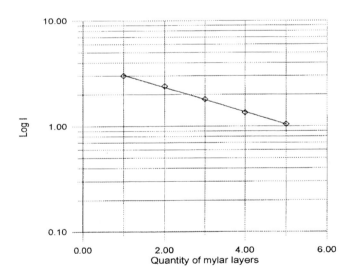

Fig.9

3.2.6 Calculation of the attenuation coefficient of mylar for 5.9 keV photons.

Install the phantom containing five mylar steps (ranging from 0.11mm to 0.55 mm) on the input window and the uncollimated source at a distance of 200-300 mm from the chamber. Make data acquisition for 3-5 minutes and process them. Plot the results as shown in fig. 9. The law of attenuation is:

$$I = I*\exp(-\mu*x)$$

Use N_{av} in the histogram as the value of I and calculate μ.

3.2.7 Phantom "black box".

Place the phantom on the input window and the uncollimated source at a distance of 200-300 mm from the chamber. Make data acquisition for 3-5 minutes and process them. What do you see on the PC monitor?

PART 3

3.3 Looking through the files from the directories.

Familiarization with typical defects of the **MWPC** and the registration system. Looking through the results obtained on the two-dimensional detector used to investigate in protein crystallography.

3.3.1 Typical defects of the MWPC and the registration system.

Looking through the files from the directory :\D.

1. NOISE D2 - map of MWPC noise events with noise spots.
2. NONHOM1 D2 - nonhomogeneity of the detector efficiency because of flowing away the gas mixture through the hole in the MWPC input window.
3. NONHOM1 D2 - nonhomogencity of the detector efficiency because of connection of neighbouring strips on the X delay line.
4. BIT-ABS D2 - no third bit of TDC.
5. MEANDR1 D2- small noncoincidence of the TDC-Y step of digitization (~2 mm) with the step of the anode wire.
6. MEANDR2 D2- significant noncoincidence of the TDC-Y step of digitization (~1 mm) with the step of the anode wires (~2 mm).

3.3.2 Two-dimensional detector for protein crystallography.

(Directory :\D for program D2.EXE or :\ARDET for program ARDK.EXE)

1. CATALAZA-example of a protein (catalaza) diffraction image.
2. LIQCRYST- example of a liquid crystal diffraction image.
3. GARNET 28-example of a garnet diffraction image.
4. COM R2 - example of a complamin diffraction image.

5. LEAFE-example of a leafe radiogramm.

SOFTWARE SKETCH.

(User's manual)

1. File structure.

D2.EXE	-	main program.
SETUP.EXE	-	parameters setting.
CONFIG.D2	-	the main parameter's values.
EXP.D2	-	exposition (hour:min:sec).
EGAVGA.BGI	-	graphic driver.

This software operates under DOS. You should include the following command line into **config.sys** file:
> device= ramdrive.sys 4096 /e

2. SETUP.EXE

At first, you should run file **setup.exe** to set the main parameters for measurement, such as:

- **Memory type**. Two kinds of memories are used in this version. That is why you should determine what type of memory you want to use. These memories have different resolutions: 512 x 512 x 16 and 1024 x 1024 x 16 bit cells.
- **CAMAC station's numbers.** You should determine CAMAC driver and memory numbers. For this driver, the number = **2**. The memory number is any from **9** to **22**.
- **RAM-drive name.** You should select the name of RAM- drive after its installation (from **A** to **I**).
- **Image resolution.** Here, you should use a resolution of **256** x **256**.
- **Zoom factor.** You can select an initial value of the zoom factor (**1 - 4**).

The button "**Save config**" - configuration is saved to disk (file config.d2).
The button "**Quit**" - exit to DOS without configuration saving.

Control keys for moving and selection - <TAB>, < ENTER> or click the mouse twice.

!!! You should **NOT** edit file "config.d2".

3. D2.EXE

This file is the main program for the investigation of detector parameters. After running, you have following command menu:
- **MEASUREMENT:**

    Start - CAMAC inialization, electronics preparation and increment mode works.

    \> - file name.

    Exp = - exposition (hour : min : sec). This time is read from file **exp.d2**.

    Accum: **OFF** - increment memory is reset before measurement.
                       ON - data accumulation in memory.

    Copy: **OFF** - increment memory data are not written to disk;
                       ON - increment memory data are written to disk.

    Sound: **OFF** - control sound is absent.
                       ON - control sound is set.

- **DISPLAY:**

    Chan X0 = displacement X

    Chan Y0 = displacement Y

    Zoom (1 -4) = integer zooming of the picture

    Max: **OFF** - data normalization for **NORM** value.
                   ON - data normalization for **MAX** value.

    Norm = normalization data.

    Palette Colour palette change.

  View video picture output.

- **FILE:**

  File: File name, which you should select from the file list by means of the marker.

  Read & Look - video picture file output.

- **TESTS:**

  Test Memory - test of increment memory is running.
  Clear Memory - control function of memory clearing.

- **Quit** - exit to DOS.

3.2 Histogramming.

  After measurement, the picture (2D histogram) outputs to a screen that depends on **Chan X0**, **Chan Y0** and **Zoom** factor. Moving the markers (vertical and horizontal), you can select the channels and get the X- (Y-) histogram. The control keys are the following:
- <Up>,<Down>, <Left>,<Right> - marker's movement with **step=1** channel;
- <Shift> + <Up> ,
  <Shift> + <Down>,
  <Shift> + <Left>,
  <Shift> + <Right>, - marker's movement with **step=10** channels;
- <F1> - *X*-histogram;
- <F2> - *Y*-histogram;
- <F4> - 2D interesting zone's statistics;
- <ESC> - exit to the main menu.
-

  You can select the zone of interest in the field of the *X-(Y-)* histogram. The control keys are the following:
- <Up>,<Down> - the marker's movement with **step=1** channel for the *Y*-histogram;
- <Shift> + <Up>,
  <Shift> + <Down> - the marker's movement with **step=10** channels for the *Y*-histogram;
- <Left>,<Right> - the marker's movement with **step=1** channel for the *X*-histogram;
- <Shift> + <Left>,
  <Shift> + <Right> - the marker's movement with **step=10** channels for the *X*-histogram;

- <F1> - integer zooming of the zone between the markers;
- <Enter> - statistics of the zone between the markers.
- <ESC> - exit to the field of the 2D picture.

REFERENCES

1. G.Charpak, R.Bouclier, T.Bressani et all

   Nucl. Inst. Meth. **62**, 262 (1968).

2. Yu.V.Zanevsky, G.A.Cheremukhina, S.P.Chernenko et al.

   Physica Medica, vol **VIII**, n.1, Jan.-Mar.,1992.

3. S.P. Chernenko, G.A.Cheremukhina, O.V.Fateev et al.

   ``X-ray detectors for structure investigations constructed at JINR.``

   Nucl. Inst. Meth. **A348**, 261 (1994).

4. E.Hell, H.J.Bosch, L.Brabetz, P.Kuhn and A.H.Walenta,

   Nucl. Inst. Meth. **A269**, 404 (1988).

5. S.E.Baru, A.G.Khabakhpashev and L.I.Shekhtman,

   Nucl. Inst. Meth. **A283**, 431 (1989).

6. Yu.Zanewsky, S.Movchan, T.Netusil

   Proceeding of the III ICFA School on Instrumentation in Elementary

   Particle Physics, Rio de Janeiro, Brazil, 16-28 July 1990. pp 327-346

   Word Scientific

7. P.Carlson, M. Danielsson, J. Solderqvist, A.Vaniachine, N.Weber,

   G. Cheremukhina, L.Smykov, Yu.Zanevsky

   X-ray Imaging with wire chamber

   ERI Technical report, KTH, Stockholm, 1996

POSTER SESSIONS

MSGC and fast neutrons

K. Bernier[1], H. Boukhal[2], J.-M. Denis[3], T. El Bardouni[2],
Gh. Grégoire, O. Grégoire[4], V. Tran[5]

Nuclear Physics Institute, Louvain-la-Neuve, Belgium

INTRODUCTION

The purpose of our work is to study the behaviour of the MSGC [1] (Micro Strip Gas Counter) which will be used for the inner tracker of the CMS [2] detector at the LHC (the future hadron collider at CERN). Since a high flux of fast neutrons ($\sim 10^6$ n cm^{-2} s^{-1}) [3] is expected in the CMS tracker, we focused our research on the effects of fast neutrons on these counters. The aim is to check that they will sustain, without degradation of their performances, the equivalent of 10 years of LHC running (i.e. a total fluence of $\sim 10^{14}$ n cm^{-2}).

THE FAST NEUTRON BEAM

To reach such a high fluence in a reasonable time, we developed an intense fast neutron beam based on the reaction Be + d → n + X using a 50 MeV deuteron beam on a 1 cm thick beryllium target. To obtain the desired intensity our test bench is located 9 cm downstream of the target.

We measured the absolute flux and the neutron energy spectrum using the activation [4] of several metallic foils. The subsequent unfolding [5] of data is estimated to give an accuracy of 20 %. The results are in fair agreement with previous measurement by Meulders et al. [6]. The production yield is 6.6 10^{11} n μC^{-1} Sr^{-1} and the mean energy is 20 MeV with a maximum of 50 MeV. The maximum intensity reached is 7 10^{10} n cm^{-2} s^{-1} which allows to reach the required fluence (10^{14} n cm^{-2}) in 23 minutes.

---

[1] Supported by the F.R.I.A., Fond pour la Formation à la Recherche dans l'Industrie et l'Agronomie.
[2] Now at University Abdelmalek Essaadi, Tétouan, Morocco.
[3] Unité de Radiobiologie et de Radioprotection, UCL-Bruxelles, Belgium.
[4] Now at Institute of Nuclear Chemistry, UCL, Louvain-la-Neuve, Belgium.
[5] Supported by the C.G.R.I., Commissariat Général aux Relations Internationales.

The dose rate is measured with PAD (Polymer Alanine Dosimeter, provided and read by the TIS-TE group at CERN) and an ionization chamber surrounded by a 20 cm thick polystyrene phantom according to the ICRU 45 protocol [7]. The results were comparable. The tissue equivalent dose rate at the build-up is 28 cGy μC^{-1} with an uncertainty of 5 %.

MSGC TESTS IN THE FAST NEUTRON BEAM

Up to know, two tests have been made with MSGC's :
- we studied the γ induced activity of two different substrates: the Desag D263 glass with aluminium strip and the conductive "Pestov" glass (provided by L. Shektman, Budker INP). The main result is that the latter gets more activated due to its higher amount of heavy metals.
- a Desag D263 glass with diamond like coating and chromium strips was irradiated with a flux of 2.7 10^{10} n cm^{-2} s^{-1} and a total fluence of 10^{14} n cm^{-2}. We measured the electrical surface resistivity before and after the irradiation and no significant effect has been observed. At the same time we saw a logarithmic increase of the leakage current when increasing the anode voltage (up to 300 volts which is lower than the working point in the gas used, Ar/DME 50/50). We interpret this as ionization close to the strips from (probably highly) ionizing particles coming from nuclear reactions of fast neutrons with the counter.

ACKNOWLEDGEMENT

We would like to thank Mr. M. Tavlet of the TIS-TE group at CERN for providing and reading the dosimeters, I. Boulogne and E. Daubie (University of Mons) for many useful discussions and for providing some MSGC substrates. We are grateful to the Cyclotron staff of Louvain-la-Neuve for the efficient operation of the accelerator.

REFERENCES

1. Oed, A., *Nucl. Instr. and Meth.* **A263** 351 (1988).
2. CMS technical proposal, *CERN/LHCC 94-38* (1994).
3. M. Huhtinen, *CERN/CMS/TN/95-198* (1995).
4. León-Florián, E., et al., *CERN/ECP/95-15* (1995).
5. Routti, J.T., and Sandberg, J.V., *Comp. Phys. Comm.* **21**, 119 (1980).
6. Meulders, J.P., et al., *Phys. Med. Biol.* **20**, 235 (1975).
7. *ICRU Report 45: Clinical Neutron Dosimetry*, Maryland: ICRU Publication, 1989.

MSGC tests with X-rays

Isabelle Boulogne* and Evelyne Daubie**

*F.R.I.A grant[1], University of Mons-Hainaut, B-7000, Belgium
**University of Mons-Hainaut, B-7000 MONS, Belgium

Abstract. Tests of MSGC detectors using an X-ray generator are reported. Results are presented for gas mixtures composed of Ar or Ne and dimethylether. The influence of the drift field and of the X-ray beam intensity is investigated.

INTRODUCTION AND TEST SET-UP

The Micro-Strip Gas Chamber (MSGC) was invented by A.Oed in 1986 [1]. Progress have been made and applications found in different physics fields [2]. The Mons laboratory is testing MSGC's using an X-ray generator as radiation source in the framework of their use in the CMS forward tracker [3].

The substrate which forms the basic element of the MSGC detector is made from diamond coated DESAG glass equipped with Au strips (10μm wide anodes, 80μm wide cathodes and 200μm pitch). In order to let the X-rays passage into the sensitive gas volume, the entrance window is a thin (50μm) Al foil and the drift electrode is made from a stainless steel mesh. The gas supply system contains stainless steel tubes, metallic filters and electronic mass-flow meters. The gas mixtures that we have used are non polymerizable mixtures composed of argon or neon and dimethylether (DME). The X-ray generator is equipped with an Fe anode emitting X-photons peaking at 6.4keV. For low rate tests an Fe absorber of 150μm is inserted between the X-ray collimator and the MSGC entrance window.

RESULTS ON OPERATING CHARACTERISTICS

Fig.1-4 show counting rate plateaux and gain curves obtained for different values of the drift field(1,3) in the Ar-DME(50-50) mixture and for different DME concentrations(2,4) in Ne based gas mixtures. The plateaux are about 100V long and they shift to lower anode potentials as the drift field

---

[1] Fonds pour la formation à la Recherche dans l'Industrie et dans l'Agriculture.

increases. The gain curves show the typical exponential behavior expected in proportional mode operation with higher values for Ne mixtures. Operating with high drift field and Ne mixtures would thus reduce the probability of occurence of discharges between anodes and cathodes. Fig.5 compares the cathode and drift currents normalized to the anode current as a function of the drift field. It should be noticed that their sum (anode current) remains practically constant and equal to 1. Fig.6 shows the rate capability of the detector showing stable operation up to rates of 10^5 X-photons/mm^2.s.

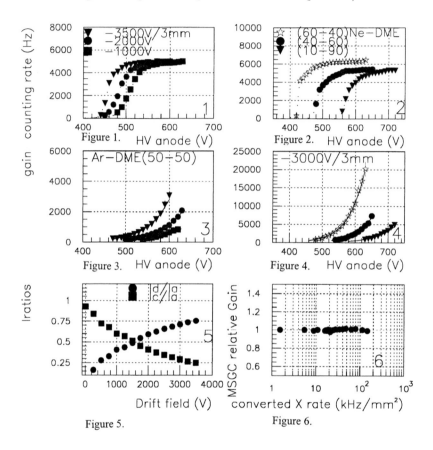

Figure 1.
Figure 2.
Figure 3.
Figure 4.
Figure 5.
Figure 6.

REFERENCES

1. Oed A.et al., *Nucl.Instr.Meth.* **A263**, 351,1988.
2. D.Contardo and F.Sauli,*Proceedings of the International Workshop on Micro-Strip Chambers*, Lyon, ed. by D.Contardo and F.Sauli,1995.
3. CMS collaboration, *Technical proposal*, CERN/LHCC/94-38, LHCC/P1,1994

High-Pressure Monitored Drift Tube Chambers for the ATLAS Detector at the Large Hadron Collider

U. Bratzler, T. Ferbel[1], A. Gabutti[2], H. Kroha, T. Lagouri, A. Manz, M. Treichel[3]

Max-Planck-Institut für Physik (Werner-Heisenberg-Institut), Föhringer Ring 6, 80805 Munich, Germany

Abstract.
The ATLAS detector is designed to exploit the full physics discovery potential offered by the Large Hadron Collider, to be built at CERN. With center-of-mass energies of 14 TeV, a proton-proton bunch crossing rate of 40 MHz, resulting luminosities of $\geq 10^{34}/\text{cm}^2/\text{s}$ and a requested operating time of 10–15 years, design, construction, and operation of this general purpose detector and its subsystems pose extraordinary challenges. High-precision measurements of final-state muons are amongst the most crucial requirements for the determination of particle signatures, both for known processes and for new discoveries alike. The ATLAS Muon System is fulfilling this requirement through the use of a system of high-pressure Monitored Drift Tube chambers. World-wide, a total of eleven chamber production sites will have to be operated for four years to produce these drift chambers for ATLAS. This article is based on work performed at the Max-Planck-Institut für Physik in Munich, one of the chamber production sites currently being set up for the ATLAS Muon Project.

INTRODUCTION

Muon measurement in ATLAS requires high-precision particle tracking through a very large volume, realized through a system of ~1,300 large-area, high-pressure Monitored Drift Tube (MDT) chambers.

The chambers, with typical active areas of 4 to 12 m², are arranged in form of an "octagonal barrel" in three layers (inner, middle, outer) about the proton-proton collision point to provide three space-points for the muon

---

[1] Home institute: University of Rochester, Rochester, NY, USA
[2] Current address: Italian Industry.
[3] Current address: CERN/PPE, CH - 1211 Geneva 23

momentum measurement. They are built from a total of 370,000 individual Al drift tubes covering an area of 5,500 m². The outside dimensions of this Muon System are ~ 22 m in diameter and 40 m in length.

The goal of the MDT chamber concept is the precise knowlege of the position of each chamber anode wire both within a chamber and throughout the Muon System. This is reached through as precise as possible MDT chamber fabrication and monitoring of any possible chamber deformations thereafter.

Precise knowledge of the positions of the chamber anode wires is the basis for the high-precision particle tracking through the magnetic field of the ATLAS Muon System and thus for the required momentum resolution ($\Delta p/p \sim 2\%$ for 100 GeV and $\Delta p/p \sim 11\%$ for 1 TeV muons). To reach this momentum resolution a particle trajectory, with typical path lengths of 6–20 m, will have to be measured to an accuracy of 50 μm which is realized with the MDT chamber concept.

In the following a brief introduction to the MDT chamber technology, along with results from the first full-size ATLAS MDT prototype chamber, built at the Max-Planck-Institut in Munich, are provided. Further information on this very current topic can be obtained from a number of recent conference contributions [1] as well as from the ATLAS Muon Spectrometer Technical Design Report [2].

MONITORED DRIFT TUBE CHAMBERS

The MDTs are fabricated from precise, industrially produced, aluminium tube extrusions with outer diameter of 30 mm, a wall thickness of 400 μm and a typical length of 4 m. The tube is closed with two "endplugs" which precisely center and hold the anode wire (50μm diameter, gold-plated W-Re), provide gas seal, and high voltage insulation between the aluminum tube (cathode) and the anode. The main challenge for MDT fabrication is the precise and reliable wire centering at affordable cost (700,000 endplugs are needed). Injection-molded plastic disks ("wire locators") produced with an accurately centered hole are used to reach the required wire centering precision of ≤ 10 μm (r.m.s).

A schematic view of an MDT chamber is shown in Fig. 1. It consists of a rectangular aluminium support structure ("spacer") and two precisely aligned multilayers of MDTs. The spacer serves as a carrier for the two, precisely aligned, tube multilayers (consisting of three tube layers each) and defines the distance between them. It also houses four infra-red alignment systems [2] to control chamber deformations. A typical chamber is fabricated from approximately 400, four meter long, pre-assembled MDTs glued layer-by-layer onto each side of the spacer. The gluing is carried out on a high-precision granite table (surface flatness ~5 μm) by means of high-precision jigs which hold the tubes into nominal position during the gluing process. With this technique

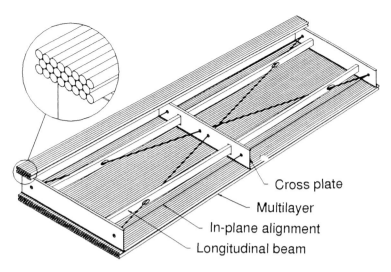

FIGURE 1. Schematic view of an MDT chamber (top multilayer only partially drawn to display spacer structure and alignment rays, enlarged view shows MDT arrangement in multilayer). Typical chamber outside dimensions are 2 m × 4 m × 0.3 m.

the wires within a chamber are positioned to ≤ 20 μm (r.m.s.).

The MDTs are operated at a gas gain of 2×10^4 with an $Ar/N_2/CH_4$ mixture (91:4:5) under an absolute pressure of 3 bars. The relatively low gas gain was chosen to minimize detector aging, the gas mixture in view of its linearity and drift velocity, while the choice for the gas pressure resulted from an optimization for spatial resolution. The typical single-tube spatial resolution reached is ≤ 80 μm. Single MDTs and full-size MDT chamber prototypes have been successfully tested in high-intensity μ-beams and with additional high-intensity background sources [2] at CERN, simulating an ATLAS-like environment. The results agreed with detector physics simulations and were satisfying the ATLAS requirements [2].

SUMMARY AND OUTLOOK

Through the introduction of high-pressure Monitored Drift Tubes chambers for large-volume, high-precision, particle tracking the ATLAS Muon System realizes muon measurement that meets the LHC performance requirements, at affordable cost. An adequately uniform momentum resolution of $\sim$2% for 100 GeV and $\sim$11% for 1 TeV muons can be expected over most of the pseudo-rapidity coverage of the system. The project is well on its way; the Technical Design Report is completed and Chamber series production will start in 1998 and proceed for four years. The years 2003-2004 are devoted for chamber installation and system integration tasks, for first muons to be measured in July 2005, the commissioning date of the LHC and its experiments.

REFERENCES

1. U. Bratzler, "The ATLAS Muon Precision Chamber System", Proceedings to *Frontier Detectors for Frontier Physics Conference*, La Biodola, Italy, May 1997 (to be published in NIM 1997).
2. The ATLAS Muon Collaboration, "ATLAS Muon Spectrometer Technical Design Report", CERN/LHCC97-22, 28 May 1997.

Order Statistics as a Tool for Analyzing Continuum Gamma Decay

Javier Cardona and Fernando Cristancho

Departamento de Física. Universidad Nacional de Colombia.[1]
Santafé de Bogotá, Colombia
Centro Internacional de Física. A.A. 4948. Santafé de Bogotá. Colombia

Abstract. Initial results of the application of the theory of Order Statistics together with the method of Energy-Ordered Spectra in the study of the nuclear continuum are presented. A first conclusion is that this method offers a possibility of distinguishing at high energies and spin different level density formulations.

In investigating the nuclear continuum it would be desirable to detect the first emmited radiation after a heavy ion fusion-evaporation reaction. However this task is nowadays not doable because of the impossibility of distinguishing a temporal sequence for the gamma rays in a cascade. To circunvent this difficulty it has been proposed [1] to use Energy Ordered Spectra (EOS): following the principle that the most energetic gamma rays are the most probable to be emitted first, we order the gamma rays in a decay sequence (event) according to their energies. We call the spectrum formed by the most energetic gamma rays, the $N=1$ EOS, the second ones the $N=2$ EOS and so on. It has been shown that in fact, by studying $N=1$ Energy Ordered Spectra, the method can yield experimental results in determining such continuum physical quantities like the level density parameter [2]. In the same work it was shown that the rather novel theory of Order Statistics [3] give a good analytical approximation to the spectral shape of $N=1$ EOS. Order Statistics give the probability $p_{N=1:n}$ for a gamma ray of energy E_γ to be the most energetic ray in a cascade of n gamma rays as

$$p_{N=1:n}(E_\gamma) = n \left\{ \int_0^{E_\gamma} f(E'_\gamma) dE'_\gamma \right\}^{n-1} f(E_\gamma), \qquad (1)$$

---

[1] This work was partially supported by the Colombian funding agency COLCIENCIAS under contract 222-96. J. C. wishes to thank the Fundación Mazda para el Arte y la Ciencia for its valuable support.

where $f(E_\gamma)$ is just the probability for that gamma ray to be emitted. In the studied circumstances this probability is just the product of the E1 gamma strength and the level density. The diamonds in Fig. 1 represent the Monte Carlo simulated $N=1$ EOS of gamma radiation originating in a region with intrinsic excitation energy $U=11$ MeV. Ref. [2] shows that using the mathematically simple dipole gamma strength and the Constant Temperature Fermi Gas level density (CTFG), the analytical CTFG curve in Fig. 1 is obtained. A very restrictive approximation in the CTFG curve is that the gamma ray is considered as a member of cascades with the lowest multiplicity n possible.

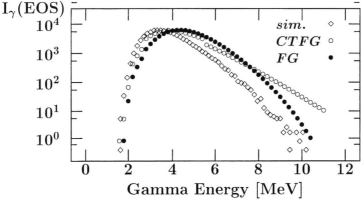

FIGURE 1. Comparison between Order Statistics results and a Monte Carlo simulated $N=1$ EOS. See text.

In the present work those cascades with all the possible multiplicities are taken into account and the Fermi Gas level density is used. Though the computations are more complicated, a decisive improvement, which is shown in Fig. 1 (FG curve), is obtained. At first it is observed that the maximum is shifted towards higher energies but the general behaviour at large energies is better reproduced. An important conclusion of this comparisons is that this method should be able to distinguish between the two level densities.

REFERENCES

1. C. Baktash et al., *Nucl. Phys.* **A520**, 555c (1990).
2. F. Cristancho, *Heavy Ion Physics* **2**, 299, (1995).
3. B. C. Arnold, N. Balakrishnan and H. N. Nagaraja, *A First Course in Order Statistics* (John Wiley and Sons, Inc. 1992).

FOCUS: a charm photo-production experiment at FERMILAB

Salvador Carrillo and Fabiola Vázquez

Physics Department, Cinvestav, I.P.N., Mexico
(for the FOCUS collaboration)

Abstract. FOCUS is designed to detect states of matter combining one or more charm quarks with light quarks (strange,up,down). The experiment aims to create 10 times as many such particles as in previous experiments and to observe rare phenomena that may shed light on fundamental interactions of the strong and electroweak forces. The experiment collected data during the 1996-97 fixed target run, and investigate several topics on charm physics including high precision studies of charm semileptonic decays, QCD studies using double charm events, measurements of D^0's absolute branching fraction, systematic investigation of charm baryons and their lifetimes, and searches for D^0 mixing, CP violation, rare and forbidden decays, and fully leptonic decays of D^+. Based on E687 experience, FOCUS (E831) expects to fully reconstruct 10^6 charm particle decays.

INTRODUCTION

FOCUS is a photoproduction experiment located at Fermilab wide band. It is an international collaboration composed of institutions from the following countries: Brazil, Italy, Korea, Mexico, Puerto Rico and USA, and it is an upgraded version of its predecessor, E687: "A High Statistics Study of States containing Heavy Quarks using the Wide Band Photon Beam and the E-687 Multiparticle Spectrometer".

DESCRIPTION OF THE BEAM

The photon beam is produced from primary protons in three stages: neutral beam production from 800 GeV protons; electron and positron beam, transport and bremsstrahlung photon beam.

800 GeV primary protons interact with a liquid deuterium target to produce an array of secondary particles which includes high energy photons from π^0 decays. The charged secondaries and uninteracted protons are swept out of

the beam and absorbed in a dump. Deuterium is chosen as the target material because it has a high ratio of radiation length to interaction length and hence will absorb a smaller fraction of the secondary photons.

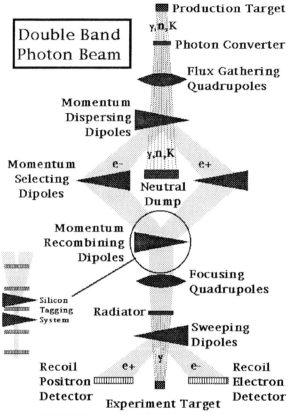

FIGURE 1. Description of secondary beam

The neutral beam produced is passed through a 60% radiation length lead convertor (fig. 1) where most of the photons will convert to an electron-positron. The secondary beamline transports electrons and positrons to a lead radiator while the remaining neutral secondaries (consisting of neutrons, kaons, lambdas and uninteracted photons) are absorbed in a neutral dump. The secondary system selects the momenta of electrons/positrons, transports them in a double dog-leg and focusses them onto the radiator. The focussed electron/positron beam is passed through a 20% radiation length lead radiator where it produces bremsstrahlung photons aimed at the experiment's target. The recoil electrons are swept out of this photon beam and are absorbed in an electron calorimeter, where their is meassured.

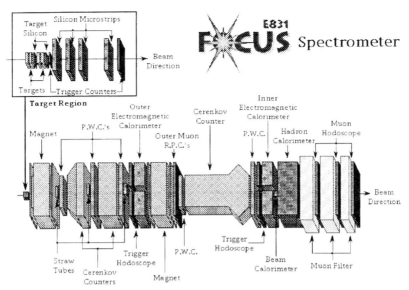

FIGURE 2. E831 Spectrometer

DESCRIPTION OF THE DETECTOR

The FOCUS Detector (fig. 2) is a large aperture fixed target multiparticle spectrometer which features excellent particle identification and vertexing for charged hadrons and leptons. The photon beam impinges on a beryllium target; charged particles which emerge from the target are tracked by two systems of silicon microvertex detectors. The first system is comingled with the experimental target, the second is just downstream of the target and consist of twelve planes of silicon microstrip arranged in three views. These detectors provide high resolution separation of primary (production) and secondary (decay) vertices. The momentum of charged particles is determined by measuring their deflections in two analysis magnets of opposite polarity with three stations of multiwire proportional chambers located in between and two stations downstream fo the second magnet. Straw tubes are used to suplement tracking in the central beam region (see paper by J. Link in these proceedings for a full description of the straw tubes). Three threshold multicell Cerenkov counters are used to identify electrons, pions, kaons and protons.

There are two electromagnetic calorimeters. The inner calorimeter, a lead glass block array, covers the central solid angle and detects particles which pass through the apertures of both magnets. The outer calorimeter covers the outer angular region described by particles that pass through the first magnet but not the second. Muons are identified in either a fine grained scintillator hodoscope with an iron filter (covering the inner region) or in an outer system that uses resistive plate chambers and the iron yoke of the second magnet as a filter. A hadron calorimeter consisting of iron and scintillating tile is used primarly in the experiment trigger but it's also used to reconstruct neutral hadrons.

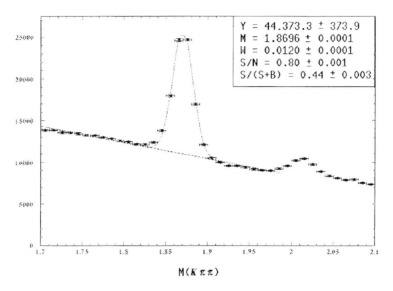

FIGURE 3. Charm golden mode: $D^+ \to K^-\pi^+\pi^+$

PRODUCTION OF CHARM PARTICLES INTEGRATED CHARM SIGNALS

In the photon-gluon fusion production mechanism the photon interacts with a gluon from the target nucleon that has fluctuated into a charm anticharm quark pair. The quarks are forced out of the nucleon and are dressed to produce the charm mesons and baryons. Charm particles live aproximately 1 ps, then dacay weakly to strange and other particles. E831 $FOCUS$ spectrometer detect these decay products. The finite lifetime of charm particles is the principal property exploted to isolate signals from copious non-charm backgrounds.

To study detector and trigger performance during the data taking we used charms "golden modes" ($D^+ \to K^-\pi^+\pi^+$, $D0 \to K^-\pi^+$, $D0 \to K^-\pi^+\pi^-\pi^+$) as shonw in fig. 3.

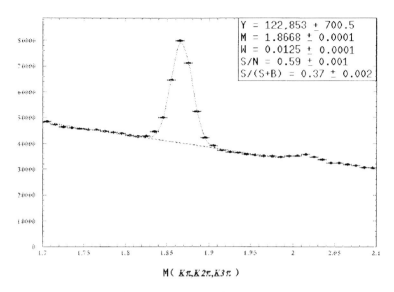

FIGURE 4. Integrated Charm Plot (all)

To have an estimated of the total charm we used an integrated charm signal see fig. 4. These doesn't represent the "best" signal we can get from the data.

ACKNOWLEDGEMENTS

We would like to thanks to FOCUS collaboration, CINVESTAV and Direccion Adjunta de Asuntos Internacionales from CONACYT (MEXICO) for supporting our participation in the data taking period at fermilab.

Assembly of Silicon Strip Detectors for the ATLAS Semiconductor Tracker

A. Cimmino[1], A. Saavedra[2], G. Taylor[1]

[1]*Physics Department, The University of Melbourne, Melbourne, Australia*
[2]*School of Physics, The University of Sydney, Sydney, Australia*

Abstract. The production of a large number of silicon micro-strip detector modules demands an assembly procedure that minimises the risk to components and error in the alignment of detectors. A novel technique, being developed to construct modules for the ATLAS SemiConductor Tracker, which employs active alignment and monitoring to assemble detector modules is discussed.

INTRODUCTION

The ATLAS detector is one of the approved experiments for the Large Hadron Collider (LHC) which will start running in the year 2005 at CERN, Switzerland. The main aim of the detector is to search for the Higgs Boson and to investigate physics such as Top and Beauty [1]. The ATLAS detector is composed of three main detectors: The Calorimetry, Muon Spectrometer and the Inner Detector. The Inner detector is the closest to the vertex and will provide the tracking of all the charged particles produced in the proton proton collision. The Inner Detector is composed of three regions: the Transition Radiation Tracker (TRT), the Pixel Detectors and the SemiConductor Tracker (SCT). The assembling technique outlined here is for detector modules to be placed in the forward region of the SCT [2].

THE SILICON MICRO-STRIP DETECTOR MODULE

A forward detector module is composed of two measuring planes, a double sided readout hybrid and a central spine. The detector modules will be supported by wheels distributed along the beam in the tracker. Hence the detectors have a fan shape. A measuring plane is composed of two fan shaped silicon detectors which are daisy chained to obtained 12cm long strips. The substrate of the silicon detector is n-type silicon, $300\mu m$ thick with implanted n-strips and aluminium readout strips. Each module has 770 strips per measuring plane with pitches depending on their position on the ring of the wheel (Outer 70.8-90.3μm, Middle 70.3-94.8μm, Inner 54.4 - 69.5 μm).

The double sided hybrid contains the binary readout electronics for both planes. The central spine or thermal/mechanical substrate provides support for the measuring planes. The measuring planes are glued to the substrate and hybrid. Detectors which are part of the same plane need to be aligned with respect to each other to provide colinearity of strips to within $\pm 5\mu m$. The strips of the measuring planes will form a stereo angle of 40 ± 0.15 mrad so that the module provides the radial position of the particle [2]. These strict tolerances are needed to minimise the error in the tracking, and any assembly procedure will need to assemble modules with the accuracy mentioned.

THE ASSEMBLY SYSTEM

Traditional assembling techniques rely on very precise machined structures to align the detectors. For example, the assembly procedure described in [2] aligns the two silicon detectors first to form a measuring plane using images from the fiducial marks. Each measuring plane is then transferred to a second vacuum chuck and the alignment is checked again. The second set of vacuum chucks will be aligned to the substrate by linear bearings and precision shafts. This procedure has an intermediate step in which the alignment has to be checked and relies on very precise mechanics to glue the two measuring planes to the substrate with the stereo angle required.

The novel technique described in this report has the following advantages:

- The alignment and gluing are done on the same support structures.

- The alignment of the detectors can be monitored before, during and after the gluing.

- The assemby system is simpler and relies on active alignment and monitoring so it only requires stability rather than high precision mechanics.

The main components of the assemby system are: the main support structure, the central frame, the detector support structure and the camera support structure (See figure 1).

The main support structure provides a stable and robust platform for the system. It has a rail which is use by the structures to slide along the z-axis. The central frame holds the substrate for detector gluing and the registration card for camera calibration. The detector support structure has two vacuum chucks on x-y stages to hold the detectors for gluing and alignment. The stage enables each detector to be translated and rotated independently. The camera support structure holds the camera which view the detectors and the substrate or registration card. There are six 752×582 CCD pixel array cameras (with $6\mu m$ square pixels). Four cameras with high magnification to view the strips of the detectors and two low magnification ones to view the reference points on the registration card. The images of the cameras are displayed on the screen of a personal computer and with the help of software developed the images are used to align the detectors.

To assemble a detector the position and magnification of the cameras is calculated using the registration card. After this a central frame containing

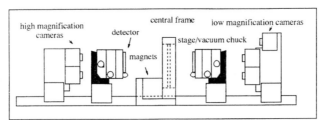

FIGURE 1. The assembly system. Once the detectors are aligned, the detector support structures will slide towards the central frame, which contains the substrate, with the cameras following to enable a continuous alignment check.

the substrate replaces the frame with the registration card. The four detectors are then placed in the vacuum chucks. The two detectors which compose a measuring plane are aligned first. The software uses the edges of the strips, which are accurate to 2μm for the alignment. It provides superimposed lines on the images for guidance and specific distances so the detector can be translated by a calibrated amount. The stereo angle between the two planes will be achieved in the same fashion but using images from both planes. While still monitoring the alignment of the measuring planes, the detectors are slowly brought to touch the substrate from both sides. After gluing, the alignment can be checked once again. Once the glue takes hold, the vacuum chucks are realeased, the detectors and the central frame are detached from the main support structure and the central frame is stored to let the glue cure. At the time of writing, the whole procedure outlined took 15 minutes for one measuring plane.

CONCLUSION

A novel technique for the assembly of silicon detectors was described and compared to a traditional technique. The main advantages being that it offers the lack of intermediate steps, active monitoring at every step of the assembly procedure and its main components do not need to be manufactured with high precision. The next step is to produce a complete detector module and quantitatively measure the performance of the assembly system.

REFERENCES

1. The ATLAS Collaboration *ATLAS, Technical Proposal*, (1994).
2. The ATLAS Collaboration *Technical Design Report* **VOL 2**, (1997).

TILECAL - The hadronic calorimeter for ATLAS

Mario David

Laboratório de Instrumentação e Física Experimental de Partículas
Av. Elias Garcia 14, 1
1000 Lisbon, Portugal

A Introduction and specifications

The TILECAL is the chosen hadronic calorimeter to be used in the Barrel and Extended Barrel regions of the ATLAS experiment at the LHC collider.

It is a sampling calorimeter with iron as passive material and scintillating tiles as active material. The scintillating tiles are readout by WaveLenght Shifter (WLS) fibres, which conduct the signal to the PMT's. The Barrel region covers $|\eta| < 1.0$, and the Extended Barrel covers $0.8 < \eta < 1.7$. There are 3 readout samples in the radial direction with 1.4, 4.0 and 1.8 λ, a segmentation in $\eta \times \phi$ of 0.1×0.1 in the first 2 radial layers and 0.2×0.1 in the last sampling layer [1].

B Goals

The major goals of the hadronic calorimetry at the LHC are to identify jets and measure their energy and direction, to measure the total missing transverse energy (E_T^{miss}), and to enhance the particle identification capability of the e.m. calorimeter by measuring quantities such as leakage and isolation [2].

C Testbeam performance

Since 1993 five prototype modules have been built and exposed to high energy beams of π's, μ's and e's. In 1994 and 1996 these modules were placed behind an electromagnetic Largon prototype in a combined setup. In 1996 a *real* size Module 0 of the Barrel region was built and tested, and in 1997 two Extended Barrel Module 0's were built and will be at testbeam this year. A brief summary of the results will be given bellow.

1 Standalone prototypes

The five prototype modules were tested with pions of energies ranging between 20 GeV and 300 GeV, with incidence angles between 0 and 45°. At 20°

an energy resolution of:

$$\frac{\sigma}{E} = \frac{(45.2 \pm 1.1)\%}{\sqrt{E}} + (1.29 \pm 0.11)\% \tag{1}$$

Was obtained after weighting corrections to acount for non-compensation. An $e/h = 1.36 \pm 0.11$ is measured, linearity and uniformity are of the order of 1% [3].

Also important, the TILECAL calorimeter was proved to detect muons due to it's good photoelectron yield. A good separation between the noise and the Landau distribution for muons in TILECAL, makes possible in the ATLAS spectrometer to correct the P_T of soft muons (2-5 GeV) with the calorimeter, in a event by event basis [4].

2 Combined testbeam

A combined test was performed in 1994 and in 1996, consisting of the 5 prototype modules behind an e.m. Largon calorimeter module. The combined calorimeter was exposed to pion energies between 20 GeV and 300 GeV, at an incidence angle of 11.3^o. An energy resolution of:

$$\frac{\sigma}{E} = \left(\frac{(38.3 \pm 4.6)\%}{\sqrt{E}} + (1.62 \pm 0.29)\%\right) \oplus \frac{3.06 \pm 0.18}{E} \tag{2}$$

Was obtained in 1994, with a H1 like weighting technic to correct non-compensation in both calorimeters, see [1,5] and references therein. Preliminary results shows that the resolution and linearity in 1996 testbeam are very similar to the one obsrved in 94.

3 Module 0's testbeam

In 1996 the first TILECAL module with the final dimensions and segmentation, was constructed and exposed to π's, μ's and e's of energies ranging between 10 GeV and 400 GeV (in the case of π's), and covering the whole η range. Preliminary results give a resolution for 100 GeV pions of 6.73% at $\eta = 0.45$ [1]. In 1997 two Extended Barrel Module 0's were already built and will be tested in August and October this year.

REFERENCES

1. ATLAS TILECAL TDR, CERN/LHCC/96-42, ATLAS TDR 3, 15 Dec. 1996
2. ATLAS Technical Proposal, CERN/LHCC/94-43, LHCC/P2, 15 Dec. 1994
3. F. Ariztizabal et al., NIM **A349** (1994) 384
4. Z. Ajaltouni et al., NIM **A388** (1997) 64-78
5. Z. Ajaltouni et al., NIM **A387** (1997) 333-351

Can we Extract Continuum Properties from Side Feeding Time Measurements?

Edgar Galindo and Fernando Cristancho

Departamento de Física. Universidad Nacional de Colombia, Bogotá
Centro Internacional de Física, A.A. 4948, Bogotá, Colombia[1]

Abstract. Monte Carlo simulated side feeding times are compared with experimental values in ^{75}Br and ^{77}Rb. The simulation predicts that $\tau \approx \tau_{SF}$. Neither of these nuclei behave in exactly this way and ^{75}Br has an extremely different behaviour. This fact should be a fingerprint of different continuum structures.

For decades the interest of gamma spectroscopy has focused on the structure of yrast states, for which life times τ have been extensively measured. Knowledge of side feeding times, τ_{SF}, has been needed just to obtain a correct state life time. Side feeding time in heavy ion fusion-evaporation reactions is produced because the nucleus has to decay along cascades in the continuum before it reaches any discrete state. The properties of the continuum should then be reflected in the τ_{SF} values. It is however extremely difficult to infere, from a single quantity those continuum properties just because of its coplexity: one has to consider different multipole gamma strenghts for statistical and collective decay, level density parameters, and several experimentally not well known quantities, for example the position of the entry states, which has in turn to do with particle decay probabilities. One tool one has available to handle that amount of different physical quantities is numerical simulation. We have simulated the gamma decay in selected reactions to obtain a theoretical side feeding time value τ_{SF}^{th} to compare with existing experimental values. We used GAMMAPACE [1], a Monte Carlo code allowing for a close handling of the statistical and collective aspects of the gamma decay. Its main input ingredients are average gamma strengths and dynamical properties in the continuum like quadrupole moment and moment of inertia. Our approach

---

[1] This work was partially supported by the Colombian funding agency COLCIENCIAS under contract 222-96. E.G. wishes to thank COLCIENCIAS for a fellowship "Jóvenes Investigadores" which allowed him to undertake this work

is to assign average experimental values obtained for the yrast states to the continuum. Fig. 1 shows experimental state lifetimes and both experimental and theoretical side feeding times in ^{75}Br and ^{77}Rb [2,3].

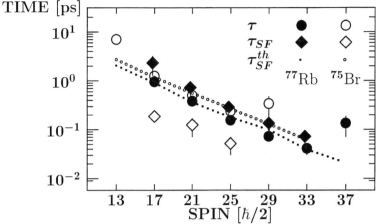

FIGURE 1. Comparison between experimental lifetimes τ and side feeding times τ_{SF} and theoretical side feeding times τ_{SF}^{th} in ^{75}Br and ^{77}Rb.

In these and in other cases studied the simulation produces curves that have the same slope as the experimental data and in general we observe that $\tau_{SF}^{t} \approx \tau$. This means that the simulation predictions are short for ^{77}Rb and too large for ^{75}Br. Two conclusions arise. 1) If nuclei would behave as schematically as simulated, the side feeding times would be as large as the corresponding lifetimes. The experimental data exhibited do not confirm this prediction. This takes us to: 2) At high intrinsic excitation energy nuclei seem to have very different structure to the one observed in the yrast. This difference produces opposite effects on τ_{SF}. In ^{75}Br: $\tau_{SF} \approx 0.1\tau$; in ^{77}Rb: $\tau_{SF} \approx 1.3\tau$. By combining the knowledge of other experimental quantities like side feeding intensity, multiplicity and simulation one could infer very reliably continuum properties as we have done very schematically here.

REFERENCES

1. F. Cristancho and K. P. Lieb, *Nuc. Phys.* **A518**, (1991).
2. L. Lühmann et al. *Phys. Rev.* **C31**, 828 (1985).
3. A. Harder, Thesis. Göttingen (1995), not published; *Phys. Rev.* **C**, in press.

SELEX Experiment
(Fermilab E781)

Fernanda G. Garcia*, Jurgen Simon[†]

*Departamento de Física
Universidade de São Paulo, USP
Caixa Postal 66318 05389-970 São Paulo,S.P., Brasil
[†]Max-Planck-Institut fuer Kernphysik
69029 Heidelberg,Germany

Abstract. The poster presents the physics goals and experimental status of Selex (Segmented Large X Spectrometer), supported by many international institutions which has been taking data since February 1997. The principal goal of this experiment is to have a high statistics sample of charm baryons from 650 GeV Σ^- and π^- beams.

INTRODUCTION

Selex is a fixed target experiment at Fermilab designed to do a systematic study of charm strange baryons. It differs from the others experiments by having a more flexible beam, a broader range of momentum coverage and a fast data acquisition system, as was described by P.Cooper's lecture. The physics goals are the following:

1. Study of the weak decay characteristics of the four stable charm baryons with high statistics.

2. Study of the strong interaction production characteristics of charm hadrons, comparing π^- and Σ^- beams. Check for leading particle effects.

3. Study the excited states of all charm baryons. Candidates for spin $\frac{1}{2}$ and $\frac{3}{2}$ states have been seen, but are not well measured.

4. Study the semileptonic decays of charm baryons.

5. Study of charm pair production and search for doubly-charmed baryon and exotic baryon production.

Besides that there will be non-charm physics topics, like Primakoff effect reactions.

I EXPERIMENTAL FEATURES AND PRELIMINARY ANALYSIS

The experiment is a three stage spectrometer with acceptance from $0.1 \leq x \leq 1.0$. Features of the spectrometer are the following:

1. High precision silicon vertex system with 30 μrad angular resolution for 100 GeV/c tracks. These silicon microstrips are located before and after the target

2. PWC's and drift chambers for measuring charged tracks;

3. Three lead glass calorimeters for photon detection and electron identification after each magnet;

4. Transition radiation detector that tags the type of beam particle;

5. Ring imaging Čerenkov counter for particle identification;

6. Forward Λ decay spectrometer to capture decay products of Ξ and Ω chain decays.

Beam composition is $\Sigma^-/\pi^- = 0.85$ at 650 GeV/c, which enhances baryon productions relative to mesons.

The trigger is based on missed distance from the primary vertex. The hardware trigger in a first level, requires

- Tracks with a momentun higher than 15 GeV/c after the second magnet and extrapolated back to the vertex detector.

- Hodoscope selection to get high multiplicity at the second spectrometer.

- Beam particle identification from beam TRD.

The raw trigger rate is 10KHz. One of the special features of E781 is that we run with a charm filter online which goal is to throw away background below charm peak as much as possible.

The run began in June of 1996 and continues until September 1997. SELEX is a new experiment and requires an extensive checkout. Data taking began in February 1997. All detectors are operational but only those used in the trigger are being fully analysed, which consist of the beam silicon, beam transition radiation detector, vertex silicon, PWC chambers after the second magnet and the RICH, in another words, just the first 2 spectrometers have been used for the preliminary analysis so far. A sample of data taken in February has been analysed offline (a total of 10^9 inelastic interactions) searching for the

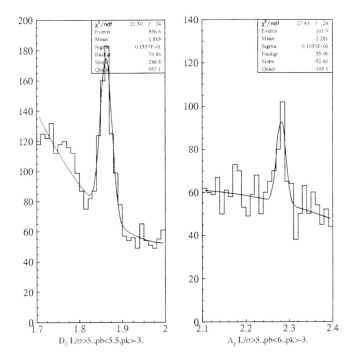

FIGURE 1. D^0 and Λ_c^+ signals from analysis of trigger detectors for 10^9 inelastic interactions.

decays $D^0 \to K^-\pi^+$ and $\Lambda_c^+ \to pK^-\pi^+$ and it was observed 395±46 for meson and 110±26 for baryon. See figure 1. These candidates were required to be separated from the primary vertex by at least 5σ and to point back to the primary with a $\chi^2 \leq 6$. These preliminary results confirm that our charm baryon signals will be comparable to our charm meson signals and verify the excellent mass and vertex resolution expected in the experiment. Further analysis using the entire acceptance of the spectrometer is underway.

Straw Tubes for Focus

Jonathan Link

University of California Davis

WHY STRAW TUBES?

Fermilab experiment 831 [1,2], known as FOCUS, uses a broadband photon beam to study particle states with open charm.

A large fraction of the FOCUS beam comes in the form of low energy photons, which create e^+e^- pairs. It was believed that the proportional wire chambers used in the preceding experiment would be unable to handle the higher pair rate in FOCUS. Thus straw tubes were constructed to cover a vertical strip down the middle of the spectrometer that is irradiated by this large pair flux. The advantage of straws is their enclosed channel design which isolates free charge in a tube. When the instantaneous rate in one straw is too high the other straw channels are not affected.

As it turned out the reconditioned PWC's were more robust than expected. Therefore, the main mission of the straws has shifted from gap coverage tracking to track parameter enhancement and track certification.

DESIGN AND CONSTRUCTION

Straw tube detectors are essentially just large arrays of Geiger tubes, where the copper coated inner tube wall is the cathode and the thin wire in the middle is the anode.

The University of California, Davis built five straw chambers for the 1996-97 run of FOCUS. Three were installed in the spectrometer and the other two were kept as spares. Each chamber has three views, and within each view there are three straw layers.

All three chambers have a vertical (x measuring) view consisting of 30 straws (3 layers by 10 rows). The most upstream chamber is built with 138 cm long straws. Its angled views ($\pm 11.33°$ from vertical) each consist of 114 straws (3 by 38). The other two chambers use 241 cm long straws and their angled views are each comprised of 222 straws (3 by 74).

The chambers were built with 5 mm diameter straws. The straw material consists of two layers of $\sim 10 \mu m$ thick mylar with a thin coating of copper on the inner surface. The anode wire is $20 \mu m$ thick gold plated tungsten. In each straw, the wire is supported at both ends by a "v" grooved feedthrough and along the straw by a helicon support [3]. The 241 cm straws have two helicon supports while the 114 cm straws require only one. The supports are placed as far from the center as possible to reduce the incidence of scattering. The supports are necessary to keep the wire from snapping over to the ground surface on the inside of the tube.

Installing the wire is a two person operation. First, thread is blown through the tube from above – which in the case of the longer straws passes four potential jam points. If the thread gets tangled between the two helicon supports the channel could be lost forever. Once the straw is threaded the person on top ties the wire to the thread and then the thread is pulled through from below. Each end of the wire is crimped off between the metal pin base and the plastic feedthrough, with a plastic collar. From below the wire is crimped with 60 grams of tension applied.

OPERATION AND READOUT

The copper-coated tube wall is held at ground and the central wire is kept at a large positive voltage ($\sim$ 1600 volts). The positive charge on the wire attracts electrons that are freed by the passage of a charged particle through the gas in the straw. As the electrons get very close to the wire there is a large signal gain due to the increased acceleration from the $1/r$ field strength and the subsequent added ionization from higher gas collison energies.

The charge is collected on the wire and a pulse forms. This pulse is amplified and shaped by the preamp/discriminator card attached to the anode at the pin on the top of the chamber. Each preamp card connects to six straw channels. The output of the preamp card is a differential ECL signal which is carried on twisted pair cables to redriver boards. The redriver boards regenerate the pulses and extend all pulse widths to a minimum of 30 ns. This is done to prevent signal loss due to attenuation in the 45 m cable run between the redrivers and the time to digital converters (TDC). When the signal arrives at the TDC it receives a rolling time stamp which is fixed when the experiment generates a trigger (common stop mode).

We are using LeCroy model 3377 TDC's with 1 ns timing resolution. The use of TDC's allows us to exploit the electron drift time in the straw to improve tracking resolution.

PERFORMANCE

The addition of the 27 straw tube planes to the existing 20 planes of PWC's has been shown to improve resolution in x by a factor of two (See figure 1a). While this incrementally improves things such as track linking to the silicon microstrip detectors, the real strength of the straw tubes is in identifying the track backgrounds.

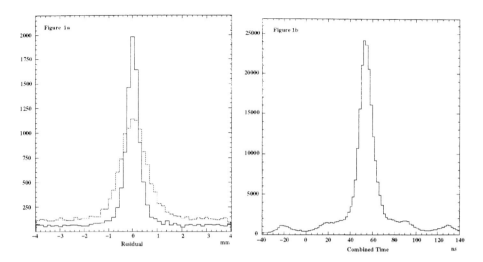

FIGURE 1. (a) Compares the straw view residuals with and without straws in the fit. In computing the residual, hits from the view in question are removed from the track fit. (solid with straws, dashed without) (b) Shows the sum of left and right drift times for straws in a track. The large peak is for the triggering rf bucket, and the smaller peaks are from tracks in other buckets.

Fermilab's Tevatron accelerator delivers beam in periodic buckets defined by the applied radio frequency (rf). Rf buckets are separated by $\sim 19ns$. The typical latched detector in FOCUS operates with a time window from 50 to 120 ns (trigger and beam tagging elements are an exception with windows of $\sim 20ns$) centered on the triggering bucket. Pair production from non-triggering buckets causes a sizable background in many detectors.

The time required for charge from the edge of the straw to drift to the wire is 53 ns. Therefore, tracks from at least 6 different buckets can leave hits that mimic a hit in the triggering bucket, but when a track causes a signal in all three layers of a straw view it is possible to, with a fair degree of accuracy, determine its rf bucket of origin (See figure 1b).

REFERENCES

1. S. Carrillo and F. Vazquez in these proceedings.
2. P.L. Frabetti, et al., *Nucl.Instrum.Meth.* **A320**, 519-547 (1992).
3. S.H. Oh, A.T. Goshaw, and W.J. Robertson, *Nucl.Instrum.Meth.* **A309**, 368-376 (1991).

Semiconductor Detectors for the ATLAS Inner Tracker

Debbie Morgan

Physics Department, University of Sheffield, England

Abstract. The ATLAS experiment currently under design for the CERN LHC contains an inner detector which tracks charged particles from the LHC beampipe to the electromagnetic calorimeter system. The main task is to reconstruct event tracks with high efficiency, to assist electron, photon and muon recognition and to reconstruct signatures of short-lived particles.

Track densities at the LHC will be extremely large, and hence high precision measurements are required. This will be achieved using semiconductor tracking detectors, making use of silicon microstrip and pixel technology.

For detectors closest to the beam interaction point the radiation levels are extremely high - up to 10 MRad. At the time of the ATLAS technical proposal, it was envisaged that gallium arsenide detectors could withstand such an environment. However, it has since become clear that GaAs is not as radiation hard as first expected, and that detectors would not perform sufficiently for the required time. In addition, progress on silicon detectors has indicated that they are able to withstand harsh radiation environments, and hence further work on silicon detectors now continues.

GALLIUM ARSENIDE DETECTORS

The radiation environment in the far forward region of ATLAS is predicted to be between 1×10^{14} and $3 \times 10^{14} n cm^{-2}$. Initial evidence that GaAs would be a suitable material came from studies with Schottky GaAs detectors that were irradiated with neutron fluences of up to $3 \times 10^{15} n cm^{-2}$. This caused the charge collection efficiency (CCE) to decrease by only 20%, with a S/N greater than 10 in the worst case. Detectors could also be operated at $T = 20^\circ C$ after irrdiation, and there was no evidence of annealing.

In order to increase the CCE of GaAs, a study was carried out to investigate the correlation between α CCE and the resistivity of a wafer. Wafers with resistivity $\leq 10^7 \Omega cm$ were obtained from various manufacturers, and two independent techniques used to verify the value. The first was a contactless technique which mapped the entire wafer and gave an indication as to variation in resistivity. The second used measurements from a forward bias I-V to calculate a Norde function, from which the resistivity could be extrapolated. Response to α and β radiation was measured on the front and back of the detector using Am-241 and Sr-90 sources. Results however were inconclusive; there was no improvement in CCE with low resistivity material, and material from different manufacturers proved to give very different results.

A pion irradiation study took place at the Paul Scherrer Institute, where detectors were irradiated to 8×10^{13} - $12 \times 10^{14} \pi cm^{-2}$. CCE measurements were taken before and after the irradiation, along with measurements of the leakage current. A severe degradation of the α signal occurred in all detectors, and no MIPs signal was seen in most detectors after the irradiation. Detectors experienced a slow breakdown, with leakage currents reaching -8 μA, compared to -0.5 μA previously.

A GaAs progress review for ATLAS took place in December 1996. It was indicated that damage from charged particles was significant, and detectors could not be guaranteed to operate efficiently for the lifetime of the LHC. Hence silicon detectors will now continue to be developed to be placed at the inner most radii, and GaAs will no longer be used.

SILICON DETECTORS

The ATLAS tracker requires one 'barrel' detector design, and five wedge shaped 'forward' detectors which will be built in discs, requiring a total of 20 000 detectors. Both barrel and wedge modules have been tested in the 1997 KEK testbeam, using binary electronics. Of the wedge detectors 1 had LBIC readout i.e. non-radiation hard , 1 had CAFE readout, which is radiation hard. The aim was to study the p-stop isolation technique, where detectors have either common, individual or combined p-stop structures. High precision telescope modules were used to localise beam hits in the different p-stop region. Analysis of the data is in progress.

Most recently,silicon detectors have been irradiated at the PS at CERN. A system has been designed and successfully implemented that has allowed detectors to be irradiated with high intensity protons, whilst being maintained at a constant temperature of $-7°C \pm 1°C$. This is done by placing the detectors in a cool box connected to a chiller which pumps glycol into a radiator. Fans in the box then ensure that the correct temperature is maintained. The cool box is placed on an x-y stage, allowing the beam to 'scan' the detector and ensure that a homogenous fluence is received over the entire surface. Detectors have been irradiated to $2 \times 10^{14} pcm^{-2}$, with both the temperature and leakage currents recorded every 10 seconds. As expected, the leakage current increases with dose and bulk damage causes the depletion voltage to increase. A further run is planned for August, where both detectors and modules will be irradiated.

The Silicon Vertex Detector of HERA-B

Basil Moshous

for the HERA-B Vertex Detector Group
Max-Planck-Institut für Physik, Werner-Heisenberg-Institut, Fhringer Ring 6,
80805 München, Germany

Abstract. HERA-B is an experiment to study CP violation in the B system using an internal target at the DESY HERA proton ring(820 GeV). The main goal is to measure the asymmetry in the "gold plated" decays of $B^0, \bar{B}^0 \to J/\Psi K_s^0$ yielding a measurement of the angle β of the unitarity triangle.
From the semileptonic decay channels of the $b, \bar{b}$-hadron produced in association with the $B^0, \bar{B}^0$ can be used to tag the flavor of the B^0. The purpose of the Vertex Detector System is to provide the track coordinates for reconstructing the $J/\Psi \to e^+e^-, \mu^+\mu^-$ secondary decay vertices and the impact parameters of all tagging particles.

THE HERA-B VERTEX DETECTOR SYSTEM

The VDS is installed between the target and the spectrometer magnet. It consists of seven superlayers of silicon strip detectors. The angular acceptance is 10-160 mrad horizontally and 10-250 mrad vertically. This corresponds to 90% of 4π in the center of mass system. Each superlayer has four quadrants consisting of two double sided silicon strip detectors each. The silicon strip detectors have sensitive areas of 5x7 cm^2. The strips on the two sides of the silicon detectors are perpendicular to each other. They are rotated by $2.5°$ with respect to the rectangular shape of the silicon detector. By placing two detectors back to back one obtains a geometry of four views where many ghost hits which appear when multiple particles cross the detector can be ruled out. This leads to a reduction of combinatorics in track finding. With a $25\mu m$ diode strip pitch and a $50\mu m$ readout pitch one can achieve a resolution of better than $12\mu m$. This leads to an impact parameter resolution for J/Ψ reconstruction of about $20\mu m$ in x and y. The B vertex resolution in z is better than $500\mu m$.

THE HERA-B 1996/97 SETUP

In 1996 three prototype silicon strip detectors were installed in the same quadrant of the last three superlayers. With these detectors which had every second strip bonded to the readout electronics we were able to verify the functionality of our detectors and the whole readout chain. Two detectors were used to reconstruct tracks coming from the target (fig. 1).

In 1997 the three old prototypes were replaced with fully bonded new prototypes. A first look on the data taken in May and June showed a very good quality of the reconstructed tracks which are in agreement with the expetations.

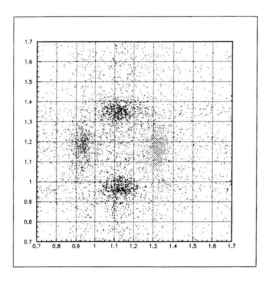

FIGURE 1. Target plane of the HERA-B detector. In this figure the impact points of reconstructed tracks using the hits from two silicon detectors in the target plane are shown. One can clearly see the 4 different targetwires and a combinatorical background.

The Mexican Participation at the Pierre Auger Observatory: Recent Results

S. Román[1], F. Alcaráz[2], E. Cantoral[1], J. Castro[3], A. Cordero[1], A. Fernández[1], R. López[1], C. Pacheco[2], M. Rubín[1], H. Salazar[1], J. Valdés[4], M. Vargas[1], L. Villaseñor[2], A. Zepeda[5]

[1] *Facultad de Ciencias Físico-Matemáticas, BUAP, Puebla, Puebla*
[2] *Instituto de Física y Matemáticas, UMSNH, Morelia, Michoacan*
[3] *Instituto Nacional de Astrofísica Optica y Electrónica, Puebla, Puebla*
[4] *Instituto de Geofísica, UNAM, México, D.F.*
[5] *Departamento de Física, CINVESTAV-IPN, México, D.F.*

Abstract. In this work we present the participations of the Mexican group at development of the Pierre Auger Observatory. We have been working in both parts of the hybrid proposed for the Auger detector, the fluorescence and the surface detectors. In the part of fluorescence, we have analyzed the resolution of the Hi-Res optical design of the fluorescence detector observatory. We have found a heterogeneus image resolution. We propose to use a lensless Schmidt camera (with spherical image surface) to duplicate the field of view to 30x30 degrees and simultaneously guarantee a resolution of one degree over of the whole field of view. By the Surface Detector, a water Čerenkov detector (WCD) prototype of reduced dimensions (cylinder 1.54 diameter filled with purified water up to 1.20 m high) is used to obtain preliminary experimental results that validate the concept of remote calibration and monitoring of WCDs. We use muons that stop and decay inside the WCD and, in a complementary way, muons that croos the WCD. We used a moun telescope trigger in order to study the charge distribution of vertical muons, their pulse amplitude decay and the Cerenkov light attenuation lenght of those secondary cosmic muons we include the bacteria population content for the four months of operation to validate the monitoring method.

INTRODUCTION

The Auger Observatory is to be a hybrid detector, employing two complementary techniques to observe extensive air showers. A giant array of particle counters will measure the lateral and temporal distribution of shower particles at ground level. An optical air-fluorescence detector (FD) will measure the air shower development in the atmosphere above the surface array. Operating together, the surface detector (SD)array and fluorescence detector characterize showers to a greater degree than either technique alone. Both methods are well established by prior experiments. The surface array resembles the array succesfully employed by the Havera Park group for over twenty years (ref. [1]), althought on much larger scale. The optical device uses the fluorescence technique pioneered by the University of Utah's Fly's Eye (ref. [2]). Measurement of atmospheric fluorescence is possible only on clear, dark nights.

The decision to use the two techniques together is based upon in this aspects, Intercalibration, Enhanced composition sensitivity, Hadronic interactions, Uniform exposure and Cost.

Auger's hybrid configuration is the most economical and robust way to obtain the the necesary data, including a subset with specially high reconstruction resolution and independent cross checks.

Each of the two surface arrays of the Auger Observatory will consist of about 1600 detectors spaced on a grid with about 1.5 km separation between individual detectors. Each array encompasses an area of 3000 km^2. The angular and energy resolution of a groun array (without coincident fluorescence data) are typically less than 1.5^0 and less than 20%, respectively. If an event trigger is assumed to require five detectors above threshold, the array is fully efficient at 10^{19} eV. New technologies are employed, making it practical to operate thousands of detectors spread over such an area. Each detector will be solar powered (consoming less than 10 watts) and will communicate via modern wireless techniques. Inter-detector relative timing is accomplished by individual Global Positionating Satellite (GPS) receivers.

The fluorescence detectors consist of many meter-sized mirrors, each of which is aquipped with a cluster of hundred or more photomultipliers. Each mirror and associated cluster will view its own segment of the sky. Together, the system of mirrors observes most of the sky above the surface array. The magnitude of the photomultipliers signals gives the number of electromagnetic particles in the shower, and hence the energy. Fast timing of the sequence of signals yield the trajectory of air showers passings in the field of view of the detector. In the hybrid mode of operation, the surface and fluorescence detectors together have a directional reconstruction resolution of about 0.3^0 for events near 10^{20} eV (ref. [3]).

In this paper we present the work done for the Mexican Auger collaboration. We will describe briefly the main objetives and principal results of our effort in the fluorescence and surface detectors development. In the second section we will detail our optical design proporsal for the FD, including the correspondent values of its optics parameters. The third section will be dedicated to the description of our results on the water Cerenkov plastic tank wich is operating since March 1997 in Mexico. Finally we present the conclusions in the Section 4.

OPTICS AT THE FLUORESCENCE DETECTOR

In the optical design in order to improve the Hi-Res telescope, the optical group (Cordero et al) to use a lensless Schmidt camera. A spherical with an aperture-stop at its center of curvature, "C", would give uniform images over a wide spherical field-surface concentric with itself, each image affected from a large amount of spherical aberration.

In table 1, the parameters of the design telescope are shown.

TABLE 1. Parameters of Schmidt Camera

Parameter	Value
Curvature Radius of the mirror	2415 mm
Diameter of the mirror from the mirror vertex	2566.4 mm
Best image plane	1161.5 mm
Effective focal distance	1207.5 mm
Detector array size	628.37 mm
Spot size	20.29 mm

The main adventage of this proposal is that the image quality can be guaranted even when the field of view is extended from 15x15 degrees to 30x30 degrees, 40x40 degrees and 45x45 degrees and so on.

This method offers several important adventages: (1) Coma aberration is elimited and the spot size image is uniform over the larger field of view. (2) The reduction in the number of buildings, electronics crates, calibration devices, etc., can reduce the cost of each eye. (3) The opening to the outside is only the diaphragm with UV transparent material (like thin mylar), it may be possible to keep dust and dirt out of the telescopes nad maintain them at nearly constant temperature, thereby eliminating important causes of sensetivity variation. However, what is the prize that we must pay for extending the field of view? If we increase the field of view and if we do not want the vignetting effect then the diameter of the mirror must to be increased. However if the field of view goes from 15x15 degrees to 30x30 degrees, then the constructed area of the mirror goes from 3696367 mm^2 to 5869603 mm^2, i.e. the area increases only 1.6 times; while the explored area on the sky grows 4 times. On the other hand we lose energy when we increase the field of because the detector area grows and it obstructs the pass of light. This is a serius problem because we must increase the diameter of diaphragm. Then we must increase the diameter mirror and must re-analyze the optical design for spherical aberration, which grows as the cube of the radius of diaphragm.

Because it is desirable to have a plane image surface instead of the spherical one, then we can make a small modification of our initial optical design. The

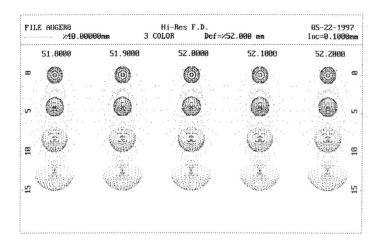

FIGURE 1. Spot diagrams of the Schmidt camera.

first step was reached locating the focal plane at the vertex of spherical image surface. We found the spot sizes grow to 5 times at the end of the field with respect to the image axis. Then, in the second step, we reached the optimal defocus and we found in this case the spot sizes only grow 1.6 times with respective to the image on axis (see fig. 1). So, we showed that even with a plane image surface we improve the Hi-Res results.

If we increase the curvature radius we find several spot sizes for one degree on the sky. We found that if the curvature radius is equal to 3500 mm then the spot size is equal to 28 mm and the image size for one degree on the sky is equal to 30 mm. In this case the diameter of the mirror is equal to 3183 mm and the new area is 4 times the area of the Hi-Res telescope. Remember that we should need four Hi-Res telescopes to cover a field of view 30x30 degrees. The very impotant difference is that the image quality is four times better than Hi-Res.

Finally we have constructed a 1/10 scaled Schmidt camera prototype at Puebla. We built a segmented spherical mirror. We obtained a uniform image over the whole field of view. With this experience in mind, we could make a full scale Schmidt lensless camera (ref. [4]).

THE MEXICAN WATER ČERENKOV DETECTOR PROTOTYPE

The mexican tank is made of a polyethylene cylinder, white on the inside and black on the outside wall, 1.54m in diameter, filled with purified water (commercially avaliable as drinkable water in 20-liter bottles) up to heght of 1.2m.

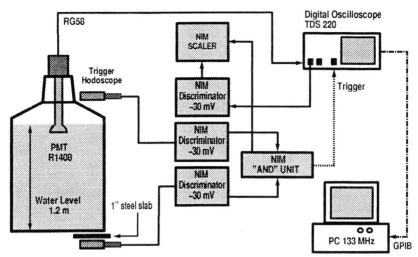

FIGURE 2. Experimental setup and data acquisition system.

A single Hamamatsu 8" PMT looking downwards was located at the center of the tank with the PMT slightly immersed in the water. The inner wall of the tank and the water surface were covered with a tyvek sheet cut to a cylindrical shape and kept in place by circular PVC hoses stretched tight against the wall of the tank. We used a Hamamatsu R5912 during the first three months of operation; then we replaced it for a Hamamatsu R1408. In the trigger system, it is consisted of two scintillation counters, one on the top and the second formed the bottom trigger. Both counters forming a vertical telescope which accepted vertical muons in a cone of almost 2 degrees with respect to the vertical with an active area of $0.073 m^2$. The hight voltage of the top (bottom) PMT paddle was 1.87 (1.60) kV., and the corresponding provided pulse rates were $3.80\,Hz$ and $3.60\,Hz$, respectively. In the data acquisition system we have that the signal from the PMT was discriminated with a commercial NIM module using a threshold of $-30\,mV$., A custom made TDC CAMAC module was used to trigger with two scintillation paddles in coincidence to measure the vertical muon pulse shape as a function of time(hundreds of nanoseconds). The signal from all the PMT's (the Hamamatsu and the small paddle's PMTs) were transmited via RG58 coaxial cables to the digital osiloscope and the NIM module. The charge was obtained integrating the pulse shape. A pentium PC running at $75\,MHz$. and a DAQ program written in LabView were used to store pulses and the charge as recorded by a Tektronix TDS220 digital scope for a large number of events(see figure 2)(ref. [5]).

By the calibration we used also a custom-made CAMAC TDC module, referred to as muon module from hereafter, was used to measure the time interval between consecutive pulses coming out of the discriminator. And the

CAMAC controller used was the LeCroy 8901; it was connected to a National Instruments GPIB port on a pentium PC running at 133 MHz(ref. [6]).

CONCLUSIONS

In the optical part is possible to build an optimized Schimidt camera without correcting plate, i.e. a spherical mirror and a diapragm located at the center of curvature plane of that mirror. This will warranty a homogeneuos image size of 20.3 mm over the detectors on all the 30 by 30 degrees field. Moreover we have analyzed flattening the image surface and we found that it is possible to use a falt or segmented flat surface in which cases the image quality is guaranteed too.

In the surface part, the average number of photoelectrons produced by the decay electron coming from stopping muons is found to be 0.18 times the average number produced by vertical muons that cross the WCD coming from the background flux of secondary cosmic ray muons. Likewise, the charge distribution of the first pulse for the events that satisfy the cut $C_2 < C_1$, which we associate with crossing muons (in the sense that this cut is the complement of the stopping-muon cuts), shows a clear peak which is slightly displaced to the right of the peak corresponding to vertical muons. The peak positions of the charge distributions of the decay electron and the crossing muons can be used to calibrate the WCD and to monitor the gain of its PMT's.

In the figure 3.a shows the charge distribution of the first pulse for events that pass the cuts $C_2 < C_1$ and time between pulses $< 8\mu s$. The solid line is a gaussian fit to the data. Figure 3.b shows the charge distribution of the first pulse for the events that satisfy the cut $C_2 < C_1$, which selects crossing-muon events. The vertical arrow indicates the position of the peak for vertical muons.

This method can be applied even in the case that the WCD is segmented in three sectors in such a way that the Čerenkov light emited in each sector is collected only by one PMT and not by the three PMT's as in the unsegmented case. In the case that the WCD's are unsegmented this method can be complemented with the requirement of double or triple coincidences among the three PMT's to reduce the background noise from the PMT's. Another important conclusion is that the calculated average attenuation length $(9.7\,m.)$ and the number of photoelectrons were practically constant during our four month data taking period. We believe that this result is very dependent on plastic tank material. From the report of the bacteria population in the inner water, we could see that the number of bacteria becomes very low with respect to the bacteria content when the tank was filled. We think that it is because of lack of air in the interior of our tank. It is also a consequence of the intrinsical hygenic properties of the plastic material. It is clear that this result is good news for the Auger Čerenkov tanks final design because we have shown that

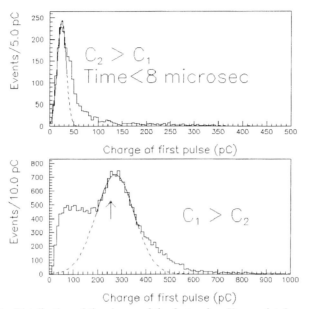

FIGURE 3. Distribution of the charge of the first pulse. Upper plot:for events satisfying charge(pulse$_2$) > charge(pulse$_1$) and time between pulses < $8\mu s$. The curve is a gaussian fit to the data. Lower plot: distribution of the charge of the first pulse for the events satisfying charge(pulse$_2$) < charge(pulse$_1$), the arrow indicates the position of the peak for vertical muons, the solid curve is a gaussian fit to the data.

the plastic material could conserve pure the inner water for a long period of time. The final conclusions will be obtained from the results of the full size plastic tank wich is already working in Puebla.

REFERENCES

1. M. A. Lawrence, R. J. O. Reid and A. A. Watson, *J. Phys.***G17**, *733* (1991).
2. R. M. Baltrusaitis et al *Nucl. Inst. Methods* **A240**, *410* (1985).
3. *"The Pierre Auger Collaboration Design Report"*, March 1997, Fermi National Accelerator Laboratory.
4. A. Cordero, et.al., " Proposal for the optical system of the fluorescence detector of the Auger Project", *Auger Project technical note (GAP- 96-045)*.
5. F. Alcaraz, et.al., "Estudies with a water Čerenkov detector for the Pierre Auger Observatory", *Auger Project technical note (GAP- 97-050)*.
6. L. Villaseñor, et. al. " Calibration and monitoring of water Čerenkov detectors with stopping and crossing muons", *Auger Project technical note, (GAP- 97-033)*.

DIRC, the Internally Reflecting Ring Imaging Čerenkov Detector for BABAR: Properties of the Quartz Radiators

Jochen Schwiening[†]

Stanford Linear Accelerator Center, Stanford, CA 94309
[†]*Representing the* BABAR DIRC *Collaboration*

A new type of detector for particle identification will be used in the BABAR experiment [1] at the SLAC B Factory (PEP-II) [2]. This barrel region detector is called DIRC, an acronym for Detection of Internally Reflected Čerenkov (light). The DIRC is a Čerenkov ring imaging device which utilizes totally internally reflecting Čerenkov photons in the visible and near UV range [3]. It is thin (in both size and radiation length), robust and very fast. An extensive prototype program, progressing through a number of different prototypes in a hardened cosmic muons setup at SLAC [4] and later on in a test beam at CERN [5], demonstrated that the principles of operation are well understood, and that an excellent performance over the entire momentum range of the B factory is to be expected.

The DIRC utilizes long, thin, flat quartz radiator bars (effective mean refractive index $n_1 = 1.474$) with a rectangular cross section. The quartz bar is surrounded by a material with a small refractive index $n_3 \sim 1$ (nitrogen in this case). For particles with $\beta = 1$, some of the Čerenkov photons will be totally internally reflected, regardless of the incidence angle of the tracks, and propagate along the length of the bar. To avoid having to instrument both bar ends with photon detectors, a mirror is placed at one end, perpendicular to the bar axis. This mirror returns most of the incident photons to the other (instrumented) bar end. Since the bar has a rectangular cross section, the direction of the photons remains unchanged during the transport, except for left-right/up-down ambiguities due to the reflection at the radiator bar surfaces. The photons are then proximity focused by expanding through a stand-off region filled with purified water (index $n_2 \sim 1.34$) onto an array of densely packed photomultiplier tubes placed at a distance of about 1.2 m from the bar end. In the present design the bars have transverse dimensions of 1.7 cm thick by 3.5 cm wide, and are about 4.90 m long. The length is achieved by gluing end-to-end four 1.225 m bars, that size being the longest high quality quartz bar currently available from industry.

Several natural and synthetic fused silica candidate materials were tested for their optical properties and radiation hardness. In a Co^{60} source, samples were exposed to doses of up to 500 krad. While natural quartz materials showed significant absorbtion in the wavelength range of the Čerenkov photons after being exposed to only a few krad, the synthetic material proved to be sufficiently radiation hard. This led to the choice of Suprasil Standard [6] and Spectrosil 2000 [7] as bar material for the DIRC.

Bars were formed from the synthetic quartz material, produced as large cylindrical ingots, using modifications of conventional optical processing techniques [8]. In order to preserve the photon angles during surface reflections, the faces and sides were nominally parallel while the orthogonal surfaces were kept nominally perpendicular. Typically, the bar's surfaces were flat and parallel to about 25 μm, while the orthogonal surfaces were perpendicular to a tolerance of 0.3 mrad. The most difficult requirements were associated with maintaining the photon transmission during reflections at the surfaces of the bar (a Čerenkov photon may be internally reflected a few hundred times before exiting the bar). This led to rather severe requirements on edge sharpness and surface finish. After polishing, the bars had an average edge radius less than 5 μm, and a nominal surface polish of better than 0.5 nm (RMS). The optical properties of the radiator bars were measured using a HeCd laser in a motion-controlled setup. The absorption of a quartz bar is typically about 1%/m at 325 nm and less than 0.2%/m at 442 nm. The coefficient of total internal reflection at 442 nm was found to be (0.99960 ± 0.00006), consistent with the expected reflectivity for the nominal surface polish.

REFERENCES

1. The BABAR Collaboration, *Technical Design Report*, SLAC-REP-950457 (1995)
2. "An Asymmetric B Factory Based on PEP: Conceptual Design Report", LBL-PUB-5303/SLAC-REP-372 (1991)
3. B.N. Ratcliff, SLAC-PUB-5946 (1992) and Dallas HEP (1992) 1889, B.N. Ratcliff, SLAC-PUB-6047 (1993), P. Coyle et.al., *Nucl. Instr. Methods* A 343 (1994) 292
4. D. Aston et.al., *"Test of a Conceptual Prototype of the Total Internal Reflection Cherenkov Imaging Detector (DIRC) with Cosmic Muons"*, IEEE Trans. Nucl. Sci. NS-42 (1995) 534
5. H. Staengle et.al., *"Test of a Large Scale Prototype of the DIRC, a Čerenkov Imaging Detector based on Total Internal Reflection for BABAR at PEP-II"*, SLAC-PUB-7428 and LBNL-40099 (1997), submitted to *Nucl. Instr. Methods*
6. Heraeus Amersil Inc., 3473 Satellite Blvd. 300, Duluth, Georgia 30136
7. Quartz Products Co., 1600 W. Lee St., Louisville, Kentucky 40201
8. Zygo Corporation, Laurel Brook Road, Middlefield, Connecticut 06455

Segmented Wire Ion Chambers (SWICs) Used as Proton Beam Position/Profile Detectors in the Fixed Target Beamlines at Fermilab

Gianni Tassotto

*Fermi National Accelerator Laboratory**
P. O. Box 500, Batavia, IL 60510

Abstract. Segmented Wire Ion Chambers have been used at Fermilab since 1972 to monitor the profile of the proton beams in the various beamlines. Many modifications and improvements have been made since that time. SWICs are presently used to display beam profiles at intensities from 10^6 to over 10^{13} particles/sec.

INTRODUCTION

After the proton beam is extracted from the Tevatron ring at 800GeV of energy it is transported through Switchyard to form the primary and secondary beams required to operate the Fixed Target Experiments. The Beams Division is responsible for transporting these beams to the various targets. The devices that have been used extensively to monitor the position and profile of the beams are Secondary Emission Electron Detectors (SEEDs), SWICs and in low intensity beamlines Proportional Wire Chamber (PWCs) (1) (2).

CHAMBER MODIFICATIONS

A detailed description of how the chambers were built can be found in Fermilab TM-1853. Figure 1 shows how the chambers are presently assembled. The modifications made take into account beam intensity and noise considerations. To avoid potential problems at high intensities the wire was changed from nickel to gold plated tungsten. A single chamber is now constructed for a certain beam spot

---

* Operated by University Research Association, Inc. under contract No. DE-AC02-76CHO300 with the United States Department of Energy.

CP422, *Instrumentation in Elementary Particle Physics:* The VII ICFA School
edited by G. Herrera Corral and M. Sosa Aquino
© 1998 The American Institute of Physics 1-56396-763-4/98/$15.00

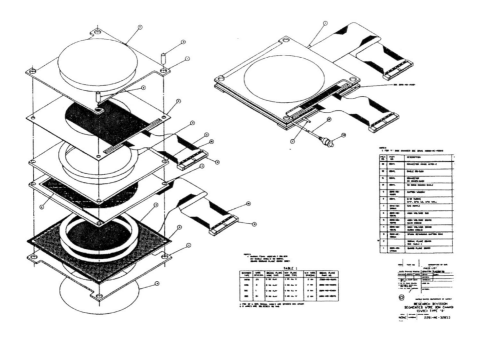

FIGURE 1. SWIC assembly.

size. Operators can optimize the profile by varying high voltage and/or scanner charge and gain parameters. The use of RG-174 coax cable was abandoned in favor of 27 twisted pairs and flat-to-round cable made by Hitachi (3). To increase the signal to noise ratio the cables are pulled directly from the scanners to the chambers located in the tunnels. The first and last wires are grounded at the scanner chassis.

ELECTRONICS

Of the 166 SWICs presently installed in the Fixed Target Beamlines 140 are still connected to a Z-80 based scanner as described in Fermilab TM-1853. The electronics for the SWICs located in the Switchyard enclosures has been upgraded. The hardware consists of a multiprocessor system that has a master CPU and multiple data collection slaves (4). The sequencer manages the integration and storage of data and stores up to 16 scans that can be recalled at will. The VME crate has access to all data.

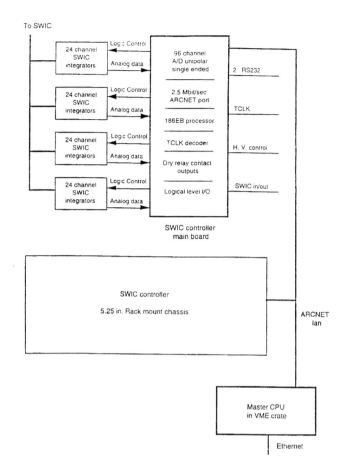

FIGURE 2. Control data schematic.

The communication between the CPU and the slaves is done via ARCNET. Ethernet supports communication between the VME crate and the computer consoles. The software has also been upgraded. Statistical calculations for the mean and the sigma now account for floating and/or dead wires. Figure 2 shows a SWIC control and data retrieval schematic layout.

BEAM POSITION/PROFILE

Of the many fixed target beamlines KTeV has the most stringent requirements for beam position stability: <100μm. The target SWIC and the target SEED were

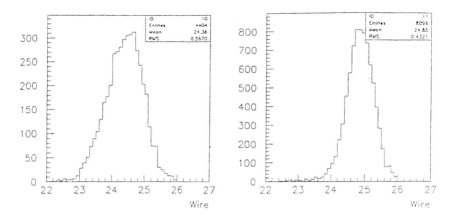

FIGURE 3. The figure on the right shows the improvement in beam stability after AUTOTUNE was implemented.

designed to have a position resolution of better than 50μm. An AUTOTUNE program was developed to maintain this stability (5). AUTOTUNE is an X-window application that displays a set of SWIC profiles relevant to a particular beamline. The program calculates the mean and the sigma for 48 wires first and then recalculates them for a smaller window i.e., ±10 wires determined empirically. The program then calculates δx, δy for spill i and changes an upstream magnet current so that if the beam does not change δx, δy will equal zero at spill i+1. KTeV target SWIC has a wire spacing of 250μm. A spill-to-spill stability of 100μm has been achieved. Work is continuing to improve beam stability particularly in long beamlines where multiple SWICs are taken into consideration. Figure 3 shows the improvement in beam stability after AUTOTUNE was implemented.

DISCUSSION

For low and medium intensity beamline we'll continue using SWICs as profile detectors; for high intensity beamlines the beam losses become intolerable due mainly to the SWIC beamline windows. To minimize the losses we are considering installing OTR screens and SEEDs particularly as target profile detectors. The detectors will be at beamline vacuum so the only material presented to the beam will be a very thin window for an OTR screen and/or a set of 0.003" AuW wires with bias foils for a SEED.

REFERENCES

1. G. Tassotto, "Beam Profile Monitors for High Intensity Proton Beams", AIP Conference Proceedings 309, Argonne IL, May 6-9, 1996, 506-509.
2. H. Fenker, "A Standard Beam PWC for Fermilab", FNAL TM-1179.
3. D. Schoo, "A Low-Cost Replacement Cable for Beamline SWIC Installation", FNAL TM-1107.
4. W. Kissel, B. Lublinsky, A. Frank, J. Smolucha, "New SWIC Scanner/Controller System", ICALEPCS Conference Proceedings, Chicago, IL, Oct. 29-Nov 3, 1995.
5. T. Kobilarcik, "Preliminary Results of Stability Study for the KTeV Bram", Work in progress.

HELIX128S-2 - A Readout Chip for the Silicon Vertex Detector and Inner Tracker Detektor of HERA-B

U. Trunk*W. Fallot-Burghardt*, E. Sexauer*, K-T. Knöpfle*,
W. Hofmann* M. Cuje[†], B. Glass[†], M. Feuerstack-Raible[†],
F. Eisele[†], U. Straumann[†]

*Max-Planck-Institut für Kernphysik, Heidelberg, Germany
[†]Universität Heidelberg, Heidelberg, Germany

Abstract. HERA-B is a fixed target experiment at the HERA proton storage ring dedicated to examine CP-violation in the B-Meson system. Based on the RD20-FElix [1] concept a readout chip has been designed in AMS's $0.8\mu m$ CMOS process for the HERA-B silicon vertex and inner tracker (MSGC) detectors. Various test chips have been submitted and successfully tested since '95, thus enabling the submission of a fully integrated 128 channel version in April '97. Design features of this chip (HELIX128S-2 [2]) and test results of its predecessor HELIX128 [3] are presented.

HELIX128S-2 SPECIFICATIONS AND DESIGN FEATURES

The chip integrates 128 channels with an overall pitch of input pads of 50 μm as imposed by the SVD's strip pitch. Major electrical specifications are a 10 MHz sampling frequency, analog pipeline with up to $12\mu s$ storage time and dead-timeless readout. Further demands are lowest possible noise, the handling of successive triggers and moderate radiation tolerance ($\approx 1kGy$). Since the inner tracker detector is included in HERA-B's Level-1 trigger decision, a discriminator is inplemented behind the frontend stage, which now features a differential amplifier design. The chip's input incorporates protection diodes for operation with MSGC detectors. The frontend uses a traditional folded cascode charge sensitive amplifier/shaper design and has been measured separately (on the HELIX 2.0 and 2.1 chips) to have an ENC of $310e^- + 39e^-/pF$ at 45 ns peaking time. The depth of the analogue storage pipeline was increased to 128×141 cells, thus providing a usable storage depth of 128 cells plus a derandomizer-buffer handling up to 8 triggered events. Via the resetable "pipeamp" and a two-stage 40MHz multiplexer, triggered events are serialized and brought offchip with a high-speed low-power current buffer. The chip's bias generator, the preamplifier and shaper control and the chips initialization circuit are programmed via a dedicated serial interface. Other features are synchronicity monitoring and daisy-chained readout of two or more chips.

HELIX128 MEASUREMENT RESULTS

HELIX128 uses the older HELIX 1.1 frontend. The bias generator, initialization circuit etc. are integrated on a separate chip ("SUFIX") [4]. The HELIX128's overall linearity range (Fig. 2) has been found to be -10 MIP to +7 MIP ($1MIP \cong 2.5ke^-$) with an amplification factor of 55mV/MIP and $\approx 60ns$ shaper time. The chip's overall noise has been measured to be $405e^- + 76e^-/pF$, suffering from a protection resistor needed for MSGC readout.

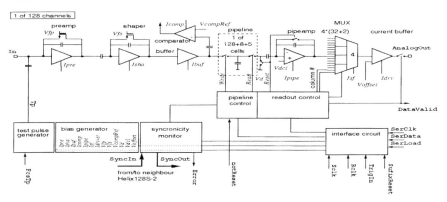

FIGURE 1. Block diagram of the HELIX 128S-2 chip

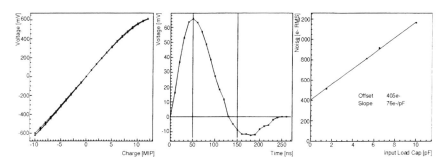

FIGURE 2. Linearity, pulse-shape and noise of the HELIX 128-1 chip

REFERENCES

1. S. Brenner et al., *Nucl. Instr. Methods* A339(1994) 477 and *Nucl. Instr. Methods* A339(1994) 564
2. W. Fallot-Burghardt, M. Feuerstack, U. Trunk et al., *HELIX128S-2 User's Manual*, HD-ASIC-33-0697
3. W. Fallot-Burghardt, M. Feuerstack, U. Trunk et al., *HELIX128 - An Amplifier and Readout Chip [...]*, HD-ASIC-18-0696
4. U. Trunk, *SUFIX 1.0 - A Support and Control Chip for the HELIX Preamplifier and Readout Chip*, HD-ASIC-13-0596

Characterisation and Preliminary Analysis of the 1996 GaAs Keystone Test Beam Detectors.

S.Walsh[*], C.Buttar[*], P.Sellin[*], D.Morgan[*], C.Hardingham[†] and J.Burrage[†]

[*]*Department of Physics, University of Sheffield, The Hicks Building, Houndsfield Rd, Sheffield S3 7RH.*
[†]*EEV, Waterhouse Lane, Chelmsford.*

Abstract. The characterisation results from GaAs microstrip detectors prior to the 1996 test beam will be presented, as well some preliminary analysis. Devices were fabricated on 200μm thick, 3 inch wafers of Semi-Insulating GaAs according to the geometry of the ATLAS[1] experiment with rear patterned ohmic contacts. The patterned rear contact was shown to have a benefit on the IV characteristics, reducing the leakage current and increasing the device breakdown voltage. CCE measurements on pad devices are also reported. Successful ac coupling is shown from the dielectric capacitance measurements, and the measurement of the Ge bias resistors implemented for the first time. Preliminary analysis of the test beam data shows a good correlation between the GaAs keystone detector and the silicon telescope. The S/N is 7.488 using tracking and the position resolution is 21μm.

INTRODUCTION

The aim of the 1996 test beam was to prototype a GaAs 6 cm AC coupled keystone detector with integrated resistors using LHC speed readout electronics. The devices were fabricated on three 3 inch wafers, 200 μm thick commercially available Semi-Insulating GaAs grown by vertical gradient freeze method. The front microstrips were a Schottky contact (Ti/Au 20:300 nm) and the rear ohmic (Ge/Ni/Au 20:20:200 nm) was continuous for one wafer and patterned for the remaining two. This was to investigate if the use of a rear patterned contact improved the breakdown behaviour of the device. The devices were fabricated by EEV using a photolithography and lift-off process[2].

A half keystone geometry detector was designed of length 6 cm with 256 strips of pitch 62 μm increasing radially to 75μm with a constant strip separation of 15μm. Also included were 2 cm AC and DC coupled version of the above and test structures for characterisation purposes.

DETECTOR CHARACTERISATION AND ANALYSIS.

IVs on pad detectors show that the rear patterned detectors have a lower leakage current of 7 nA/mm^2 with a maximum bias voltage of -250 V before slow breakdown occurs while the non patterned pads have a higher leakage current of 16 nA/mm^2 with an abrupt breakdown between 150-200 V. The IV for a single strip of a keystone detector with rear patterned strip was biased to 180 V while the non patterned broke down at 100 V. The lower breakdown and higher leakage current of the detector, compared to the pad devices, is probably due to the poor lift-off of the lower metal layer which caused surface breakdown.

Charge collection efficiency for a rear patterned pad was measured for α irradiating on the front (rectifying) and rear (ohmic) contact, and with mips (Sr-90). The device is fully active at 250 V since back alpha irradiation was observed. The CCE for α irradiating the front contact in the plateau region is 36% and a maximum of 38 % corresponding to a collected charge of 485247e. The maximum CCE for back alphas is 49% and for mips 47%. The low CCE is believed to be due to charge trapping in the wafer and a material limitation rather than due to the detector fabrication.

The silicon nitride capacitance was measured across a 128 (2 cm keystone geometry) strip detector and the average value found to be 0·1508 nF which corresponds to a thickness of 476 nm.

Problems were encountered bonding to the detector. To successfully bond a high force had to be applied which cracked the silicon nitride destroying the dielectric layer, hence an external capacitor had to be introduced. The bond pad regions were later examined by a scanning electron microscope and atomic force measurements which revealed a very smooth surface with no surface contamination. This problem can be avoided in future by depositing a layer of aluminum to the bond pad to which the wire has no problem adhering to.

Analysis to date shows a good correlation between the GaAs strips and Si telescope. However, the data is noisy so that a signal-to-noise of 7.488 has only been achieved. From the residual of single strip hit the position resolution[3] is 21μm. At present there appears to be limited evidence of charge sharing within the detector, but the analysis continues.

REFERENCES

1. ATLAS Technical Proposal, **CERN/LHCC/94-43** (1994)
2. S.Walsh et al, Como Conf Proceedings to be published.
3. P.J.Sellin et al, *Nucl. Inst. Meth.* **A381** (1996) pp57.

LIST OF PARTICIPANTS

Lecturers
Peter Cooper	USA
Tord Ekelöf	Sweden
Claus Grupen	Germany
Helmut Marsiske	USA
Aurora Savoy-Navarro	France
Helmut Spieler	USA
Jaroslav Va'vra	USA
Peter Weilhammer	Switzerland

Review Talks
Muzaffer Atac	USA
Carlos Escobar	Brazil
Dan McCammon	USA
Kenzo Nakamura	Japan
Lev Shekhtman	Russia
Arnulfo Zepeda	Mexico

Lab. Instructors
José Rubén Alfaro Molina	Mexico
Muzaffer Atac	USA
Alessandro Brez	Italy
Sergei Chernenko	Russia
Marcus French	England
Paolo Giubellino	Italy
Geoff Hall	England
Marvin Johnson	USA
Idzik Marek	Italy
Arnulfo Martínez Dávalos	Mexico
Alan Rudge	Switzerland
Marleigh Sheaff	USA
Lev Smykov	Russia
Jonathan Streets	USA
Luis Villaseñor	Mexico

Students
Duccio Abbaneo	Italy
Francisco Javier Alcaraz Ayala	Mexico
Serkant Ali Cetin	Turkey
Carlos Avila	Colombia
Betzaida Batalla García	Puerto Rico

Tom Beckers	Belgium
Kim Bernier	Belgium
Uwe Bratzler	Germany
Isabelle Boulogne	Beligium
Javier Cardona	Colombia
Salvador Carrillo	Mexico
Edgar Casimiro	Mexico
Eleazar Cuautle Flores	Mexico
Eva Dahlberg	Sweden
Mario David	Portugal
Galileo Domínguez Zacarias	Mexico
Jorge de la Torre	Guatemala
Mattias Ellert	Sweden
Emilio Esparza Coss	Mexico
Edgar Emilio Galindo León	Colombia
Fernanda Gallinucci García	Brazil
Jaime Arturo García Garcia	Mexico
Mario Ranferi Gutiérrez Morales	Guatemala
Erick Donaldo Guzmán Ramírez	Guatemala
Moshe Hanlon	England
Leandro Hernández de la Peña	Cuba
Francisco Javier Hernández Moreno	Mexico
Raúl Alejandro Hernández Montoya	Mexico
Luis Herrera Valadez	Mexico
Angeyo Kalambuka Hudson	Kenya
Muge Karagoz	Turkey
Gregor Kramberger	Slovenia
Lena Leinonen	Sweden
Jonathan Link	USA
Ricardo López Fernandez	Mexico
Andrew Maier	Switzerland
Javier Magnin	Argentina
Fernando Martínez Pulido	Mexico
Alejandro Mirles	Puerto Rico
Luis Manuel Montaño Zetina	Mexico
Enrique Montiel Piña	Mexico
Debbie Morgan	England
Anibal Armando Morán Valdez	Guatemala
Tania Moulik	India

Basil Moshous	Germany
Claudia Oliva Mendoza Barrera	Mexico
Miguel N. Mondragón	Mexico
Juan Adolfo Ponciano	Guatemala
Cuauhtémoc Pacheco Díaz	Mexico
Armando Perdomo Almeida	Cuba
Gilberto Perea	Mexico
Fco. Javier Ramírez Jiménez	Mexico
Petra Riedler	Austria
Carlos J. Rivera	Puerto Rico
Sergio Román López	Mexico
Piyali Roy	India
Aldo F. Saavedra	Australia
Humberto Salazar	Mexico
Rosa Elena Sanmiguel Domínguez	Mexico
Francesco Sciacca	Switzerland
Jürgen Simon	Germany
Damijan Skrk	Slovenia
Jochen Schwiening	Germany
Jürgen Sjolin	Sweden
Kevin Stenson	USA
Mauricio Suárez	Puerto Rico
Gianni Tassotto	USA
Rolando D. Ticona Peralta	Bolivia
Luis Adolfo Torres González	Mexico
Jan Troska	England
Ulrich Trunk	Germany
Zaida del Rosario Urrutia del Cid	Guatemala
Andrea del Rocío Vargas Treviño	Mexico
Elsa Fabiola Vázquez	Mexico
Román Villanueva Aboytes	Mexico
Susanne Walsh	England
Adolfo Zarate Morales	Mexico

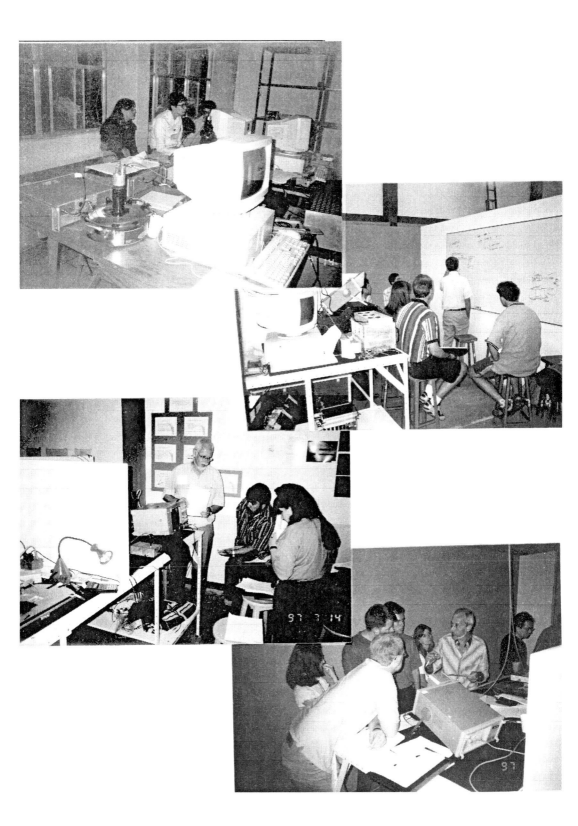

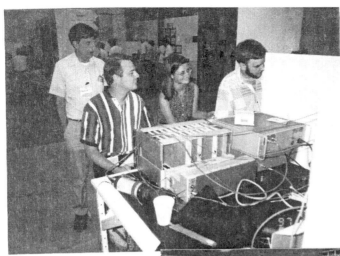

Author Index

A

Alcaráz, F., 399
Alfaro-Molina, R., 324
Atac, M., 191, 251

B

Bernier, K., 367
Boukhal, H., 367
Boulogne, I., 369
Bratzler, U., 371
Burrage, J., 417
Buttar, C., 417

C

Cantoral, E., 399
Cardona, J., 375
Carrillo, S., 377
Castro, J., 399
Cheremukhina, G., 347
Chernenko, S., 347
Cimmino, A., 382
Cooper, P. S., 3
Cordero, A., 399
Cristancho, F., 375, 387
Cuje, M., 414

D

Daubie, E., 369
David, M., 385
Denis, J.-M., 367

E

Eisele, F., 414
El Bardouni, T., 367

F

Fallot-Burghardt, W., 414
Ferbel, T., 371
Fernández, A., 208, 399
Feuerstack-Raible, M., 414

G

Gabutti, A., 371
Galindo, E., 387
Garcia, F. G., 389
Giubellino, P., 257
Glass, B., 414
Grégoire, Gh., 367
Grégoire, O., 367
Grupen, C., 14

H

Hall, G., 283
Hardingham, C., 417
Hofmann, W., 414

I

Idzik, M., 257

J

Johnson, M., 313

K

Knöpfle, K-T., 414
Kroha, H., 371

L

Lagouri, T., 371
Link, J., 392
López, R., 399

M

Manz, A., 371
Martínez-Davalos, A., 324
McCammon, D., 225
Morgan, D., 395, 417
Moshous, B., 397

N

Nakamura, K., 235

P

Pacheco, C., 399
Payne, R. G., 283

R

Román, S., 399
Rubín, M., 399
Rudge, A., 257

S

Saavedra, A., 382
Salzar, H., 399
Savoy-Navarro, A., 47

Schwiening, J., 407
Sellin, P., 417
Sexauer, E., 414
Sheaff, M., 313
Simon, J., 389
Smykov, L., 347
Straumann, U., 414
Streets, J., 251

T

Tassotto, G., 409
Taylor, G., 382
Tran, V., 367
Treichel, M., 371
Trunk, U., 414

V

Valdés, J., 399
Vargas, M., 399
Va'vra, J., 117
Vázquez, F., 377
Villaseñor, L., 333, 399

W

Walsh, S., 417
Weilhammer, P., 257
Wilcer, N., 251

Z

Zanevsky, Yu., 347
Zepeda, A., 208, 399